19th Drive Train Technology Conference (ATK 2021)

IOP Conference Series: Materials Science and Engineering
Volume 1097

Aachen, Germany
9 - 11 March 2021

ISBN: 978-1-7138-3352-9
ISSN: 1757-8981

Printed from e-media with permission by:

Curran Associates, Inc.
57 Morehouse Lane
Red Hook, NY 12571

Some format issues inherent in the e-media version may also appear in this print version.

Institute of Physics
Dirac House, Temple Back
Bristol BS1 6BE UK

Phone: 44 1 17 929 7481
Fax: 44 1 17 920 0979

techtracking@iop.org

Additional copies of this publication are available from:

Curran Associates, Inc.
57 Morehouse Lane
Red Hook, NY 12571 USA
Phone: 845-758-0400
Fax: 845-758-2633
Email: curran@proceedings.com
Web: www.proceedings.com

TABLE OF CONTENTS

IOP Conf. Series: Materials Science and Engineering 1097 (2021) 011001 doi:10.1088/1757-899X/1097/1/011001

Preface

This present volume contains the Proceedings of the 19th Drive Train Technology Conference (ATK) in the year 2021. The ATK 2021 is organized by the Institute for Machine Elements and Systems Engineering of RWTH Aachen University, Germany, and virtually delivered through an online event on March 09-11, 2021.

The traditional focus of the ATK on drive train innovations, including machine elements and tribology, this year is extended the first time on **Model Based Systems Engineering.** The ATK is an interdisciplinary platform that brings together experts in the field of Drive Technology and Systems Engineering to discuss the latest findings in research as well as to provide the transfer of research into industries. To identify practical needs, the conference's contributions from industry experts and academic researchers are balanced. By providing access to the 5th Conference for Wind Power Drives, taking place at the same time, the multidisciplinary idea is raised further. Both Conferences together have hosted over 100 live talks in 29 expert sessions.

We really appreciated the support provided by our friends and colleagues, leading to the great success of ATK 2021 and would like to thank:

- Our Program Committee for providing deeper insights into the industry's challenges, assessing numerous contributions to design our conference program and hosting our expert sessions.
- Our keynote speakers Prof. Dr.-Ing Christian Guist (Bauhaus-Universität Weimar), Andreas Mitschke (Airbus Defence and Space GmbH) and Dr. Georg Strobl (BMW AG) for introducing our guests and shaping the framework of our conference.
- All contributors providing an extensive and high-quality base for the ATK conference program.

We would like to thank all attendees for this fruitful and enjoyable gathering.

Sincerely,
Prof. Dr.-Ing. Georg Jacobs
Chairman of ATK

IOP Conf. Series: Materials Science and Engineering 1097 (2021) 011001 doi:10.1088/1757-899X/1097/1/011001

Committees

Organisation Committee:
Prof. Dr.-Ing. Georg Jacobs
Sebastian Stein, M.Sc.

Program Committee and Scientific Board:
Dr.-Ing. Ralf Dinter (Flender GmbH, Germany)
Prof. Dr.-Ing. Christian Guist (Bauhaus-Universität Weimar, Germany)
Prof. Dr.-Ing. Gunter Knoll (IST GmbH, Germany)
Dr.-Ing. Oliver Koch (Schaeffler Technologies AG & Co. KG, Germany)
Dr.-Ing. Lutz Lindemann (Fuchs Petrolub SE, Germany)
Dr.-Ing. Christian Muggeo (Contact Software GmbH, Germany)
Dr.-Ing. Nadine Nagler (Bosch Rexroth AG, Germany)
Dr.-Ing. Marco Ramm (Dr. M. Ramm, CLAAS Industrietechnik GmbH, Germany)
Dr.-Ing. Ernst-August Werner (Isatec GmbH, Germany)
Dr.-Ing. Dorothee Wieczorek (SEW-Eurodrive GmbH & Co. KG, Germany)
Dr.-Ing. Alexander Hassis (SEW-Eurodrive GmbH & Co. KG, Germany)

Extended Scientific Board:
Dr.-Ing. Jörg Berroth
Dr.-Ing. Francisco Gutierrez Guzmann
Dr.-Ing. Florian König
Dipl. Ing. Christian Konrad
Stephan Neumann, M. Eng.
Dr.-Ing. Christopher Sous

Editorial Board:
Prof. Dr.-Ing. Georg Jacobs
Sebastian Stein, M.Sc.

IOP Conf. Series: Materials Science and Engineering 1097 (2021) 011002 doi:10.1088/1757-899X/1097/1/011002

Peer review declaration

All papers published in this volume of *IOP Conference Series: Materials Science and Engineering* have been peer reviewed through processes administered by the Editors. Reviews were conducted by expert referees to the professional and scientific standards expected of a proceedings journal published by IOP Publishing.

- **Type of peer review:**
  Single-blind

- **Describe criteria used by Reviewers when accepting/declining papers. Was there the opportunity to resubmit articles after revisions?**
  The Reviewers have assessed all submitted papers in accordance to the IOP's peer review policy, as published on 20 July 2020. All Authors received the Reviewer's anonymised comments and were offered the opportunity to revise their papers before their final assessment for acceptance or rejection.

- **Conference submission management system:**
  Converia

- **Number of submissions received:**
  26

- **Number of submissions sent for review:**
  26

- **Number of submissions accepted:**
  22

- **Acceptance Rate (Number of Submissions Accepted / Number of Submissions Received X 100):**
  85%

- **Average number of reviews per paper:**
  2,38

- **Total number of reviewers involved:**
  17

- **Any additional info on review process:**
  n/a

- **Contact person for queries:**
  Georg Jacobs, Institute for Machine Elements and Systems Engineering, RWTH Aachen University, Germany. Email: georg.jacobs@imse.rwth-aachen.de

Mechanical concept development using principle solution models

Thilo Zerwas[1], Georg Jacobs[1], Kathrin Spütz[1], Gregor Höpfner[1], Imke Drave[2], Joerg Berroth[1], Christian Guist[3], Christian Konrad[1], Bernhard Rumpe[2] and Jens Kohl[3]

[1] Institute for Machine Elements and Systems Engineering, RWTH Aachen University, Templergraben 55, 52062 Aachen, Germany

[2] Software Engineering, RWTH Aachen University, Templergraben 55, 52062 Aachen, Germany

[3] BMW Group AG, Petuelring 130, 80788 München, Germany

thilo.zerwas@imse.rwth-aachen.de

Abstract. Cyber-physical systems (CPS) are characterized by the interaction of mechanical, electronic and information technology subsystems. Model-Based Systems Engineering (MBSE) is an approach for the cross-domain development of CPS and requires compatible methods and models for a function-oriented collaboration of the domains. However, the mechanics operate mainly component-oriented and thus cannot participate in the function-oriented development process. We propose a new modeling method that allows mechanics to develop a consistent and function-oriented system model from requirements and functions to principle solutions. The principle solution formally specifies the physical effect, active surfaces and material through which a function is fulfilled. Since principle solutions are consistently parameter-based, they can be calculated by linked external models and checked against requirements. This enables to examine possible solutions for functions at an early stage without having to develop components. Since principle solutions consist of recurring elements, we also propose a modeling concept for a solution libraryso that proven models can be efficiently reused and the modeling effort is reduced. Modeling method, test and solution library are explained using the example of an electric water pump of an automotive cooling circuit.

1. Introduction

The objective of product development is to transform the customer's requirements and wishes as efficiently as possible into a functionally fulfilling product. In order to meet today's customer requirements, more and more cyber-physical systems (CPS) are being developed. These are characterized by the interaction of mechanical, electronic and information technology subsystems [1,2]. Model-Based Systems Engineering (MBSE) is an approach for the cross-domain development of CPS. It is based on a common, parameter-based system architecture, which structures the system to be developed in a function-oriented way. In order to successfully develop CPS in the context of MBSE, the methods and models of all participating domains must be compatible and suitable for collaboration [1,3,4]. While the other domains operate function-oriented, mechanical engineers usually develop component-oriented: Requirements are translated directly into components without any detours. Since

the development of components is very time-consuming compared to the small-step procedure of the other domains, the mechanics have to invest a lot of effort in order to show first development results. Thus, parallel and agile developing is not possible. Instead, the mechanics must also start to develop function-oriented so that it can collaborate adequately with the other domains.
For this purpose, the mechanics must be able to derive a functional architecture from its requirements that is solution- and domain-neutral and can thus structure the development in a function-oriented way. A function can often be technically realized with different solutions from different domains. Therefore it is important for the mechanics to be able to define their possible solutions with little effort and test them as early as possible against requirements. One possible approach from design methodology are principle solutions. These describe the essential properties of a possible solution by specifying a physical effect, effective surfaces and material [5].
However, up to now there is no suitable way to describe principle solutions in the sense of [5] in a system architecture and test them against requirements. Hence, the contributions of this paper are

1. a modeling method for the function-oriented and continuous development of requirements, functions and principle solutions,
2. a formalization of principle solutions that allows testing against requirements, and
3. a modeling concept for a solution library for efficiently reusing principle solutions.

The paper is structured as follows: Section 2 provides an overview of the state of research concerning function-oriented development with principle solutions in the mechanical domain. Section 3 defines the research question and hypothesis of the paper at hand. Section 4 presents a method for the function-oriented and continuous modeling of requirements, functions and principle solutions, while Section 5 illustrates how principle solutions can be tested against requirements. Section 6 focuses on the efficient storage and reuse of principle solutions with a solution library before Section 7 concludes.

2. State of research

The Systems Modeling Language (SysML) has become established for modeling CPS in the context of MBSE. SysML is a language family that extends a subset of the Unified Modeling Language for the integrated representation of several domains in systems engineering [6]. For this purpose, the SysML comprises multiple language elements and diagrams for modeling behavior, structure and requirements [6]. Since these elements often result in several options for modeling the same information in the mechanics, it is important to define which SysML elements are used for which development models.

For the domain of mechanics it was noted in the introduction (cf. section 1) that principle solutions are essential for a function-oriented development of CPS, since they efficiently describe a possible solution by specifying effect, active surfaces and material without losing the functional orientation [5,7]. The basic idea is to derive possible physical effects for the realization of a function from the function itself. For this purpose, the functional architecture must have so-called elementary functions on its leaves, which describe a concrete physical relationship between the incoming and outgoing function flows (e.g. increase torque). The set of elementary functions is finite and allows the description of any overall technical function by its combinatorics. For each of these elementary functions the Koller catalog documents which physical effects are basically suitable for their realization [8]. If the selected physical effect is supplemented by active surfaces and material specifications, a principle solution is obtained. In the state of research of [5,8] this principle solution is only represented by a sketch and description. This representation does not fulfill the demands of developing CPS with SysML in the mechanics: it is neither formal or parameter-based, nor testable or efficiently reusable.

There are several research approaches to describe principle solutions or mechanical realizations of a function with SysML. One approach is the description of concepts by manual sketches [9]. Although this approach allows the integration of sketches into SysML, it is not a parameter-based description of a physical behavior that can be tested against requirements. Other approaches describe the realization of a function by components [1,3,10]. Thus, components have to be elaborately detailed before the realization can be tested against requirements and the mechanical domain can only collaborate with a time delay. Another proposal for modeling principle solutions in the sense of [5] principle solution is provided by [11]. However, the parameters are neither linked within a principle solution nor externally. Hence, the principle solution cannot be tested directly with CAD or simulation models [12–15]. All in all, there is no integrated approach to model, test and efficiently reuse principle solutions in the sense of [5] in a function-oriented development process with SysML.

3. Research question and hypotheses

In the introduction (cf. section 1), the challenges for mechanics in the development of CPS were described, which have not yet been overcome by the current state of research (cf. section 2). Therefore, the research question of this publication is:

How can principle solutions be described in a function-oriented, testable and efficient way?

To answer this question, three research hypotheses are formulated:

1. *Principle solutions can be formalized as SysML principle solution models in an object-oriented way.*
2. *SysML principle solution models can be tested due to their formalization.*
3. *SysML principle solution models can be captured in SysML libraries and efficiently reused.*

The following three chapters each address one of the three hypotheses. In section 4 we present a method for modeling requirements, functions and principle solutions, which allows a seamless connection of the individual artifacts and parameters. Section 5 illustrates how principle solution models can be tested against requirements and thus allow early, function-oriented validation independent of components. Finally, section 6 shows how functions and principle solution models can be stored in a model-based solution library and can be efficiently reused.

4. Modeling method for function-oriented development

In this section a modeling method is presented, which allows to describe requirements, functions and principle solutions of a development process in a system model. An essential characteristic is the consistent linking of artifacts and parameters. In this way, the function orientation can also be consistently maintained for the mechanical domain and cross-domain collaboration is improved. The modeling method is based on the SysML profile SysML4FMArch, which was developed as a linguistic basis for modeling functional architectures in the mechanical domain [16].

The modeling method is explained on an automotive cooling system, which will be briefly introduced at this point. The main function of a vehicle is locomotion. For this purpose, the drive system provides mechanical energy that is conducted to the wheels and then transferred to the road. In vehicles with combustion engines, mechanical energy is obtained from the chemical energy of a fuel. For this purpose, the physical effect of combustion is used in the cylinders of the engine, resulting in a thermal expansion of the fuel-air mixture. The sudden increase in pressure accelerates the piston, which transfers the mechanical energy to the rest of the drive system. During combustion of the fuel-air mixture, not all the chemical energy is converted into mechanical energy for propulsion: A part of the energy is conducted out of the system via the escaping exhaust gas and another part is induced into the engine components as thermal energy. Since the engine is often unable to release all of this thermal energy via its outer surfaces, its internal energy and temperature rise.

The rising component temperature is becoming increasingly critical for the component material as well as the combustion process and endangers the functional reliability of the engine. For this reason, combustion engines are usually kept within an optimum temperature window by a liquid-based cooling system. A cooling medium circulates in this cooling system, which absorbs heat from the engine and releases it to the cooler. At the cooler the thermal energy is emitted to the environment. The cooling medium is accelerated by a pump so that it can absorb and release sufficient heat by convection and remains in motion despite the pressure losses. In our example system, the coolant pump is not operated mechanically but electrically and can thus be set to a certain rotational speed by a control unit and based on the current temperatures of the engine. In addition, the engine is simplified and consists of the components cylinder head (CH) and crankcase (CC), which both require different target temperatures.

The following subchapters each describe the methodical modeling of this example system with regard to its requirements, functions and principle solutions.

4.1. Requirements

Requirements are demands and wishes that customers, manufacturers, legislators and many other stakeholders have for the product to be developed. A successful product must not only fulfill the wishes of the customer, but must also be able to be produced efficiently (e.g. factory standards) and comply with legal requirements. Therefore it makes sense to document these requirements and to continuously check their compliance during the development process [5]. Until today, requirements in many companies are still formulated in textual sentences (unstructured text) and stored as a requirements list in unlinked documents or software tools. Furthermore, it is not uncommon to try to define the requirements finally at the beginning of a project and not to change them afterwards. This endeavor is understandable, but usually not compatible with the dynamic development processes of CPS. The essential disadvantages are the often-ambiguous formulation of requirements and the missing link to the development models based on them, which are thus cut off from requirement changes.

Instead, for some time now, there have been a wide range of suggestions on how requirements can be modeled using MBSE approaches. Many of them have two common features that make a significant difference to document-based and informal requirements. One is a clear formalization and explicit description that leaves no room for interpretation. On the other hand, the requirements are expressed as far as possible by (physical) quantities with concrete values. These can be linked to other development models so that changes in requirements directly reach all relevant models.

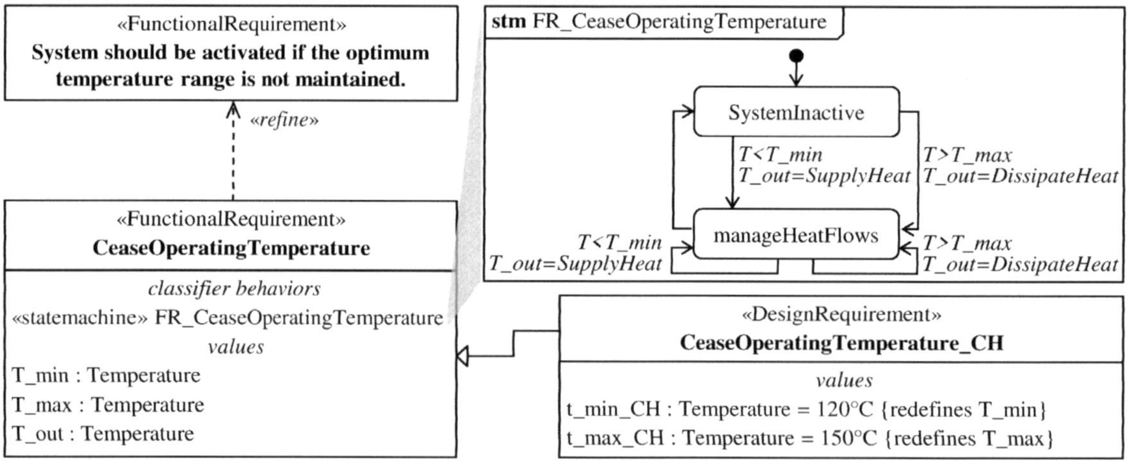

Figure 1. Requirements of the example system.

The requirements are differentiated into two categories. A «FunctionalRequirement» specifies the desired functionality of a technical system and is formalized as state machine or activity diagram. In the example system the superior behaviour of the whole cooling system is described as a state machine, which is always in one of two states: Either it is idle or active (figure 1). The transitions between the states depend on the temperature states of cylinder head and crankcase. If, e.g., the cylinder head exceeds its permissible maximum temperature, the system switches to the active state. This modeling allows for example the automated generation of test cases. This way it can be checked whether the system fulfills the prescriptive behavior and is always in the expected state [17,18].

The second category are design requirements («DesignRequirement»), which limit the value range of a parameter occurring in the system. For example, the optimum range for the operating temperature of the cylinder head can be specified as 120 °C to 130 °C. Since this temperature range is decisive for the described transition in the state machine, this design requirement is modeled as a specialization of the functional requirement. Thus, the modeling of behavior and the restriction of parameter values are clearly separated, but can be used for common statements.

4.2. Functional architecture

Functions describe the specific behavior of a product without specifying which components, effects, etc. physically implement this behavior. The concept of functions is based on the idea that physical flows enter and leave a system over a given system boundary. These *function flows* are quantified by concrete parameters values and can be categorized as flows of energy, material, and signal. Functions describe not only which function flows enter and exit, but also which *operation* takes place [5]. The decomposition of the overall function into subfunctions results in a functional architecture [7]. According to the SysML profile SysML4FMArch, the functions can be distinguished into decomposed functions («Architecture») and elementary functions («ElementaryFunction») [16]. Each elementary function describes an elementary mathematical relationship between the input and output flows [5]. Functions can be linked to the requirements they fulfill through function calls or satisfy relationships.

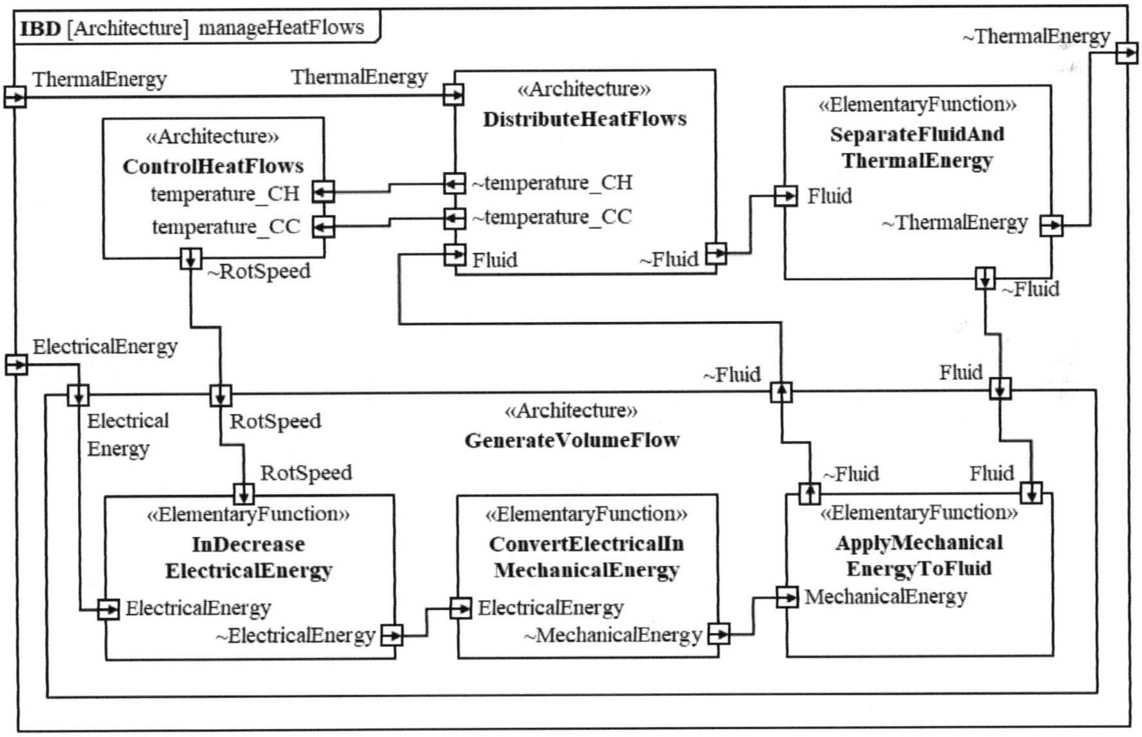

Figure 2. Functional architecture of the example system.

In the example system, the function manageHeatFlows is decomposed into four functions (figure 2). The function *generateVolumeFlow* generates a volume flow of the coolant according to the specified signal from the function *controlHeatFlows*. This volume flow is directed to the function *DistributeHeatFlows*, where heat is absorbed (at the cylinder head and crankcase). The heated coolant leaves this function and releases the thermal energy to the environment in the elementary function *SeparateFluidAndThermalEnergy*, before it circulates back into the *GenerateVolumeFlow* function. This function can be divided into three elementary functions: *InDecreaseElectricalEnergy* transforms the incoming electrical energy so that the following function *ConvertElectricalInMechanicalEnergy* generates mechanical power according to the specified rotational speed. This mechanical power is used in the subsequent function *ApplyMechanicalEnergyToFluid* to pressurize and accelerate the coolant.

The realization of an elementary function is often possible in multiple domains [19]. For example, cooling circuits can be controlled by mechanical thermostats or software-based controllers. Often the flows of a function can give a hint, in which domain this function can be realized. At the latest by defining a physical effect, the further development of this elementary function is assigned to a certain domain. Therefore, the transition from functions to principle solutions also entails a shift from cross-domain to domain-specific development.

4.3. Principle solution models

While functions only describe the desired changes of function flows, so-called principle solutions concretize how this change is physically realized. Therefore, elementary functions and principle solution models are linked with a generalization relationship. Thus, the principle solution inherits all functional flows as ports from the elementary function and can use them to describe more precisely how the incoming flows are transformed into the outgoing flows using a physical effect, active surfaces and material. As elements of the principle solution, physical effect, active surfaces and material can be modeled with the corresponding stereotypes defined in [16] in the internal block diagram of the principle solution. Physical effects can usually be described with mathematical equations which are modeled as constraints. The parameters of such equations can depend on function flows, active surfaces and material and are linked to them accordingly. Physical quantities, which refer to function flows (e.g. volume flow, pressure, torque), are linked to the corresponding values of the incoming and outgoing function flows. Parameters that refer to geometric or material-related quantities are linked to the value properties of an «ActiveSurface» or a «Material». If the equation contains natural constants, these are modeled as value properties directly into the principle solution and linked to the constraint parameters.

The upper section of figure 3 shows the continuous modeled path from a requirement to the associated function and its principle solution for the example system. The requirement that a volume flow of 7 l/s should be generated is met by the elementary function *ApplyMechanicalEnergyToFluid*. This elementary function is now specialized by the principle solution *CentrifugalPumpWheel* that describes how mechanical energy is applied to the fluid.

The lower section of figure 3 illustrates the internal block diagram of the principal solution. Several possible physical effects for the elementary function *ApplyMechanicalEnergyToFluid* can be found in the Koller catalog [8]: conservation of momentum, friction, Boyle's law, adhesion, Coulomb's law, Bernoulli's principle and others. Here the physical effect *CentrifugalForce* is chosen, which is modeled as «PrincipleEffect» with all relevant parameters in the principle solution model. The active surfaces are selected to match this effect: The *PumpWheel* rotates and conveys the fluid outwards against the *Cylinder*, where the induced kinetic energy is converted into static pressure and the fluid can exit through a radial opening. These active surfaces are described by a few parameters for their geometry (e.g. outer diameter of the *PumpWheel*) and design parameters (e.g. optimal volume flow), which are essential for their physical behavior. In the example shown, it is not the material parameters of the active surfaces that are relevant for the modeled physical effect, but the material parameters of the fluid flow. Therefore, the density parameters of the incoming and outgoing fluid flow are linked to the density parameter of the physical effect. This is also the reason why *PumpWheel* and *Cylinder* do not contain any material parameters here, as it is basically enabled by [16]. In addition to the essential physical effect, the principle solution contains another «EffectElement» representing the pressure difference. Finally, the parameters of the «PrincipleEffect» are linked to the corresponding counterparts of the active surfaces and function flows.

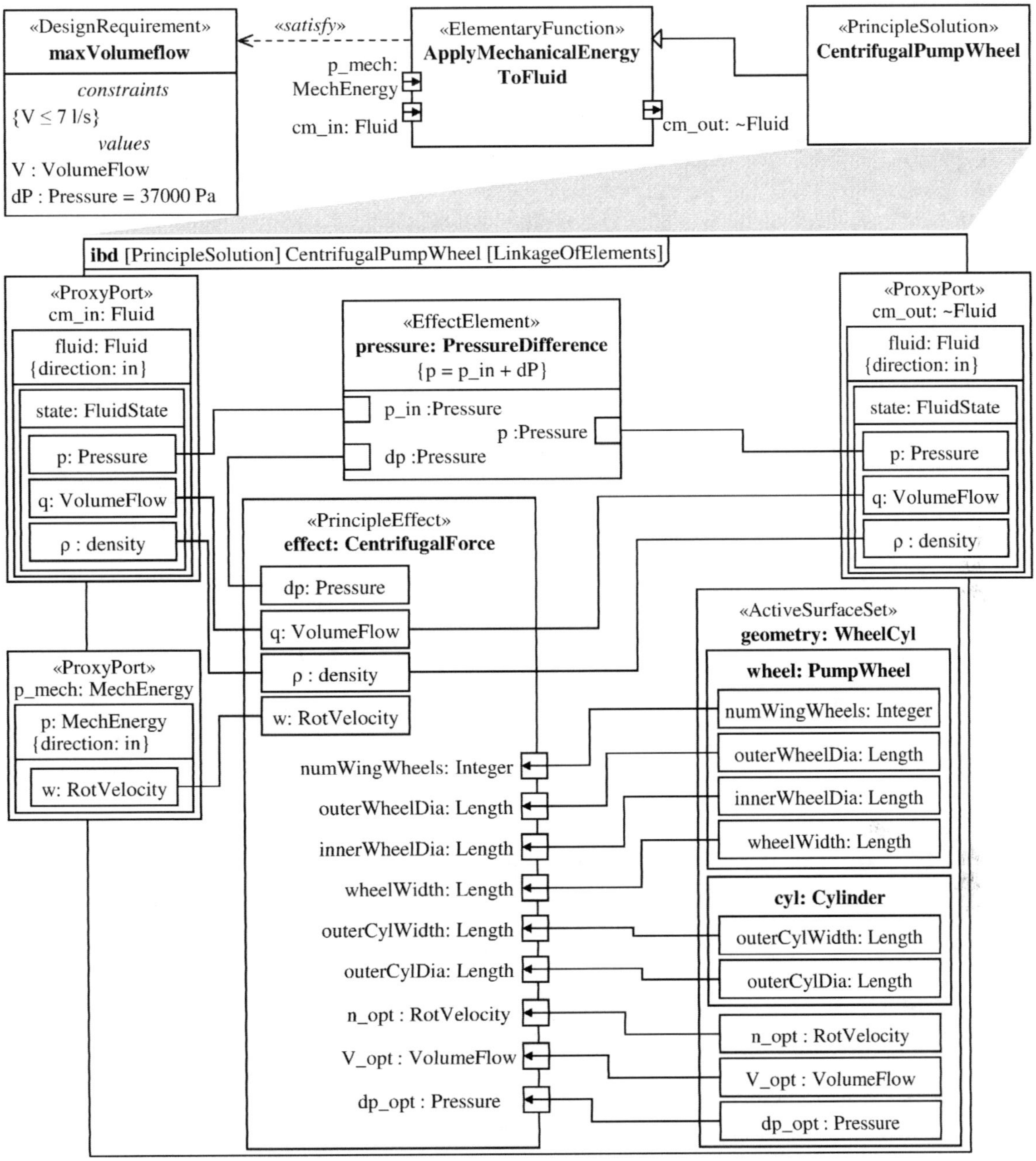

Figure 3. Requirements are satisfied by elementary functions and their principle solutions whose elements, parameters and relationships are visible in the internal block diagram.

4.4. Interim conclusion

The modeling method presented demonstrates the function-oriented development in mechanics by consistently linking requirements via functions to principle solutions. The modeling of the latter transfers the concept of Koller [5] into a formalization for SysML and, in contrast to other research approaches [20], uses a consistently parameter-based representation of effect, geometry and material [16]. This is a key advancement that enables initial performance testing of principle solutions (cf. section 5) and linkage to more detailed behavior models (cf. section 6). And this confirms the first research hypothesis "Principle solutions can be formalized as SysML principle solution models in an object-oriented way".

5. Test of principle solution models

The basis for testing a principle solution is the previously presented formalization (cf. section 4). Due to the parameter-based representation, physical effect and active surfaces can be linked to external models, see figure 13.7. In our example system, the *CentrifugalForce* of the principle solution *CentrifugalPumpWheel* is linked to a MATLAB model that calculates the hydrodynamic behavior of a pump wheel. Similarly, the parameters of the «ActiveSurfaces» are linked via an Excel table to the corresponding CAD models of these active surfaces. With these links the described principle solution (figure 3) can be executed by an engine that handles the execution of the single solvers. In our case, we used the integrated simulation engine of a system modeler (e.g. Cameo Systems Modeler). For this purpose, parameter values are applied externally via function flows and read from other external models (e.g. the CAD model). All values are passed to the MATLAB function for effect calculation, which returns the calculated results. Those can be further processed in the principle solution or passed on to the subsequent principle solution via a function flow.

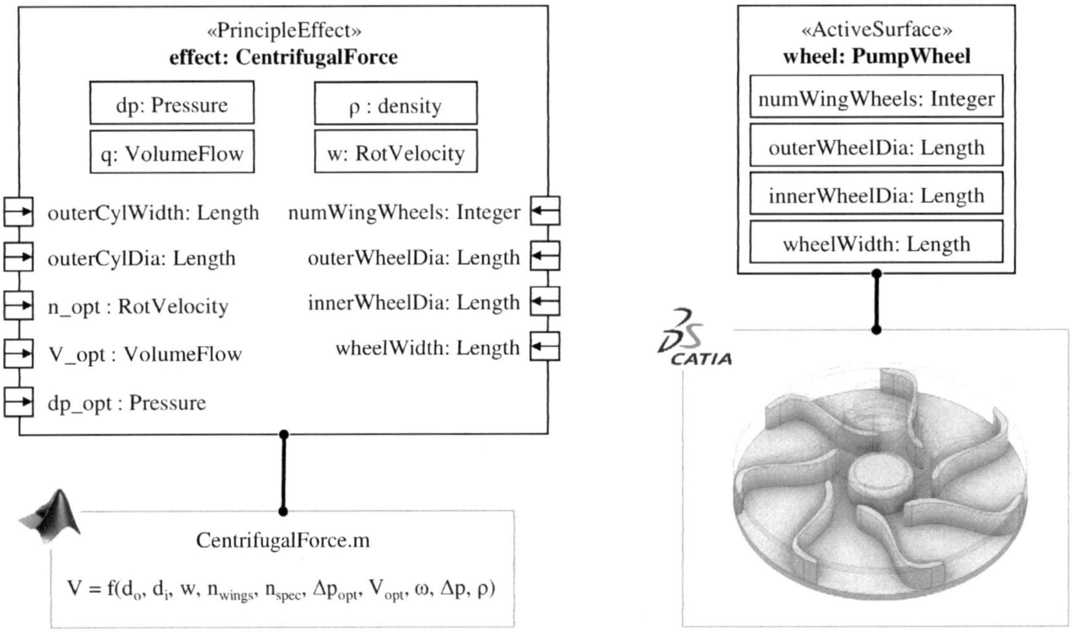

Figure 4. The «PrincipleEffect» *CentrifugalForce* is connected to an external MATLAB model and the «ActiveSurface» *PumpWheel* is connected to a corresponding CAD model.

In our example system it was required that the defined principle solution must overcome a pressure difference of 37 kPa (from the cooling circuit) and generate a volume flow of at least 0.007 m^3/s to ensure sufficient cooling of the motor. The execution of the principle solution results in a volume flow of 0.0074 m^3/s and confirms that the requirement can be met with this principle solution (table 1).

This confirms the second research hypothesis. With the described procedure, principle solutions can be tested without having to develop complex components. This means that the mechanics can collaborate with the other domains about the function realization earlier than before and rely on objective test results. In accordance with the described procedure, not only individual but also several principle solutions can be interconnected to create system tests. In addition, it is possible to define design processes as activities, for example to parameterize principle solutions for optimal functional fulfillment [21].

Table 1. Parameter values for the calculation of the principle solution *CentrifugalPumpWheel.*

Parameters of the functional flows	Parameters of the principle solution	Resulting parameters of the effect calculation
$\omega = 15$ 1/s $\rho = 1000$ kg/m^3 $\Delta p = 37000$ Pa	$V_{opt} = 0.00606$ m^3/s $\Delta p_{opt} = 40000$ Pa $n_{spec} = 40{,}7$ $n_{wings} = 7$ $d_o = 115$ mm $d_i = 57{,}5$ mm $w = 32{,}1$ mm	$V = 0.0074$ m^3/s

6. Solution Library

Besides the described advantages of parameter-based modelling of principle solutions, it is unmistakable that this involves a certain modelling effort. One approach to reduce the modelling effort is to reuse the models created once. Since elementary functions and physical effects are not only a finite but also a known quantity, their reuse offers high potential.

Koller has structured elementary functions and physical effects with non-formalized descriptions in a document-based catalog [5]. Since, according to Roth [22], catalogs should basically fit the method used and enable efficient use, it is necessary to develop a new concept for a digital library. Therefore, we use SysML as conceptual design language to develop the *Solution Library* and the interfaces to the system architecture.

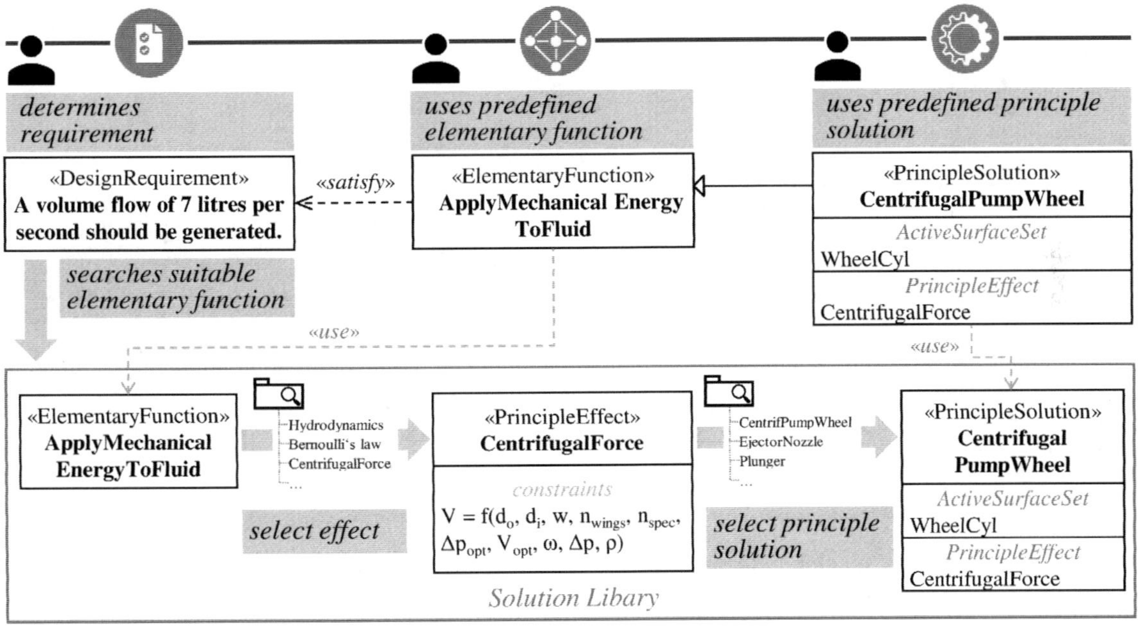

Figure 5. The solution library (lower part) supports the development process (upper part) by identifying possible physical effects and principle solutions for a selected elementary function.

The first use of the solution library takes place during the modelling of the functional architecture. Here, the user can reuse exactly those elements from the finite and predefined pool of elementary functions that he needs to fulfill the requirements (figure 5 left). By selecting the elementary function (here: *ApplyMechanicalEnergyToFluid*) the solution library is automatically filtered and only those physical effects are displayed, which can realize the chosen elementary function (figure 5 center). After a physical effect is also selected (here: *CentrifugalForce*), the set of stored principle solutions is

filtered so that only those containing the selected physical effect are listed (figure 5 right). All these listed principle solutions are suitable as technical realization of the elementary function and are able to fulfill the initial requirement (chosen here: *CentrifugalPumpWheel*).
The described approach is feasible because we have extended the well-known Koller catalog [8] by central elements: Our solution library contains not only elementary functions and physical effects (like Koller), but complete principle solutions including frequent active surfaces and materials (figure 6 left). Thus, our principle solutions can be varied not only with regard to the physical effect, but also with other active surfaces and materials.

With principle solutions, initial functional tests (cf. section 5) can be carried out, but more specialized models often have to be used in order to validate further requirements (e.g. service life or noise propagation). Therefore, as a second major enhancement compared to Koller, we store each principle solution together with suitable behavior models and workflows in a so-called solution element (figure 6 right). In this way, behavior models are clearly assigned to concrete principle solutions in our solution library and can be used efficiently for virtual behavior testing of evolving solutions based on the chosen purpose [21]. This confirms the third research hypothesis.

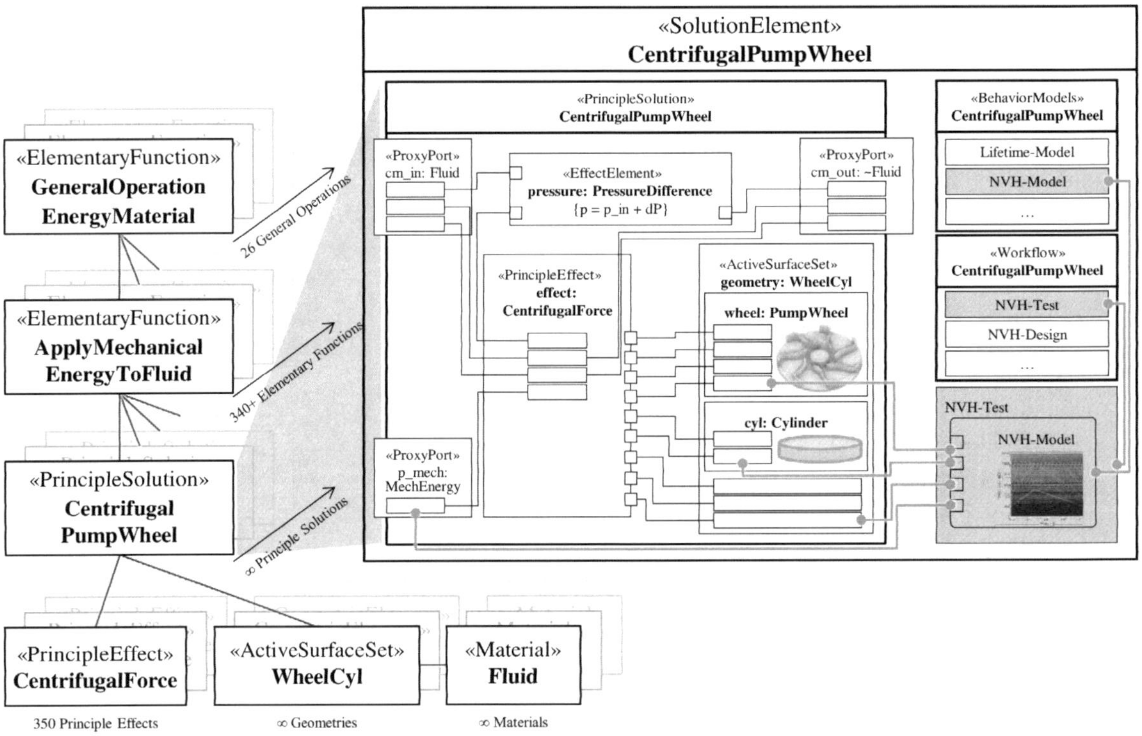

Figure 6. In addition to elementary functions, physical effects, active surfaces and materials, the solution library primarily contains a set of predefined principle solutions. These principle solutions are stored together with more detailed behavior models and workflows in a so-called solution element. Here it is shown for the example system how the principle solution *CentrifugalPumpWheel* is classified in the variant structure of the solution library.

7. Conclusion

In this paper a SysML-based modeling method for the mechanical domain in the development of CPS was presented. This method enables the mechanical domain to consistently link its requirements, functions and principle solutions with each other, so that the function orientation and thus the possibility of collaboration with other domains is maintained. The principle solution was formalized based on the concept of Koller [8] for SysML and consists of physical effect, active surfaces and

IOP Conf. Series: Materials Science and Engineering 1097 (2021) 012001 doi:10.1088/1757-899X/1097/1/012001

material, which are linked to each other via their parameters. This parameter linking makes the principle solution executable so that it can be calculated by external models and the results can be compared against requirements. In contrast to the Koller catalog [8], the modeling concept for the solution library contains not only elementary functions and effects, but entire principle solutions with more sophisticated behavior models and their corresponding test workflows. The solution library thus makes principle solutions efficiently reusable and paves the way for virtual behavior testing of evolving solutions based on the chosen purpose.

References

[1] Eigner M, Gilz T and Zafirov R 2012 Proposal for functional product description as part of a PLM solution in interdisciplinary product development *DESIGN 2012* ed Marjanović D, Storga M, Pavkovic N, Bojcetic N (Zagreb: UNIZAG-FSB) pp. 1667–76.
[2] Alur R 2015 *Principles of cyber-physical systems* (Cambridge: The MIT Press).
[3] Gausemeier J, Dorociak R, Pook S, Nyßen A and Terfloth A 2010 Computer-aided cross-domain modeling of mechatronic systems *Design 2010* ed Marjanović D, Storga M, Pavkovic N, Bojcetic N (Zagreb) pp. 723–32.
[4] Broy M 2010 *Cyber-Physical Systems* (Berlin Heidelberg: Springer).
[5] Koller R 1998 *Konstruktionslehre für den Maschinenbau* (Berlin: Springer).
[6] OMG OMG Systems Modeling Language Version 1.6, https://www.omg.org/spec/SysML/1.6/PDF.
[7] Feldhusen J and Grote K-H (eds.) 2013 Pahl/Beitz Konstruktionslehre Methoden und Anwendung erfolgreicher Produktentwicklung 8th ed. (Berlin: Springer).
[8] Koller R and Kastrup N 1994 *Prinziplösungen zur Konstruktion technischer Produkte* (Berlin: Springer).
[9] Moeser G, Albers A and Kumpel S 2015 Usage of Free Sketches in MBSE *2015 IEEE International Symposium on Systems Engineering (ISSE)* ed IEEE) pp. 50–55.
[10] Moeser G, Kramer C, Grundel M, Neubert M, Kümpel S and Scheithauer A et al. 2015 Fortschrittsbericht zur modellbasierten Unterstützung der Konstrukteurstätigkeit durch FAS4M *Tag des Systems Engineering* ed Schulze S-O, Muggeo C (München: Hanser) pp. 69–78.
[11] Wölkl S and Shea K 2009 A Computational Product Model for Conceptual Design Using SysML *Volume 2: 29th Computers and Information in Engineering Conference* ed ASMEDC pp. 635–45.
[12] Pasch G, Jacobs G, Höpfner G and Berroth JK 2019 *Multi-Domain Simulation for the Assessment of the NVH Behaviour of a Tractor with Hydrostatic-Mechanical Power Split Transmission*: RWTH Aachen University.
[13] Golafshan R, Jacobs G, Wegerhoff M, Drichel P and Berroth JK 2018 Investigation on the Effects of Structural Dynamics on Rolling Bearing Fault Diagnosis by Means of Multibody Simulation.
[14] Andary FS, Berroth JK and Jacobs G 2019 An Energy-Based Load Distribution Approach for the Application of Gear Mesh Stiffness on Elastic Bodies *Journal of mechanical design* **141** 95001.
[15] Berroth J, Jacobs G, Kroll T and Schelenz R 2016 Investigation on pitch system loads by means of an integral multi body simulation approach.
[16] Drave I, Rumpe B, Wortmann A, Berroth J, Höpfner G and Jacobs G et al. Modeling Mechanical Functional Architectures in SysML *MODELS '20: Proceedings of the 23rd ACM/IEEE International Conference on Model Driven Engineering Languages and Systems* ed pp. 79–89.
[17] Rumpe B 2012 *Agile Modellierung mit UML* (Berlin, Heidelberg: Springer).
[18] Drave I, Hillemacher S, Greifenberg T, Kriebel S, Kusmenko E and Markthaler M et al. 2019 SMArDT modeling for automotive software testing *Softw: Pract Exper* **49** pp. 301–28.

[19] Albers A and Zingel C 2011 Interdisciplinary Systems Modeling using the Contact and Channel-Model for SYSML *Impacting society through engineering design* ed Culley SJ (Glasgow: Design Society) pp. 196–207.
[20] Munker F and Albers A 2015 SystemSketcher – Entstehung eines anwenderorientierten Ansatzes zur interdisziplinären Systemmodellierung *Tag des Systems Engineering* ed Schulze S-O, Muggeo C (München: Hanser) pp. 291–300.
[21] Höpfner G, Jacobs G, Zerwas T, Drave I, Berroth J and Guist C et al. Model-Based Design Workflows for Cyber-Physical Systems Applied to an Electric-Mechanical Coolant Pump *Antriebstechnisches Kolloquium 2021*
[22] Roth K 1994 *Konstruieren mit Konstruktionskatalogen* (Berlin: Springer).

IOP Conf. Series: Materials Science and Engineering 1097 (2021) 012002 doi:10.1088/1757-899X/1097/1/012002

Optimization-based Component Sizing Method for Electrified Heavy-Duty Powertrain Concepts

D. Roth[1], C. Habermehl[1], G. Jacobs[2], S. Neumann[2], B. Juretzki[1], D. Bayer[3]

[1] IME Aachen GmbH, Aachen
[2] RWTH Aachen University, Institute for Machine Elements and Systems Engineering, Aachen
[3] Liebherr Hydraulikbagger GmbH, Kirchdorf an der Iller

david.roth@imse.rwth-aachen.de

Abstract. The demand for battery-powered, exhaust emission-free mobile machinery is increasing. However, the high energy requirements of mobile machines result in large, heavy and expensive battery storage systems. Driven by continuously decreasing but still fairly high battery costs battery costs, the efficiency of the drive system is becoming the focus of development activities, even in an industry where robustness, service lifetime, productivity and power density have been the driving factors up to now. Interactions between topology, component sizing, and control strategy require a system-level approach for the development of drive systems. In this paper, a novel optimization-based sizing method for electrified heavy-duty powertrain concepts is presented and applied to subsystems of a wheeled excavator. Scalable component cost and efficiency models are used in a system simulation to calculate objective function values for a multi-objective optimization. The result of the optimization, a trade-off between energy consumption and costs provides the OEM with a decision support in the development process. The modular structure of the approach allows a subsequent extension of the design parameter space as well as the detailing of the models used.

1. Introduction

In reaction to the ongoing climate change and the existential threat for Europe and the world, the European Union has agreed on the Green Deal. Part of the Green Deal is the goal of becoming the first climate-neutral continent and releasing no net greenhouse gas emissions by 2050 [1]. In addition, ongoing urbanization contributes to increasing exceedance of current emission limits in inner-city areas. In order to comply with current and future emission limits, local and regional authorities are planning and already implementing action plans that include the establishment of low- and zero-emission zones in inner-city areas. In this context, the Committee for European Construction Equipment expects increasing restrictions on the use of diesel-powered mobile machinery [2]. In the future, the demand for alternative drive concepts will increase not only in the automotive sector, but also in the field of mobile machinery. One possibility for operating construction machinery in zero-emission zones is the use of battery-powered electrical machines. The use of passenger cars in urban areas is characterized by short distances, short periods of use, low power requirements and numerous opportunities for charging a battery. In comparison, the use of construction machinery requires high power over a long period of time (e.g. an eight-hour workday) without the possibility of recharging in between. As a result, comparatively large amounts of energy have to be stored, which leads in combination with the low energy density of electric accumulators to exceedingly large, heavy and expensive batteries. Therefore, in the development of battery-electric mobile machines, the energy efficiency of the drive system is becoming much more important than it was the case in the development of conventional Diesel driven machines. Mobile machines are characterized by the interaction of several consumers (e.g. travel drive, cylinder drive and swing drive in hydraulic excavators). The system efficiency and thus the energy requirement of the machine therefore depends not only on the sizing of the components, but also on the choice of a suitable topology and an efficient control strategy. In order to be able to take into account the mutual influence of topology, component sizing and control strategy on the energy requirement of the machine in early phases of the development, it is necessary to develop and evaluate on a system level [3].

For the development of hybrid and battery electric vehicles, optimization-based approaches that focus on minimizing system characteristics, such as fuel consumption or manufacturing costs, have proven beneficial in the automotive sector [4]. Multiple approaches for finding an optimal vehicle topology [5–9], an ideal component sizing [9–16] and an appropriate control strategy [11–14, 17] using several different kinds of optimization strategies, such as Genetic Algorithms, Particle Swarm Optimization, Nested Optimization or extended coordination methods can be found in the literature [4, 18].

Optimization algorithms are also used for development tasks in the field of mobile machinery. For example, approaches for optimizing the component sizing and control strategy of hybrid energy storage systems consisting of batteries and fuel cells or capacitors can be found in the literature [19–21]. Furthermore, there are existing approaches that address topology selection [22] and component specification of electrical components as well as internal combustion engines for hybrid mobile machinery [23–28].

Compared to an automotive application, in which the travel drive is the main consumer, the energy consumption of mobile machinery is composed of the parallel operation of numerous travel and working drives. Therefore, methods from the automotive sector focusing on one single consumer cannot be used directly for the development mobile machine drive systems. However, individual component models used in the automotive sector, such as efficiency and cost models for electric motors and power electronics, can be used for the development of mobile machines. Still, further models are required for the efficiency and cost calculation of hydraulic components and heavy-duty gearboxes, which are not used in the automotive sector. Most studies from the field of mobile machinery have only focused on specific domains, such as the optimization of energy storage systems. Furthermore, there has been no detailed investigation of the optimal hydraulic component or heavy-duty gearbox sizing. To the best of our knowledge, there are no optimization-based development approaches for battery-electric drive concepts (travel and work drive) of mobile machinery. To answer the question of how an efficiency- and cost-appropriate drive concept consisting of topology, component sizing and control strategy can be found for battery-electric mobile machines, a novel method is required that uses a system model capable for the parallel operation of multiple consumers as well as performant models for heavy-duty components, which can be used by well-established optimization algorithms.

In this paper, a novel method for optimization-based component sizing is presented and exemplarily applied to a wheeled excavator using a preselected topology. For this topology, a multi-objective optimization using scalable component models is performed to identify Pareto-optimal component design parameters. The questions of topology synthesis and an optimal control strategy for the specified components will be covered in a future work. The presented method is characterized by a combination of a computationally intensive and therefore for optimization methods unsuitable transient dynamic simulation, which makes it possible to consider the interaction of parallel operated consumers in the component sizing, and a performant quasi-static simulation, which is necessary to apply known optimization methods for the component sizing. The qusi-static simulation uses interchangeable component models that can be extended in the level of detail. The model level of detail can be adapted according to the data availability, so that the method can be used by OEMs in early phases of the product development for decision making.

2. Use Case and Scope

2.1 *Considered System*

This paper investigates an electrified drive concept for a wheeled excavator. It is based on a conventional drive system (Figure 1, left) for a wheeled excavator with an operating weight of 12.5 t to 17.5 t. A 105 kW diesel engine drives a variable displacement pump that supplies a state-of-the-art load-sensing (LS) system. The LS system supplies five consumers: one hydraulic motor each for the travel drive of the undercarriage and the slewing drive of the superstructure (SS), as well as hydraulic cylinders for the bucket, stick and a boom. The motion of the superstructure is transmitted by a swing reducer (SR). At each consumer valve, the load pressure p_L is signaled to the pump. In the model the two cylinders which actuate the boom are combined into one consumer assuming symmetrical operation. The travel drive is designed as a central drive with a variable displacement hydraulic motor. The working drive comprises the cylinders and the slewing drive. In practice, the travel and the working drive are usually not operated simultaneously. The diesel engine speed remains largely constant during the execution of a work task. Its level is selected by the operator depending on the required power.

The electrified drive topology (Figure 1, right) was conceptualized in a study using discursive methods such as the morphological analysis. Among the defined evaluation criteria, it represents a preferred solution, therefore it was selected for the subsequent optimization which is presented in this paper. The hydraulic travel drive as well as the slewing drive are replaced by electric motors (EM). The travel drive is implemented as four single-wheel drives with planetary hub gearboxes (G). Both drives can thus utilize the high recuperation potential when decelerating the vehicle or the superstructure. In addition to the two rotary consumers, the boom drive has a high recuperation potential. This means that the potential energy can be converted into electrical energy when the boom is lowered. Accordingly, the boom has its own energy supply that enables recuperation. The recuperation potential was derived from measured duty cycles. For reasons of confidentiality, no detailed description of the technical implementation can be given. The stick and bucket drive have a low recuperation potential, so that the drive of these consumers is kept as a conventional LS drive. In addition to recuperation, the new drive concept offers the advantage of dividing the LS system up. In conventional LS systems, consumers operated in parallel with different loads cause throttling losses at the low-load consumers. In the conventional system, the slewing drive and the boom drive are typically the highest-load consumers. In the electrified concept, these throttling losses are thus avoided.

For the sake of brevity, this paper focuses on the component sizing for the travel drive as well the stick and bucket drive of the electrified drive concept (Figure 1, right, highlighted).

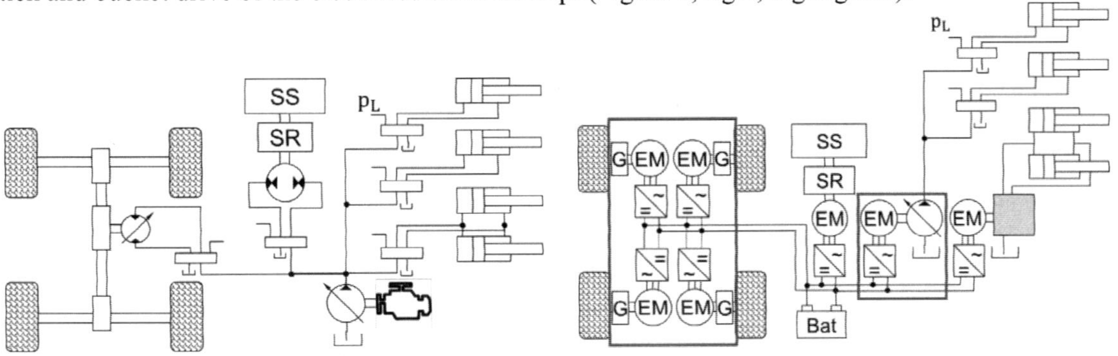

Figure 1. Conventional drive system (left) and electrified drive concept (right) of a mobile excavator. The subsystems dimensioned in this paper are highlighted.

2.2 *Duty Cycle*

To ensure consistent performance of the electrified drive concept compared to the conventional drive system, representative operating scenarios of the conventional drive system are used for component dimensioning. These were obtained from measurements for the travel and working drive. For the travel drive, a combined cycle with driving and grading sections is adopted in this paper. Figure 2, left depicts examples of the excavator speed v in $km \cdot h^{-1}$ and track gradient in ° of a driving section. For the working drive, a 90° full load working cycle is used as a representative cycle. In Figure 2, right the exemplary stick cylinder stroke s_S in mm and chamber pressures p_S in bar are shown. In this cycle, the excavator picks up soil at one point and, after a 90° rotation of the superstructure, dumps it onto a truck, for example. The dumping point can be higher or lower than the pickup point.

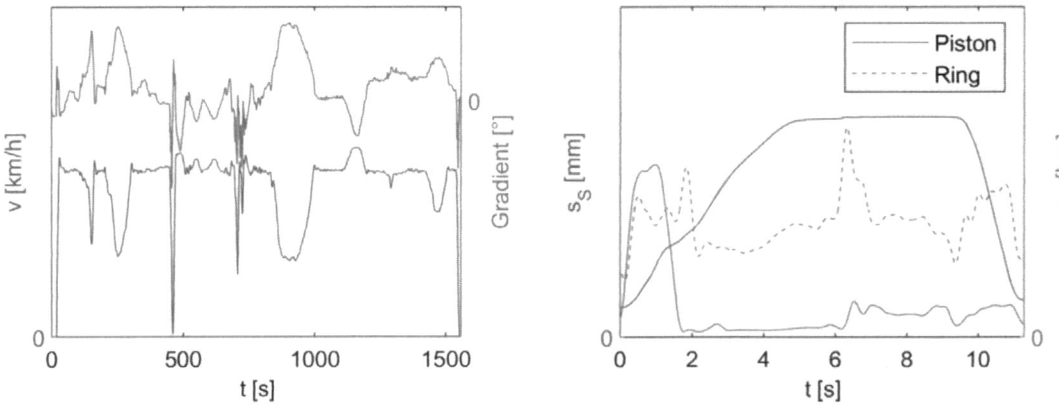

Figure 2. Excavator velocity and track gradient of a driving section (left) and stroke and pressure of the stick cylinder in the 90° working cycle (right)

2.3 *Design Parameters and Objectives*

The approach of this paper is applied to the optimal sizing of the outlined subsystems in Figure 1 while respecting the defined load and motion cycles of the consumers. This concerns the optimal selection of the supply design parameters. Here, the design parameters of the travel drive are the installed EM and power electronics (PE) power as well as the gear ratio of the planetary hub gearboxes. The design parameters of the working drive are the installed EM and PE power and the maximum displacement of the pump. The constant speed operation of the pump is adopted from the conventional system, however, the speed is defined as a design parameter. System costs and energy consumption are considered as the optimization objectives.

3. Materials and Methods

In this section, the proposed approach for optimal component sizing is presented. The workflow of the approach as well as the necessary steps are shown in Figure 3.

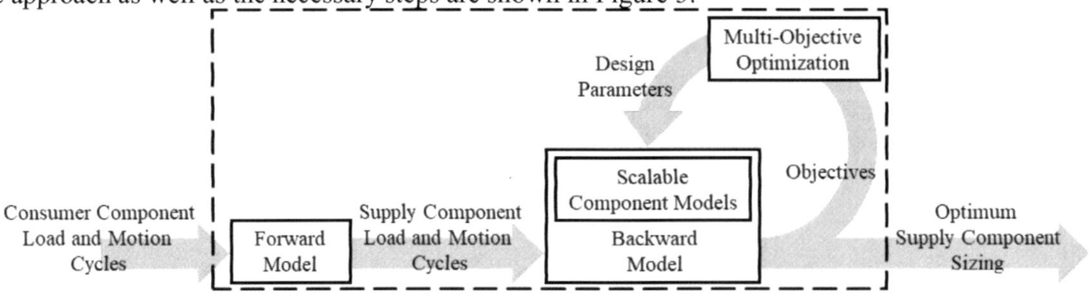

Figure 3. Approach for optimization-based supply component sizing.

On the input side, load and motion cycles (Figure 2) of the consumers are fed into a forward model. These cycles contain the design-relevant operating scenarios of the system and were obtained from measurements in the framework of the presented investigations. Alternatively, consumer cycles can be scaled from similar applications or fully synthesized. The forward model is a transient simulation model that represents the topology to be investigated. It is used to calculate the loads and motions of the supply components to be dimensioned while respecting the consumer cycles. The supply component cycles along with the design parameters form the input variables of the backward model. In contrast to the forward model, it is quasi-static and thus more computationally efficient. This property significantly reduces the computational effort as it is evaluated numerously during the following optimization process. It consists of scalable supply component models that can be varied in their dimensions by means of the design parameters. These models describe the costs and the energy consumption of the components based on the design parameters considering the supply cycles. The costs and the energy consumption are the objective functions that are minimized in the following multi-objective optimization. The optimization algorithm iteratively evaluates the values of the objective functions and selects new design parameter sets. After this iteration, optimal component specifications are available.

3.1 *Forward Model*

The forward model is a 1D dynamic simulation model built in the MATLAB/Simscape [29] environment from the hydraulic and mechanical library. This model of the electrified wheeled excavator assuming ideal power sources is used to determine the supply component load and motion cycles, which are shown in Figure 4.in the form of mechanical and hydraulic loads, from the specified consumer cycles.

In the case of the bucket and stick cylinder drive, an ideal pressure source supplies the system. The pressure of this source is set by the current maximum consumer load and offset by an LS pressure. The cylinder strokes are controlled by proportional valves which are modeled by variable orifices. An operator model, implemented as a PI-controller, transforms the difference between the actual and the desired cylinder stroke into a valve actuation. The so released volumetric flow leads to a pressure buildup in the cylinder chambers, which is considered by the pressure buildup equation. To calculate the cylinder motion, a force balance between pressure and external forces is performed at the cylinders. The external forces consist of inertial forces of the cylinder piston and rod as well as its attachments, friction, digging and weight forces. The sum of the external forces is derived from the cylinder chamber pressure difference in the reference vehicle measurements. The pressure and flow rate provided by the ideal power source is used in the following backward model to identify optimal component specifications of the hydraulic pump, EM, and PE.

In the case of the travel drive the excavator is modeled as a single mass under the influence of propulsion torques from ideal torque sources and resistances from the track gradient, rolling resistances, air drag resistance and external forces (e.g. soil forces during grading). Here the operator model, likewise implemented as a PI-controller, transforms the difference between the actual and the desired excavator velocity into propulsion torques. The wheel torque and speed are used for the component sizing of the single-wheel drive. The supply component load and motion cycles for the travel and working drive determined in the forward model are shown in Figure 4.

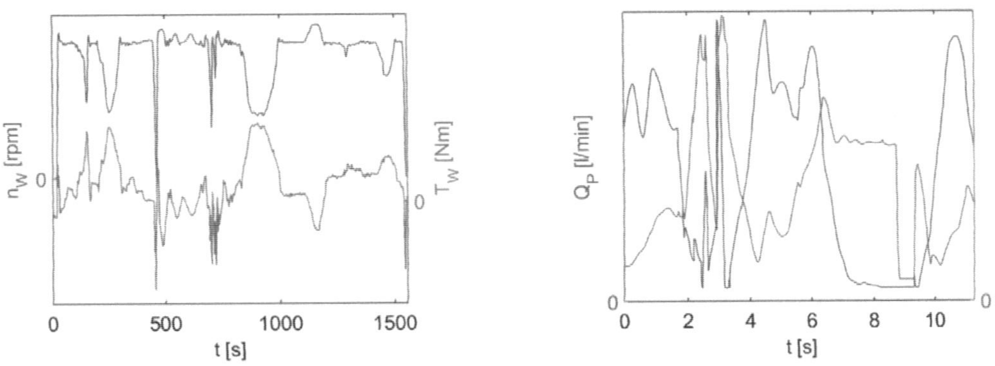

Figure 4. Supply component load and motion cycles. Left: Wheel torque T_W and speed n_W. Right: Pump flow rate Q_P and pressure p_P.

3.2 *Backward Model*

A backward model consists of scalable component models. Based on the design parameters, it describes the costs and, considering the component loads, the energy consumption of a component taking into account the energy losses. In the optimization process, backward models are evaluated numerously and are therefore modeled in a quasi-static approach to reduce computational effort. Losses are thus described by constant factors, maps or analytical equations. The level of detail of the models is selected according to the authors' available database (parameters) of the components.

3.2.1 *Scalable component efficiency models*

Electric motor and power electronics

To evaluate the energy efficiency of the electrical system consisting of EM and PE, a model is used that calculates the electrical energy demand (objective) at a given mechanical load cycle (input from forward model) as a function of the installed electrical power (design parameter). Therefore, a precalculated reference efficiency map for a power class of 60 kW, which takes into account the operation point dependent losses of both components, is used. The map is available for motor operation in the first quadrant and is adopted to the remaining quadrants under the assumption of symmetrical loss behavior. The design parameter of the EMs and PEs is their power. For the scaling of the electrical system, the torque axis of the reference map is stretched or squeezed and thus the loss behavior is adapted to the required power class. By evaluating the mechanical operating points in the efficiency map, the efficiency of the electrical system $\eta_{EM,PE}$ and ultimately electrical power is obtained. The reference map used is shown in Figure 5. The highest overall efficiency is achieved in the low power range.

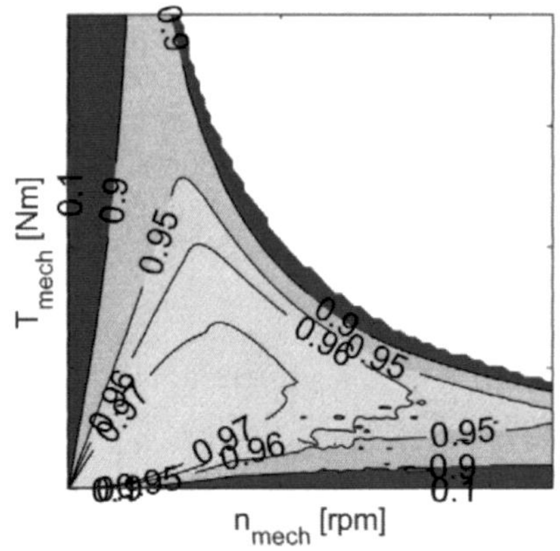

Figure 5. Reference efficiency map of the electrical system consisting of EM and PE depending on the mechanical operating point of the EM.

Hydraulic pump

To describe the efficiency of the hydraulic pump, efficiency maps of a complete product series were available to the authors. On the basis of the known efficiency maps, parameters were extracted that enabled the use of a more detailed analytical efficiency model, which is described in the following. According to Equation (1) the total efficiency of a hydraulic pump η_{tot} is the product of the volumetric η_{vol} and the hydraulic-mechanical efficiency η_{hm}.

$$\eta_{tot} = \eta_{vol} \cdot \eta_{hm} \quad (1)$$

The volumetric efficiency is the ratio of the effective Q_{eff} and theoretical volumetric flow rate Q_{th} (Equation (2)). The effective volumetric flow is smaller than the ideal volumetric flow by a leakage flow Q_{leak}. Here, α is the relative displacement, V_0 the maximum displacement in cm^3 and the speed of the pump n_P in rpm.

$$\eta_{vol} = \frac{Q_{eff}}{Q_{th}} = \frac{Q_{th} - Q_{leak}}{Q_{th}} = \frac{\alpha \cdot V_0 \cdot n_P - Q_{leak}}{\alpha \cdot V_0 \cdot n_P} \quad (2)$$

The leakage flow Q_{leak} in $m^2 \cdot s^{-1}$ is calculated using Equation (3). K_{HP} is the Hagen-Poiseuille coefficient in $m^3 \cdot s^{-1} \cdot Pa^{-1}$ and Δp the pressure difference of the pump ports in Pa.

$$Q_{leak} = K_{HP} \cdot \Delta p \tag{3}$$

The hydraulic-mechanical efficiency is the ratio of theoretical T_{th} and effective torque T_{eff} at the pump shaft (Equation (4)). The effective torque is higher than the theoretical torque by a friction torque T_{fric}.

$$\eta_{hm} = \frac{T_{th}}{T_{eff}} = \frac{T_{th}}{T_{th} + T_{fric}} = \frac{\alpha \cdot V_0 \cdot \Delta p}{\alpha \cdot V_0 \cdot \Delta p + 2 \cdot \pi \cdot T_{fric}} \tag{4}$$

The total frictional torque T_{fric} in Nm is composed of a constant component T_0 and a pressure-dependent component T_p (Equation (5)). Here, K_{TP} describes the proportionality factor for determining the pressure-dependent component in $N \cdot m \cdot Pa^{-1}$.

$$T_{fric} = T_0 + T_p = T_0 + K_{TP} \cdot \Delta p \tag{5}$$

A parameter estimation approach is used to evaluate five efficiency maps of a pump model series and fitted the parameters of the loss model described above to satisfy the efficiency maps. The parameters obtained are depicted in Figure 6 as a function of the corresponding maximum displacement. In the optimization, parameters of pump sizes which are located between the available data are determined by interpolation using neighboring points. Pump sizes beyond the boundary values are not considered. The model allows the loss behavior of hydraulic pumps to be continuously scaled using the displacement as a design parameter.

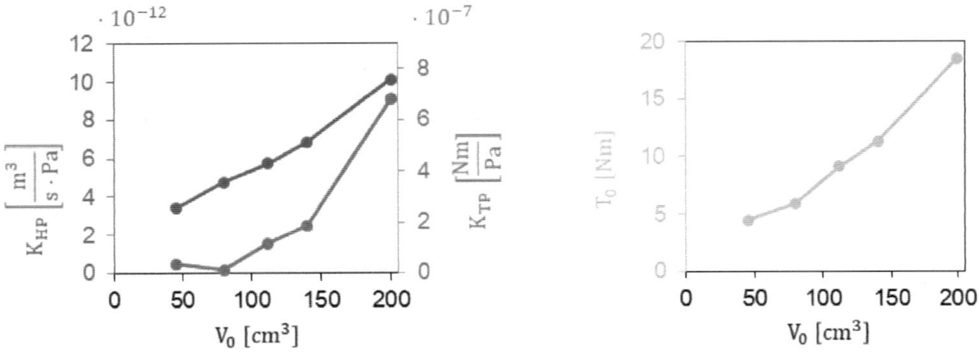

Figure 6. Parameters of the efficiency model of a pump series as a function of the maximum displacement volume. Left: K_{HP} and K_{TP}. Right: T_0.

The efficiency model is used to calculate the mechanical operating points of the connected electric motor based on the hydraulic operating points determined in the forward model. Figure 7 shows an exemplary overall efficiency map of a 200 cm^3 pump at 1,800 rpm which was derived from the analytical model. The highest overall efficiency is achieved at maximum displacement and a mid-range pressure level.

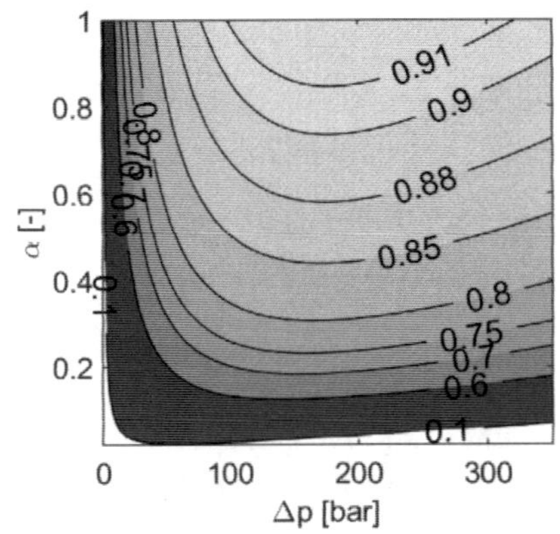

Figure 7. Exemplary overall efficiency map of a 200 cm^3 pump at 1,800 rpm

Planetary hub gearbox

The design parameter of the gearbox is the gear ratio. Thus, a model is required that determines the efficiency of the gearbox over the selected ratio. In a first approach, the efficiency of the gearbox is modeled by a fixed efficiency depending on the selected gear ratio. For an economic design of gearboxes, Niemann, Winter [30] identified a maximum ratio of $i_{St} = 6$ for a single-stage, $i_{St} = 35$ for a two-stage and $i_{St} = 150$ for a three-stage gearbox. For each gear stage, an efficiency of $\eta_{Pl} = 0.981$ is assumed for a planetary gearbox at its rated operating point [30]. This results in the equation shown in Equation (6) for the efficiency of the gearbox η_G. More detailed models to calculate the gearbox component and operating point specific losses such as bearing and gear losses both load dependent and independent can be used in future works [31].

$$\eta_G = \begin{cases} \eta_{Pl} & \text{if} \quad i_{St} \leq 6 \\ \eta_{Pl}^2 & \text{if } 6 < i_{St} \leq 35 \\ \eta_{Pl}^3 & \text{if} \quad i_{St} > 35 \end{cases} \tag{6}$$

3.2.2 *Scalable Cost Models*

Electric motor, power electronics and hydraulic pump

The costs of EM C_{EM}, PE C_{PE} and hydraulic pumps C_{Hyd} are described in € using linear approaches according to Equations (7), (8) and (9). The costs of EM and PE result from the product of their powers P_{EM} and P_{PE} in kW and the respective constant cost factors k_{EM} and k_{PE} in $€ \cdot kW^{-1}$. The costs of hydraulic pumps scale with the constant cost factor k_{Hyd} in $€ \cdot cm^{-3}$ and the maximum displacement V_0 in cm^3. The cost factor of EM and PE is assumed as $k_{EM} = 15\,€ \cdot kW^{-1}$ and $k_{PE} = 16\,€ \cdot kW^{-1}$ respectively according to [22]. For the combination of EM and PE, other sources give similar numerical values ($30\,€ \cdot kW^{-1}$ [11], [12]). Due to confidentiality, the cost factor for the hydraulic pumps is omitted.

$$C_{EM} = k_{EM} \cdot P_{EM} \tag{7}$$
$$C_{PE} = k_{PE} \cdot P_{PE} \tag{8}$$
$$C_{Hyd} = k_{Hyd} \cdot V_0 \tag{9}$$

Planetary Hub Gearbox

For the calculation of the gearbox costs, there are approaches in the literature that are based on the gearbox mass [32] as well as approaches that are based on the number of required components [33]. In order to be able to consider different gearbox costs within a selected number of gear stages depending on the transmitted torque, a mass-based approach is chosen in the context of this contribution. To estimate the gear mass, a mass factor $f_{G,mass}$ is determined which allows the mass of a reference gearbox to be scaled to other gear ratios. This mass factor was derived in [34] and used in [25] for a single-stage planetary gearbox. It can be determined according to Equation (10) where N is the number of planets and i_{Pl} is the overall gear ratio of a single planetary gear stage. The derivation of the equation is based on the assumption of a maximum allowable surface durability factor of the gears [34].

$$f_{G,mass} = \frac{1}{N} + \frac{2}{N \cdot (i_{Pl} - 2)} + \frac{i_{Pl} - 2}{2} + \left(\frac{i_{Pl} - 2}{2}\right)^2 + \frac{0{,}8 \cdot (i_{Pl} - 1)^2}{N \cdot (i_{Pl} - 2)} + \frac{0{,}4 \cdot (i_{Pl} - 1)^2}{N} \tag{10}$$

For a reference gearbox, the mass factor can be used to scale the gearbox' mass according to Equation (11), where i_{Ref} is the overall gear ratio of the reference gearbox.

$$M_{G,mass|i=i_{Pl}} = \frac{f_{G,mass|i=i_{Pl}}}{f_{G,mass|i=i_{Ref}}} \cdot M_{G,mass|i=i_{Ref}} \tag{11}$$

The mass estimation of a single-stage planetary gearbox is not sufficient for a planetary hub gearbox and was therefore extended to two- and three-stage planetary gearboxes. For the extension to a two- or three-stage gearbox, the total gear ratio is assumed as equally divided among all stages. A separate mass factor is determined for each stage. The additional ratio per stage and the additional mass per stage are used from catalogue references of a planetary hub gearbox manufacturer [35]. The parameters used for the reference gearboxes are summarized in Table 1.

Table 1. Parameters of the reference planetary hub gearboxes [35].

	First Stage	Second Stage	Third Stage
Additional Gear Ratio	6.09	6.19	4.51
Additional Weight	30 kg	56 kg	39 kg

A relative weight cost of 14.25 € · kW^{-1} according to [32] is assumed. The nonlinear shape of the cost function for two- and three-stage planetary hub gearboxes is shown in Figure 8. The costs of a gearbox C_G are determined by an evaluation of the cost function for a given gear ratio.

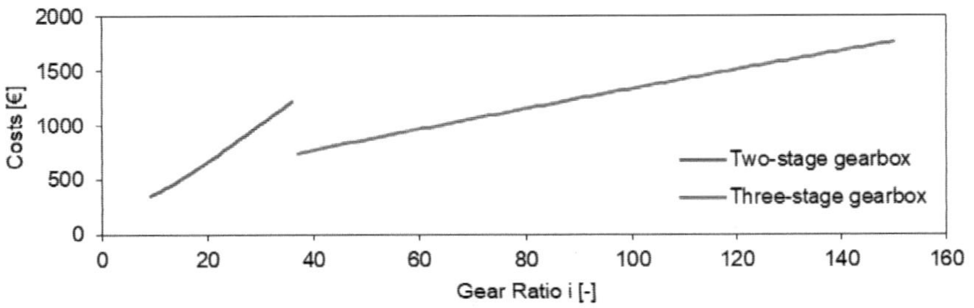

Figure 8. Cost function for planetary hub gearboxes

Gearbox costs are calculated by estimating the gear mass. In addition to the pure gearbox mass, other factors such as the number of gear box components installed have an influence on the costs. Taking into account the number of gearbox components would result in a different cost leap between a two-stage and a three-stage gearbox design. The three-stage gearbox has less gear mass due to smaller gear diameters, but more components. The interchangeability of the component models allows a more detailed cost model for the gearbox to be added at a later point in time and the influence on the component specification to be investigated.

3.2.3 *Model Structure*

The scalable component efficiency and cost models are combined to backward models of the travel and working drive subsystems (Figure 9). The model inputs are the precalculated supply component cycles ($T_W(t), n_W(t)$ and $p_P(t), Q_P(t)$) and the design parameters (i, V_0, P_{EM}, P_{PE}). To expand the solution space the constant pump speed n_P is treated as a design parameter as well. The model calculates and outputs the costs (C) and the electric energy consumption (E_{El}) of the subsystem.

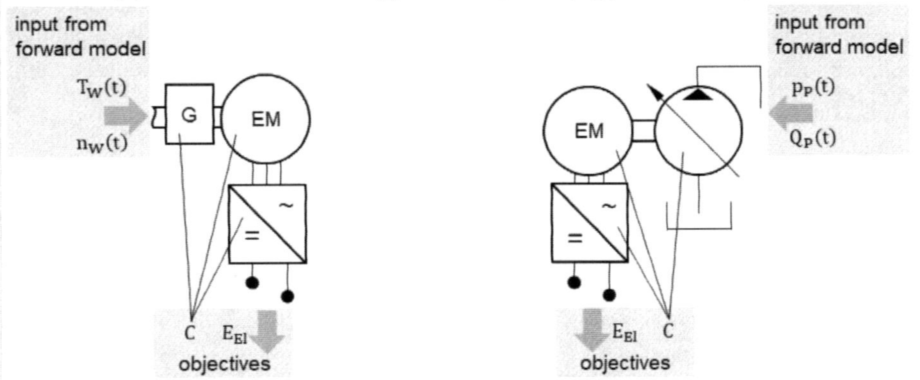

Figure 9. Structure of travel (left) and working drive (right) backwards models.

The costs of a subsystem are the sum of the component costs (see Section 3.2.2) and calculated using Equation (12).

$$C = \sum_i C_i \tag{12}$$

The consumption of electrical energy is determined by integration of electrical power, which is given by Equation (13).

$$E_{El} = \int P_{El}(t)dt \tag{13}$$

The electrical power is evaluated along the efficiency path (see Section 3.2.1) of the respective subsystem (Equations (14) and (15)). T_P and n_P are the torque and speed of the pump shaft respectively.

$$P_{El,travel}(t) = \eta_{EM,PE}(T_W, n_W) \cdot \eta_G \cdot 2\pi \cdot T_W \cdot n_W \tag{14}$$

$$P_{El,working}(t) = \eta_{EM,PE}(T_P, n_P) \cdot \eta_P(p_P, Q_P) \cdot p_P \cdot Q_P \tag{15}$$

3.2.4 *Multi-Objective Optimization*

To find an optimal component specification, a multi-objective optimization problem is formulated. J is the multi-objective optimization function consisting of the costs and the electrical energy consumption of the respective drive.

$$J := [C, E_{El}] \tag{16}$$

$$\min_x J \tag{17}$$

x is the optimization vector. For the travel drive it is $x_{travel} = (P_{El}, P_{PE}, i)$ and for the working drive $x_{working} = (P_{El}, P_{PE}, V_0, n_P)$.

In preference to combining the objective functions for costs and energy consumption Pareto optimization is used to find trade-offs between the two objectives. As the algorithm a state-of-the-art implementation `paretosearch` in MATLAB is used [29].

Stopping criteria are defined for the optimization, which discard unsuitable solutions. In the case of the working drive, for example, these are solutions that exceed the comparatively low maximum speeds of the pump or cannot provide the required flow rate due to low pump speed and maximum displacement. For both the travel drive and the working drive, exceeding the maximum torque of the EM is not permitted.

4. Results

In the following section, the results of the multi-objective optimization are presented and discussed. The results are divided into travel and working drive. Since no weighting of costs and energy consumption was made, the results are presented as Pareto fronts.

4.1 *Travel Drive*

Figure 10 shows the optimization results of the single-wheel drive system. Each circle represents one set of design parameters whose energy consumption and costs are displayed against each other. The size of the circles reflects the installed electrical power, with larger circles representing a higher installed power. The color of the circles represents the required gear ratio.

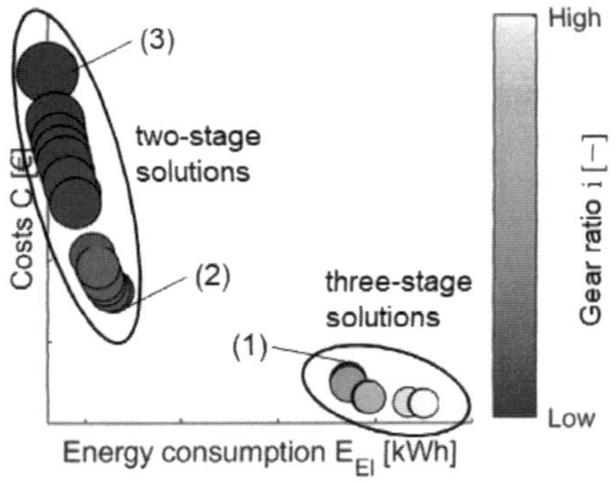

Figure 10. Result of the multi-objective optimization visualized as a Pareto front for the travel drive. The size of a circle is proportional to the installed electrical power of a solution. The color represents the gear ratio.

For the travel drive, two separate Pareto fronts are obtained, which result from the discontinuous cost and efficiency functions of the planetary hub gearboxes. The first Pareto front at the top left of the diagram is characterized by high relative costs and low energy consumption. Within this Pareto front, only design parameters of two-stage planetary hub gearboxes are present. Variants with high installed electrical power lead to higher costs. Due to more efficient operating points in the efficiency map of the EM and PE, these variants have a lower energy consumption. The second Pareto front, which is characterized by low costs and a comparatively high energy consumption, shows only variants that utilize a three-stage planetary hub gearbox. The current gearbox cost model considers the material mass only, so that if more extensive influences are taken into account in the future, the results may vary. The results indicate that the costs for providing the required high torque by means of a third gear stage are lower than implementing a more powerful electric motor. However, the lower energy efficiency due to the additional gear stage is disadvantageous. A permission of additional system costs of 26% can reduce the energy consumption of the most efficient three-stage solution (1) to the most cost-effective two-stage solution (2) by 3%. A further 36% increase in cost between the most cost-effective two-stage solution (2) and the most efficient two-stage solution (3) improves energy consumption by only 1%.

4.2 *Working Drive*

Figure 11 shows the results of the multi-objective optimization for the working drive. In the left plot the size of the circles represents the installed electrical power while the color of the circles represents the maximum displacement of the hydraulic pump. In the right plot the circle color represents the speed of the hydraulic pump while their size provides no further information.

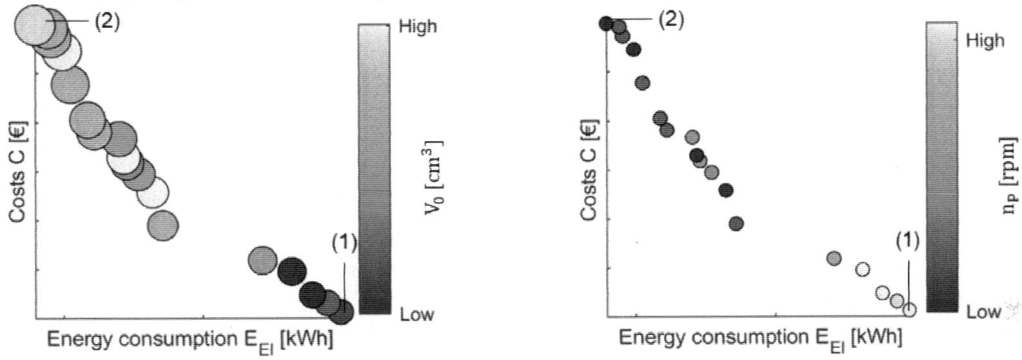

Figure 11. Result of the multi-objective optimization visualized as a Pareto front for the working drive. Left: The size of a circle is proportional to the installed electrical power of a solution. The color represents the required pump displacement. Right: The circle color represents the required pump speed.

Within the Pareto front, a gap is apparent, but unlike the travel drive, it is not due to a discontinuous component model. In fact, it shows a region where no solution satisfies the boundary conditions of the system. In the left plot of Figure 11 a clear correlation can be seen between the installed electrical power and the objectives cost and energy consumption. As the installed electrical power increases, the relative costs of the working drive increase and the energy consumption decreases due to more efficient operating points of the EM, PE and hydraulic pump. More efficient solutions utilize higher maximum pump displacements. In addition, they are operated at lower speeds (Figure 11, right). If 68% higher system costs of the most cost-effective solution (1) shown are permitted, the energy consumption can be decreased by 3.2% to the most energy efficient solution (2) shown.

5. Conclusion and Outlook

Due to stricter legislation, the demand for zero-emission drive systems in the field of mobile machinery will increase today and in the future. One possibility for operating construction machinery in zero emission zones is the use of battery-electric drive systems. Compared to the automotive sector, however, the high power requirements of mobile machinery over long periods of time lead to high energy demands, which are reflected in heavy, large and expensive batteries. The efficiency of the drive concept consisting of topology, component sizing and control strategy thus moves to the center of the development.

In order to answer the question of how an efficiency- and cost-appropriate drive concept can be found, this paper presented a novel optimization-based method that allows the identification of an optimal component sizing for a given topology. The novelty of the presented method results from the combination of a computationally intensive and therefore for optimization methods unsuitable transient dynamic simulation, which makes it possible to consider the interaction of parallelly operated consumers in the component specification, and a performant quasi-static simulation, which is necessary to apply known optimization methods for the component sizing. The component models used in the simulations are characterized by scalability in the design parameter space and are arbitrarily interchangeable and extendable in the level of detail. Using a mobile excavator as an example system, the design parameter space was defined first. Corresponding to the selected design parameters, suitable models for the description of the objective criteria energy consumption and costs for the components EM, PE, hydraulic pump and planetary gearbox were presented. In a transient dynamic simulation, component loads were derived from measured load and motion cycles. These component loads serve as input for backward models, which implement the objective functions for a multi-objective optimization using scalable component models. Based on the results of the multi-objective optimization, which are presented in the form of Pareto fronts, a quantifiable trade-off of the objective criteria energy consumption and costs was derived. This trade-off helps the OEM to identify suitable component specifications in early stages of development by determining the additional costs they are prepared to accept in order to achieve an increase in efficiency. Using the example of the travel drive, it was possible to quantify the influence of the number of gear stages on the manufacturing costs and the energy consumption. By using a two-stage planetary gearbox paired with a more powerful EM, a 3% reduction in energy consumption can be achieved with additional costs of 26%. Using the example of the working drive, a reduction in energy consumption of 3.2% for additional costs of 68% could be demonstrated.

In the future, the presented approach and given framework should be extended and detailed. For this, the cost and efficiency models can be refined to provide an expansion of the current design parameter space. Using the example of the of the planetary gearbox a predimensioning can be added to the approach. This would expand the parameter space with quantities such as number of teeth, tooth module as well as shaft (diameter, length) and bearing (diameter, type) properties. In addition to higher accuracy in the mass estimation, analytical loss models can be used to calculate the gearbox component and operating point specific losses such as bearing and gear losses both load dependent and independent. The sizing of the EM can be extended to the specification of the number of pole pairs as well as the length and diameter of the rotor and stator. The efficiency model can be expanded to more fidelity using the well-known Willans approach. The presented method thus can be adapted to the specific development task as needed and can be used in different phases of the product development.

6. Acknowledgements

The authors thank Liebherr-Hydraulikbagger GmbH for providing data, insight and expertise. The project is funded by the Federal Ministry for Economic Affairs and Energy (BMWi).

Supported by:

Federal Ministry
for Economic Affairs
and Energy

on the basis of a decision
by the German Bundestag

IOP Conf. Series: Materials Science and Engineering 1097 (2021) 012002 doi:10.1088/1757-899X/1097/1/012002

Abbreviations

EM	Electric motor
GB	Gear box
OEM	Original equipment manufacturer
PE	Power electronics

Nomenclature

α	Relative displacement	–
η_{EM+PE}	Efficiency of the electric system	–
η_{G}	Efficiency of the gearbox	–
η_{hm}	Hydraulic-mechanical efficiency	–
η_{Pl}	Efficiency of a single planetary gear stage	–
η_{tot}	Total efficiency	–
η_{vol}	Volumetric efficiency	–
C	Costs of a subsystem	€
C_{EM}	Costs of electric motor	€
C_{G}	Costs of gear box	€
C_{Hyd}	Costs of hydraulic pumps	€
C_{i}	Costs of a component	€
C_{PE}	Costs of power electronics	€
E_{El}	Energy consumption	kWh
$f_{G,mass}$	Gear mass factor	–
i_{Pl}	Overall gear ratio of a single planetary gear stage	–
i_{St}	Gear ration of a gear stage	–
J	Multi-objective optimization function	–
k_{EM}	Costs factor of electric motor	$€ \cdot kW^{-1}$
K_{HP}	Hagen-Poiseuille coefficient	$m^3 \cdot (s \cdot Pa)^{-1}$
k_{Hyd}	Costs factor of hydraulic pumps	$€ \cdot cm^{-3}$
k_{PE}	Costs factor of power electronics	$€ \cdot kW^{-1}$
K_{TP}	Proportionality factor for pressure-dependent frictional torque	$Nm \cdot Pa$
$M_{G,mass}$	Gear mass	kg
N	Number of planets	–
n_{mech}	Rotational speed of the electric engine	rpm
n_{P}	Rotational speed of the pump	rpm
n_{W}	Rotational speed of the wheel	rpm
p	Pressure	bar
Δp	Pressure difference	bar
P_{El}	Power demand	kW
P_{EM}	Installed power of the electric motor	kW
p_{L}	Load pressure	bar
p_{P}	Pressure at the pump	bar
p_{S}	Stick cylinder pressure	bar
P_{PE}	Installed power of the power electronics	kW
Q_{eff}	Effective volumetric flow rate	$l \cdot min^{-1}$
Q_{leak}	Leakage flow rate	$l \cdot min^{-1}$
Q_{P}	Volumetric flow rate of the pump	$l \cdot min^{-1}$
Q_{th}	Theoretical volumetric flow rate	$l \cdot min^{-1}$
s_{S}	Stick cylinder stroke	mm
T_{0}	Constant frictional torque of the pump	Nm
T_{eff}	Effective torque of the pump	Nm

T_{fric}	Frictional torque of the pump	Nm
T_{mech}	Output torque of the electric engine	Nm
T_p	Pressure-dependent frictional torque of the pump	Nm
T_{th}	Theoretical torque of the pump	Nm
T_W	Torque at the wheel	Nm
v	Vehicle velocity	$km \cdot h^{-1}$
V_0	Maximum displacement	cm^3
x	Optimization vector	–

References

[1] Communication from the Commission to the European Parliament, the European Council, the Council, the European Economic and Social Committee and the Committee of the Regions 2019 *The European Green Deal*

[2] Committee for European Construction Equipment 2017 *CECE Position on European Low Emission Zones and retrofitting*

[3] Kunze G 2010 Mobile Construction Machinery - Trends and New Developments *ATZ Offhighway* **3** 6–22

[4] Silvas E, Hofman T, Murgovski N, Etman P and Steinbuch M 2016 Review of Optimization Strategies for System-Level Design in Hybrid Electric Vehicles *IEEE Trans. Veh. Technol.* 1

[5] Alparslan Emrah Bayrak, Yi Ren and Panos Y. Papalambros 2016 Topology Generation for Hybrid Electric Vehicle Architecture Design *Journal of Mechanical Design*

[6] Alparslan Emrah Bayrak, Yi Ren, Panos Y. Papalambros Design of Hybrid-Electric Vehicle Architectures Using Auto-Generation of Feasible Driving Modes *Proceedings of the ASME 2013 International Design Engineering Technical Conferences andComputers and Information in Engineering Conference*

[7] Bertram C and Herzog H-G 2013 - 2013 Optimization method for drive train topology design and control of electric vehicles *2013 World Electric Vehicle Symposium and Exhibition (EVS27) 2013 27th International World Electric Vehicle Symposium and Exhibition (EVS27) (Barcelona, 17.11.2013 - 20.11.2013)* (IEEE) pp 1–8

[8] Silvas E, Hofman T, Serebrenik A and Steinbuch M 2015 Functional and Cost-Based Automatic Generator for Hybrid Vehicles Topologies *IEEE/ASME Trans. Mechatron.* **20** 1561–72

[9] Zhou X, Qin D and Hu J 2017 Multi-objective optimization design and performance evaluation for plug-in hybrid electric vehicle powertrains *Applied Energy* **208** 1608–25

[10] Desai C and Williamson S S 2009 - 2009 Optimal design of a parallel Hybrid Electric Vehicle using multi-objective genetic algorithms *2009 IEEE Vehicle Power and Propulsion Conference 2009 IEEE Vehicle Power and Propulsion Conference (VPPC) (Dearborn, MI, 07.09.2009 - 10.09.2009)* (IEEE) pp 871–6

[11] Dimitrova Z and Maréchal F 2016 Techno–economic design of hybrid electric vehicles and possibilities of the multi-objective optimization structure *Applied Energy* **161** 746–59

[12] Ebbesen S 2012 Optimal sizing and control of hybrid electric vehicles ETH Zurich

[13] Finesso R, Spessa E and Venditti M 2016 Cost-optimized design of a dual-mode diesel parallel hybrid electric vehicle for several driving missions and market scenarios *Applied Energy* **177** 366–83

[14] Millo F, Zhao J, Rolando L, Cubito C and Fuso R 2015 Optimizing the design of a plug-in hybrid electric vehicle from the early phase: an advanced sizing methodology *Computer-Aided Design and Applications* **12** 22–32

[15] Zhou Q, Zhang W, Cash S, Olatunbosun O, Xu H and Lu G 2017 Intelligent sizing of a series hybrid electric power-train system based on Chaos-enhanced accelerated particle swarm optimization *Applied Energy* **189** 588–601

[16] Pourabdollah M, Egardt B, Murgovski N and Grauers A 2018 Convex Optimization Methods for Powertrain Sizing of Electrified Vehicles by Using Different Levels of Modeling Details *IEEE*

Trans. Veh. Technol. **67** 1881–93
[17] Guzzella L and Sciarretta A 2013 *Vehicle Propulsion Systems* (Berlin, Heidelberg: Springer Berlin Heidelberg)
[18] Gao W and Mi C 2007 Hybrid vehicle design using global optimisation algorithms *IJEHV* **1** 57
[19] Alan C, Ali S, Alaa H and Eric B 2014 - 2014 Optimal sizing of an energy storage system for a hybrid vehicle applied to an off-road application *2014 IEEE/ASME International Conference on Advanced Intelligent Mechatronics 2014 IEEE/ASME International Conference on Advanced Intelligent Mechatronics (AIM) (Besacon, 08.07.2014 - 11.07.2014)* (IEEE) pp 775–80
[20] Sim K, Vijayagopal R, Kim N and Rousseau A 2019 Optimization of Component Sizing for a Fuel Cell-Powered Truck to Minimize Ownership Cost *Energies* **12** 1125
[21] Yi H-S, Jeong J-B, Cha S-W and Zheng C-H 2018 Optimal component sizing of fuel cell-battery excavator based on workload *Int. J. of Precis. Eng. and Manuf.-Green Tech.* **5** 103–10
[22] Verbruggen F J R, Silvas E and Hofman T 2020 Electric Powertrain Topology Analysis and Design for Heavy-Duty Trucks *Energies* **13** 2434
[23] Kolodziejak DPH, Pham T H, Hofman T and Wilkins S 2019 An Optimization and Analysis Framework for TCO Minimization of Plug-in Hybrid Heavy-duty Electric Vehicles *IFAC-PapersOnLine* **52** 484–91
[24] Wei D, He H and Cao J 2020 Hybrid electric vehicle electric motors for optimum energy efficiency: A computationally efficient design *Energy* **203** 117779
[25] Kim D-M, Benoliel P, Kim D-K, Lee T H, Park J W and Hong J-P 2019 Framework Development of Series Hybrid Powertrain Design for Heavy-Duty Vehicle Considering Driving Conditions *IEEE Trans. Veh. Technol.* **68** 6468–80
[26] Fries M, Lehmeyer M and Lienkamp M 2017 *Multi-criterion optimization of heavy-duty powertrain design for the evaluation of transport efficiency and costs* (Piscataway, NJ: IEEE)
[27] Swagata Borthakur and Shankar C. Subramanian 2018 Optimized Design and Analysis of a Series-Parallel Hybrid Electric Vehicle Powertrain for a Heavy Duty Truck
[28] Wang E, Di Guo and Yang F 2015 System design and energetic characterization of a four-wheel-driven series–parallel hybrid electric powertrain for heavy-duty applications *Energy Conversion and Management* **106** 1264–75
[29] Mathworks *Entwickler von MATLAB und Simulink* https://de.mathworks.com/ (accessed 2 Feb 2021)
[30] Niemann G and Winter H (eds) 2003 *Maschinenelemente: Band 2: Getriebe allgemein, Zahnradgetriebe - Grundlagen, Stirnradgetriebe* 2nd edn (Berlin, Heidelberg: Springer Berlin Heidelberg)
[31] Habermehl C, Jacobs G and Neumann S 2020 A modeling method for gear transmission efficiency in transient operating conditions *Mechanism and Machine Theory* **153** 103996
[32] Naunheimer H, Bertsche B, Ryborz J and Novak W 2011 *Automotive Transmissions: Fundamentals, Selection, Design and Application* 2nd edn (Berlin, Heidelberg: Springer Berlin Heidelberg)
[33] Reick B Methode zur Analyse und Bewertung von stufenlosen Traktorgetrieben mit mehreren Schnittstellen *Dissertation* Karlsruher Institut für Technologie (KIT)
[34] Rejman E. 2011 Gears Weight Equations - Gear Chain Weight Calculation Methodology *Scientific Proceedings VIII International Congress "Machines, Technologies,Materials"*
[35] Messerer H 2020 *Reggiana Riduttori General Catalogue: Wheel Gears* http://www.reggianaridutt.it/pdf/c310.pdf

Role model of model-based systems engineering application

Iris Gräßler, Dominik Wiechel and Jens Pottebaum

Paderborn University – Heinz Nixdorf Institute, Fürstenallee 11, 33102 Paderborn

dominik.wiechel@hni.upb.de

Abstract. The transition to model-based systems engineering (MBSE) is essential to meet the demands of increasing system complexity, productivity, quality and shorter design cycles. Necessary perspectives, associated roles and competencies for the application of MBSE are not sufficiently defined. The paper at hand presents a role model for the practical application of MBSE based on three major steps: (1) identification, (2) categorization and (3) association of stakeholders, which implement and apply MBSE within an enterprise. The use of the role model is demonstrated and evaluated in a test case based on personas within a fictional scenario which illustrates the implementation of MBSE. The roles, which are adapted to today's conditions, serve as a practical guide comprising all basic conditions that have to be considered when applying MBSE. Practitioners can easily tailor the illustrated results to their own use case.

1. Introduction

Trends like digitalization and increasing system complexity lead to changes in working environment, engineering approaches and required competencies [1–4]. Systems Engineering (SE) practically represents the most relevant approach to handle these challenges. Combining SE with the transition from a document-centric to a model-based approach leads to Model-based Systems Engineering (MBSE). It is essential for Systems Engineering to meet the demands of increasing system complexity, productivity by the engineering team and within the desired quality of the system model [5]. The extending demand for MBSE implies the requirement of specified roles to define the activities and tasks, which have to be performed to implement and apply MBSE. Without corresponding guidelines, the implementation and application of these theoretical approaches fail due to the lack of necessary environmental conditions. An essential element of guidelines to support the implementation is an applicable Role Model. Roles represent and define activities and tasks to be taken into account for performing Systems Engineering throughout the entire system life cycle [6]. Two crucial Lessons Learned from the application of MBSE are [7]: (a) The best way to start modeling is to hire people who already know how to do it and (b) everyone needs training, but not to the same depth. The research question to be raised is stated as follows: Which participating actors and competencies can be defined to enable a successful MBSE application and adaption?

The aim of this research paper is to answer the research question and provide guidance to practitioners in implementing MBSE by defining a Role Model for Model-Based Systems Engineering adapted to the specific circumstances of MBSE. Answering the state problem is important to bridge the gap between the theoretical power of MBSE and the challenges of implementation. For this purpose, results from literature as well as from research projects and industrial applications were analyzed and supplemented to define and adjust a Role Model for Model-Based Systems Engineering Application. The Role Model is an incremental part for defining the theory and supporting the application of MBSE.

2. State of the Art

2.1. Systems Engineering
According to Broy [8], a system is divided in Sub-Systems which interact with one another and can be viewed as a whole that interacts with its external environment to achieve an objective. A practically relevant structured approach to handle the complexity of the development process is Systems Engineering. Systems Engineering is defined as a structured cross-disciplinary approach for the development of complex technical systems to achieve a cross-disciplinary optimum in a specified time frame and budget [9]. The practice of Systems Engineering is supported by SE frameworks (for instance, specified in the INCOSE Handbook [6]) and standards (for instance, IEEE 1220 [10] and ISO/IEC 15288 [11]). The application of SE varies across industries, organizations and system types [12]. In the Systems Engineering Handbook, 30 processes are defined to apply Systems Engineering. These processes are divided into four categories: Technical processes, agreement processes, organizational project-enabling processes and technical management processes. Each process is described by relevant inputs, activities and outputs arising from the application. In order to implement SE successfully, companies need support for the implementation of Systems Engineering to transform their processes, organizational structure, role definition and project responsibilities [13].The results of the application of SE are document-based artefacts. Several documents include information about the system, which are spread across all related disciplines [14]. This results in the difficulty to assess all relevant information about the relationship between requirements, design, analysis and tests [8].

2.2. Model-Based Systems Engineering
Model-Based Systems Engineering (MBSE) is the formalized application of modelling to support system requirements, design, analysis, verification, and validation activities beginning in the conceptual design phase and continuing throughout development and later lifecycle phases [6]. MBSE is expected to replace the document-centric approach that has been practiced by systems engineers in the past and influence the future practice of Systems Engineering by being fully integrated into the definition of Systems Engineering process [15]. MBSE is characterized by the synergetic integration of model based engineering, systems engineering, and – inherently – digital engineering. Mental models are transferred to digital models to create, manage, review and advance product designs from system to component level across all disciplines.
Various modelling languages, methods and tools can be used to apply MBSE. In particular, a major focus for model-based systems engineering approaches is the ability to understand a system from multiple perspectives and to ensure integration across these perspectives. Thus, activities of both individuals in specific tasks as well as teams in collaborative tasks are supported. Advantages of MBSE are early defect detection, model and data reuse, technical and programmatic risk reduction, improved communication, supply chain efficiency, product line definition, and standards conformance [16]. Beside the implementation into the process and the availability of tools, competent people with the right skills are a fundamental aspect to implement MBSE [15]. Compared to Systems Engineering, additional skills are needed for the successful application of MBSE. Examples are the required understanding of the abstraction of an overall system model, the awareness of the "single source of truth" paradigm and the understanding of the separation between view and model [17]. So far, required perspectives, associated roles and competencies are not sufficiently defined in literature. The resulting challenge for the application of MBSE is the lack of guidelines and support to reconcile different perspectives and enable the multidisciplinary development. Thus, potentials of the approach can rarely be transferred.

2.3. SE and MBSE Role Models
A role is the user-neutral description of a function, a task or an activity to be executed [18]. A role includes a role description that summarizes the objectives, tasks, responsibilities and competences of the role within the specific topic [17]. One person can take on several roles and perform them simultaneously. Interdependent roles that carry out coherent activities can be combined in a role model.

IOP Conf. Series: Materials Science and Engineering 1097 (2021) 012003 doi:10.1088/1757-899X/1097/1/012003

The level of detail in descriptions of roles varies depending on the scope. By defining competencies, tasks and activities, the role descriptions imply additional information which increase the added value of the approach. Competencies are defined by several organizations in context of systems engineering, for example by the INCOSE Systems Engineering Competency Framework [19], the NASA's Systems Engineering Competencies [20] and the MITRE Systems Engineering Competency Model [21]. Task, on the one hand, are described within other Role Models descriptions and, on the other hand, are derived from the competencies and activities from other literature-based sources. Key activities are described within the processes of the INCOSE Handbook [12]. Sheard [22] defines twelve well established roles of Systems Engineering in 1996. Sheards roles were updated twenty-three years later by Gräßler et al. who define the Role Model for Systems Engineering Application (RM-SEA) [13]. The RM-SEA includes fifteen roles that describe possible and frequent activities and tasks in Systems Engineering projects and is meant to be used as a reference for Systems Engineering application. For MBSE, it needs to be reviewed and extended for dedicated roles regarding model creation and management.

2.4. Narrative methods in product development

The rising complexity in the development of systems and the increasing number of system-elements and disciplines involved lead to a high amount of uncertainties [23]. Researchers in the field of product development explore approaches to handle both complexity and uncertainties. One way to find solutions is to adopt and customize approaches from other domains, for example by adopting narrative methods [24]. As a relevant example, personas are fictional characters built upon available information about a considered group of people. Information is stored in templates structured by different facets, for example name, age, educational achievements, profession and goals [25]. Personas enhance cross-team collaboration, synergy, and communication as well as supporting the decision-making [26]. Another example are user scenarios, which are written in natural language and focus on a user within a considered context [27]. These two approaches can be combined resulting in a Persona-scenario. Here, define elements of persona-scenarios are considered according to Madsen and Nielsen [28] (see **Table *1***).

Table 1: Narrative elements of a persona-scenario according to Madsen and Nielsen [28]

Narrative elements	Narrative elements in Persona scenario
Character(s)	Protagonists represented by Personas
Time	Fixed time horizon
Problem	The problem which is investigated (and solved) within the story
Setting	Presentation of the Personas, the problems and the place of action
Opening episode	Definition of Personas goal and start of action
Episodes	Sequences of episodes that concern the problem
Resolution	Reaching the goal / solving the problem
Plot	Mostly linear manner
Overall story	Overall context of the episodes

3. Methodology

The methodology to define a Role Model for the practical application of MBSE is divided into three major steps: (1) identification, (2) categorization and (3) association of stakeholders. The term stakeholder indicates people and roles who implement and apply MBSE within an enterprise. The steps are performed sequentially in the logical order which is illustrated in **Figure *1***.

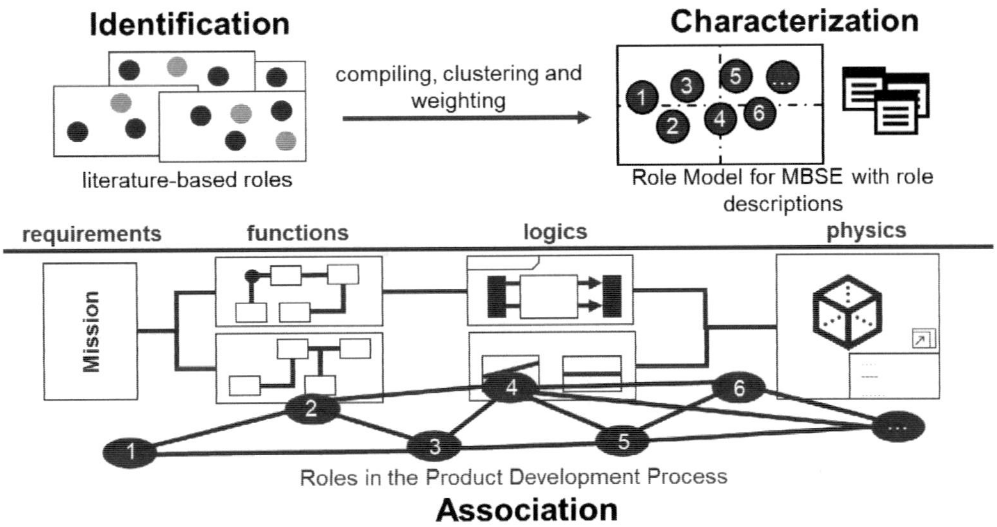

Figure 1. Methodology for deriving Role Model of Model-Based Systems Engineering Application

In the first step (1), Role Models are identified and analyzed using management layers of [29] as search areas: Systems Engineering, enterprise management structures (organizational management) and project management. The literature-based analysis is executed in accordance to Petersen et al. [30]. In the second step (2), the resulting roles are categorized in terms of homonyms and synonyms. Afterwards, an analysis of the semantic similarity is conducted to find taxonomic relationships and clear relation to MBSE. Finally, roles are clustered based on relevance for MBSE key activities. In the third step (3), roles are associated with an implementation process. Personas are used according to Pruitt and Grudin [26], combined with persona-scenarios to create synthetic test cases for validation purposes.

4. Elaboration of the Role Model for MBSE

Different Role Models are analyzed in the first step (1) to identify roles that need to be represented with competent employees in order to ensure successful application of MBSE. Based on the management layers of [29], the search areas are compiled to: Systems Engineering, enterprise management structures and project management. The literature based analysis is executed in accordance to Petersen et al. [30]. Strings and keywords are derived for each area to ensure that all relevant literature is identified. The results are illustrated in

Table *2* and 3. Due to the large number of results a reading scheme is applied to select relevant literature [31]. Within the reading scheme, literature is selected that does not provide definitions of roles, only describes the topic MBSE incidentally or exclusively addresses a certain domain.

Table 2. Results of the string in two different research platforms.

		Results	Selected
Web of Science	TS=("Role-Model" OR "Role Model" OR "Roles") AND "Project Management"	518	11
	TS=("Role-Model" OR "Role Model" OR "Roles") AND ("Systems Engineering" OR "SE")	111	6
	TS=("Role-Model" OR "Role Model" OR "Roles") AND ("Organizational Management") // TS=("Role Model" OR "Role-Model") AND "Organization"	153	5
Google Scholar	allintitle: Role Model "Project Management" // allintitle: Role-Model "Project Management" // allintitle: Roles "Project Management"	116	10
	allintitle: Role Model "Systems Engineering" // allinititle: Role-Model "Systems Engineering"	41	4
	allintitle: "Role Model" "Organizational Management" // allinitite: "Role-Model" Organizational // allintitle: Role Model Organization	183	5
		1.194	41

Table 3: Representation of considered Role models and roles of the three categories.

	Systems Engineering	Project Management	Organizational Management
Number of literature considered	8	9	5
Number of included roles	93	90	30

In the second step (2), the resulting 213 roles are categorized. In the beginning, homonymous roles are combined, reducing the number of roles to 166. Afterwards, an analysis of the semantic similarity is conducted. Roles that are semantically close to each other and have a large overlap of the information in their role descriptions are summarized to a superior role, for example the requirements engineer and the request analyst. 91 roles are semantically different. Thereafter, roles are rejected that do not show a clear assignment to MBSE or are described generically, for example "Influencer" [32]. The remaining 66 roles are transferred into a matrix in which they are compared with relevant MBSE key activities. These key activities are derived from the processes of the Systems Engineering Handbook [6]. The processes are categorized into key processes which can be transferred to MBSE. The activities of the key processes are adjusted from an MBSE perspective to cover the specifics of MBSE. The results are 23 key activities which have to be performed for successful MBSE. The roles and the key activities are transferred into a fulfilment matrix (F_{ra}). By comparison, it is determined whether a certain role can perform the corresponding MBSE key activities. For evaluation, a five-part rating scale is introduced, ranging from "0" (the role does not fulfil an activity) to "4" (the role fulfils the activity completely). The remaining roles are classified, resulting in the fulfilment matrix F_{ra}, where each value represents the fulfilment of an activity a by the role r.

The similarity of two roles was calculated by the formula

$$dist(i,j) = \sum_k |F_{ik} - F_{jk}| \quad (1)$$

The similarity matrix $dist$ is a symmetric matrix where the value $dist(a,b)$ denotes the difference (i.e. distance) in processes between role a and role b. To identify similar roles, the distance of all roles was visualized and clustered. The 68-dimensional matrix $dist$ was first reduced to 23 dimensions by classical multidimensional scaling. Therefore a non-classical multidimensional scaling approach is chosen with presenting the dimension as 2. The resulting reduction was clustered with an average linkage algorithm with a distance-similarity measure, resulting in 14 clusters.

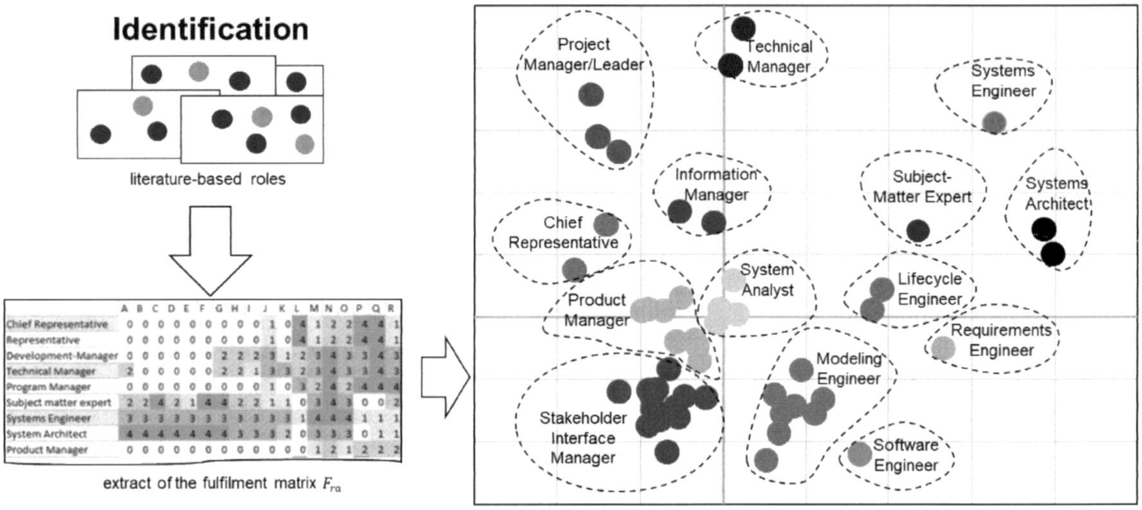

Figure 2. Compiling clustering and weighting of the roles for the application of MBSE

For characterization, existing role descriptions from literature are adapted for MBSE, which imply suitable competencies, tasks and activities based on literature-based competency profiles. Due to the strong correspondence of the roles to the RM-SEA, it is adapted as a baseline for overarching SE roles. The remaining roles are classified in three categories, based on the three management layers: Organizational Management, Project Management and technical/Specialty [29]. The adjusted roles as well as the added role are defined in the following:

(Project) Leader *based on references* [13, 17, 32]: The role is responsible for the operational management of the engineering project. Additionally, he is responsible for the achievement of objectives in the project. He provides leadership, coordinates the project tasks and project team members and manages the project-relevant activities. He may be responsible for one or more projects tasks, other tasks could be delegated to team members. He ensures the optimal distribution of resources. The system architect helps the project leader to break down the MBSE activities in projects into manageable tasks.

Systems Analyst *based on references* [6, 13, 33, 34]: A Systems Analyst analyzes, designs, and develops information systems to support the MBSE activities. He is responsible for developing, analyzing and modeling the architecture under the guidance of the *Systems Architect.* His key MBSE activities include collecting and analyzing objective data, measure the performance of the system model and generating and disposing information. Models are crucial input for all activities. Output is generated which is linked into different partial models. The System Analyst requires explicit and formal connections between system models domain specific models.

Subject-Matter expert *based on references [6, 13, 35]:* The Subject-Matter expert is an expert in his domain. He is responsible for the modeling of the sub-system or system-element architectures in specific areas and disciplines. Furthermore, the role has to establish traceability and implement the required systems elements within his tasks. Compared to the Systems Engineer, the subject-matter expert has the detailed view of a specific subject or discipline. In coordination with the project manager, the knowledge of various subject-matter experts is bundled. Typically, subject-matter experts use domain-specific languages in modeling tasks.

In addition to these roles, the following role is identified but not included in the Role Model, because the responsibilities of the role depend on the firm size and the enterprise organization. It is assumed that this decision-making role exists in every company, which want to implement and apply MBSE.

Chief Representative *based on references* [6, 13, 36]: The Chief Representative is involved in management decisions and has the necessary responsibilities for overall business decisions. The role has to decide, if a project is aligned with the enterprise strategy. The role also receives relevant systems engineering reports (e.g. implementation reports or transition reports), represents the opinion of the company leadership and takes care of MBSE key activities like assessing a feasible project plan. Additionally, the role interacts with the *Leader* and the *Entrepreneur* to identify the best possible solutions for the further strategic plan.

Fehler! Verweisquelle konnte nicht gefunden werden. depicts which roles are adopted or adjusted from the RM-SEA [13]. Furthermore, two essential roles for MBSE are added, called the Subject-matter Expert and the Systems Analyst. Compared to the Modeling Engineer, the Subject-Matter experts addresses (a) the need of linking models from different disciplines within an overall system model and (b) the detailed decomposition of the overall system into sub-systems and system elements. This requires specific knowledge of a discipline and the understanding of dependencies between partial models. The systems analyst addresses the increasing focus of digital simulations and test cases within an overall system model, for example a change impact analysis in engineering change management [37]. Thus, these essential roles have to be considered additionally when applying MBSE compared to applying the document-based SE.

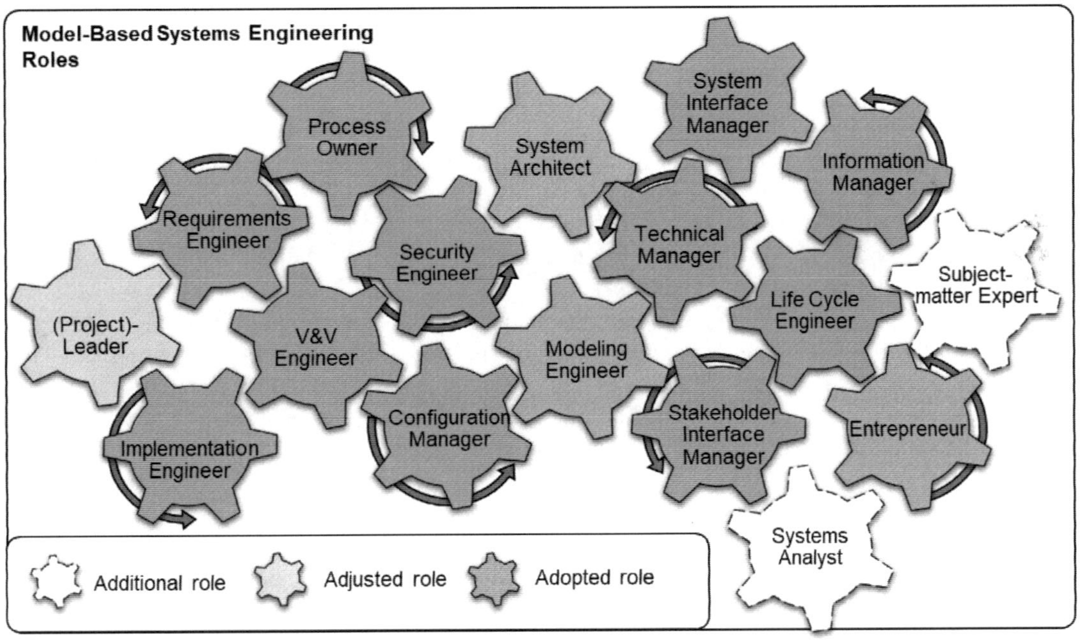

Figure 3. Role Model for Model-Based Systems Engineering, based on [13]

Lastly, the roles are associated with an implementation process (3). The association indicates situations in which different roles interact with each other. These interactions are illustrated in a test case, created by a persona-scenario which depicts implementation process. Therefore, the information within the defined role descriptions (2) are extended by additional data for each persona.

The elaboration of personas is done based on a four step approach proposed by Pruitt and Grudin [26]:

- The first step is to choose a manageable amount of personas (3-5 personas).
- Afterwards, a common template with relevant facets is developed.
- In the following step, this template is filled by internal and external resources.
- The final step is the presentation of the personas within the intended context.

By allocating the roles to Personas, it is demonstrated how the role model solves challenges of MBSE described in literature.

5. Validation by narrative test case

The use of the role model is demonstrated and evaluated in a test case based on personas and a scenario within an organizational structure. The narrative elements from Madsen and Nielsen [28] are the foundation of the scenario . By allocating the roles to Personas, it is demonstrated how challenges of MBSE described in literature can be solved using the role model. According to current survey results [38], typical challenges of MBSE are: upfront investment, the purpose and scope definition, the awareness and change resistance as well as the tool dependency.
Four personas are determined as *characters* of a typical application scenario. Challenges define the *setting* of the scenario. The *time* is illustrated by the implementation phase and application phase, which connects the *problem* with the *resolution*.

The generic product development process of Figure 1 is deposited in the laboratory environment at the HNI (Chair of Product Creation) of Paderborn University. Within a company entitled HNI Inc. (represented by product portfolio, historic manufacturing data etc. specified for research purposes), a project team is established to implement MBSE, including the project leader, a technical manager and a subject-matter expert. In the beginning, the project team has to convince the Chief Level Managers to implement MBSE.

- The representative of the Chief Level Managers is called *Paul*. He is 56 years old and was promoted to CEO after many years as CFO. His high integrity, his ethical principles, his proactive approach, but most of all his respectful manner, has helped him to work his way to the top of the company. As CEO, he is characterized by his critical and constructive thinking, but also by his creative, visionary and holistic approach. Paul interacts on the one hand with the board members of the organization, but also directly with the CTO / CIO and CFO. His connection to the other employees is mostly indirect.
- The Project Leader is called *Susanne*. Prior to her position as Project Leader, she distinguished herself through her very good results as Project Team Member. Additionally, she holds a Master of Business Administration. Through the coordination of the MBSE project there is a direct contact to C-level roles like Paul. Their tasks include the definition of roles and responsibilities, the communication of management expectations, and the management and review of results. Personally, she is committed to the development of her employees, positive project results and continuous improvement.
- The Technical Manager is called Natasha. She has a Master of Science in Mechanical Engineering Technically, she distinguishes herself above all by her logical thinking, her problem-solving skills and her general knowledge in the field of engineering sciences. She is in close contact with your team, but also forms the bridge to the C-Level she reports to.
- Kyle is enterprise-internal Subject-Matter engineer of the discipline Software. His statement about MBSE was: "We (i.e. the developers) have wanted an MBSE solution for a long time. The other departments just don't understand how important the implementation is." His IT skills mainly include UML and SysML. He cooperated directly with the project team and sees himself as a representative of the users.

The objective of HNI Inc. is change from a document-centric to a model-based development environment and therefore to implement and apply MBSE. When he was CFO *Paul* has supervised the ERP implementation five years ago and remembers how many unforeseeable difficulties arose during the implementation, which delayed the project by 6 months. He also regards the upfront investment critically, as the benefit of MBSE is not sufficiently high. Paul's decisions are influenced by the role of the Project Leader. Susanne assigns her MBSE implementation team to identify the potential of MBSE. The team prepares a set of slides which Susanne uses to convince Paul to implement MBSE in a meeting. The advantage becomes especially clear to him when Susanne presents how a comprehensive analysis of the required modeling depth is performed before each modeling. The issue of modeling depth was identified by Kyle, who was able to identify an MBSE method that considers setting the modeling depth as the first step of the method. The purpose and scope of MBSE becomes more transparent for all involved. Natasha remains skeptical. From a technical perspective, she thinks that the switch to model-based development will lead to resistance from her employees. She also fears that the investment will make the company become too dependent on an IT vendor. Due to the increasing relevance of Kyle's role and his involvement in the MBSE team, there is a conversation between the Technical Manager and the System Architect. With an insight into a system model that he developed in a freeware software environment, Kyle explains the benefits and highlights the increasing system quality, the reduction of errors and defects and the improved traceability. Additionally, Kyle's previous experience with SysML means that he knows the different vendors of the tools and can discuss with Natasha which vendor would be most suitable for a pilot project. With the new insights, Natasha improves in arguing the case to her staff, making the added value of MBSE clearer and reducing resistance within her own ranks.

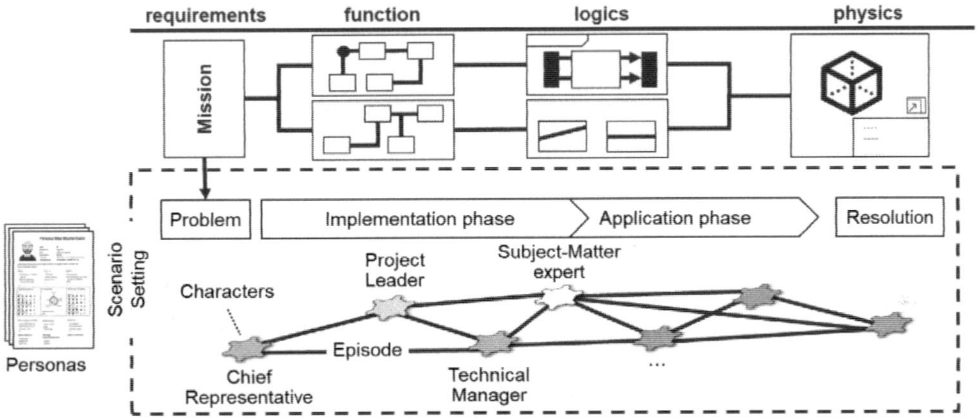

Figure 4. Persona-scenario of the MBSE implementation process within the HNI Inc.

The narrative test case represents an example for the application of the role model. Roles can be adapted and transferred to individual needs of a company. Based on the Role Model, the MBSE Project Leader can infer job profiles from role descriptions to receive decision support for the selection of suitable employees. In addition, actions required can be derived in order to fill the roles that are necessary for the application of MBSE. One example is the additional qualification of employees.

6. Conclusion

The paper at hand contributes a role model for the implementation and application of Model-Based Systems Engineering. Due to the specific circumstances of the MBSE application, actors and competencies have to be defined to enable successful MBSE application and adaption. Therefore, a comprehensive literature analysis is conducted to identify roles from the areas systems engineering, enterprise management structures and project management. 213 roles are identified and then condensed to 66 semantically different roles. Additionally, key activities of MBSE are derived from the processes of the INCOSE Handbook to benefit from MBSE advantages. The identified roles and key activities are transferred to a similarity matrix. The matrix enables the clustering and weighting the identified roles due to their field of activity. Based on the results, the existing Role Model for Systems Engineering Application (RM-SEA) is adjusted and expanded to satisfy the specialities of model-based development.

An adaptable role model for MBSE is presented. The application of the role model is demonstrated in a synthetic test case. Narrative methods are combined to create the test case and apply a scenario, where personas are defined, which have MSBE roles within the fictional enterprise. Industrial application demands are considerd, which increases the acceptance and the neccessity to implement the model-based approach. The test case illustrates that the definition of roles is crucial for the successful implementation of MBSE. The role model helps the development team to define relevant responsibilities and key activities to implement and apply an appropriate MBSE approach. The approach supports the mutidimensional development and recounciles different perspectives on the system model.

References

[1] Gräßler I, Hentze J and Oleff C 2018 Systems engineering competencies in academic education *System of Systems Engineering Conference* **13** p. 542–547.
[2] Gräßler I, Hentze J and Bruckmann T 2018 V-Models for interdisciplinary systems engineering *International Design Conference* **15** p. 747–756
[3] Boston consulting group 2015 Industry 4.0 the future of productivity and growth in manufacturing industries *Die Zukunft der Produktivität und des Wachstums in verarbeitenden Industrien*
[4] BMWi 2015 Industrie 4.0 und digitale wirtschaft. impulse für wachstum *Beschäftigung und Innovation* Berlin
[5] Object management group Systems modeling language (SysML®) v2 request for proposal (RFP) *OMG*
[6] Walden D D, Roedler G J, Forsberg K, Hamelin R D, Shortell T M 2015 Systems engineering handbook *INCOSE-TP-2003-002-04* Hoboken
[7] Fosse E 2019 Model-based systems engineering (mbse 101) *NASA*
[8] Friedenthal S 2015 A practical guide to SysML. the systems modeling language *Morgan Kaufman* Waltham
[9] Gräßler I 2015 Umsetzungsorientierte synthese mechatronischer referenzmodelle *VDI Mechatronik* p. 167–172
[10] IEEE 1220-2005 - IEEE Standard for application and management of the systems engineering process
[11] ISO/IEC/IEEE 15288 2015 Systems and software engineering - System life cycle processes
[12] Systems Engineering Vision 2025 *INCOSE* 2014
[13] Gräßler I, Oleff C and Hentze J 2019 Role Model for Systems Engineering Application *International Conference on Engineering Design* **22** Delft
[14] VDI 2020 VDI 2206. Entwicklung cyber-physischer mechatronischer systeme, Düsseldorf
[15] Holt J and Perry S SysML for Systems Engineering A Model-Based Approach *The Institution of Engineering and Technology*, Stevenage
[16] Madni A, Purohit S 2019 Economic Analysis of Model-Based Systems Engineering *Systems* **7** p. 12
[17] Weilkiens T, Lamm J G, Walker M, Roth S 2016 Model-based system architecture *John Wiley & Sons Inc* Hoboken, New Jersey
[18] Feldhusen J and Gebhardt B 2008 Product lifecycle management für entscheider. *Springer* Berlin
[19] Delicado A B 2019 INCOSE systems engineering competency framework (ISECF) Madrid
[20] NASA's Systems engineering competencies *NASA*
[21] Metzger L and Bender L 2007 Mirte systems engineering competency model
[22] Sheard S A 1996 Twelve Systems Engineering Roles *INCOSE Sixth Annual International Symposium*
[23] Deuten J J and Rip A 2000 Narrative infrastructure in product creation processes *Organization* **7** p. 69–93

IOP Conf. Series: Materials Science and Engineering 1097 (2021) 012003 doi:10.1088/1757-899X/1097/1/012003

[24] Riessman C K 2007 Narrative methods for the human sciences *SAGE Publications, Inc* Thousand Oaks
[25] Campos J and Pavia A 2011 A personal appraoch the personas technique in a companion's design lifecycle *Human-computer interaction - INTERACT* Heidelberg
[26] Pruitt J and Grudin J 2003 Personas: practice and theory *Designing for User Experience*
[27] Nielsen L 2004 Engaging personas and narrative scenarios. - a study on how a user-centered approach influenced the perception of the deisgn process in the e-business group at AstraZeneca, Copenhagen
[28] Madsen S and Nielsen L 2010 Exploring persona-scenarios - using storytelling to identify requirements *IFIP Advances in Information and Communication Technology* p 57-66
[29] Murdoch J and Astley K 2004 Role-based planning for systems engineering and project management *IEEE International Conference on Systems, Man and Cybernetics* p. 5447–5454
[30] Petersen K, Feldt R, Mujtaba S and Mattson M 2008 Systematic mapping studies in software engineering *EASE' 08:Evaluation and Assessment in Software Engineering* p. 68–77
[31] Blessing L T and Chakrabarti A 2009 DRM, a design research methodology *Springer* London
[32] Pádár K, Pataki B and Sebestyén Z 2017 Bringing project and change management roles into sync *Journal of Organizational Change Management* **30** p. 797–822
[33] Broy M 2013 Modellbasiertes software und systems engineering als element eines durchgängigen systems lifecycle managements *Industrie 4.0.* Berlin
[34] Crawford J K 2004 Project management roles & responsibilities *Center for Business Practices* Havertown
[35] Borky J M and Bradley T H 2019 Effective Model-Based Systems Engineering *Springer International Publishing*
[36] Menz M 2012 Functional top management team members *Journal of Management* **38** p. 45–80
[37] Graessler I, Oleff C and Scholle P 2020 Method for Systematic Assessment of Requirements Change Risk in Industrial Practice *Applied Science* **10** (23)
[38] Chamni M and Bruel J M 2018 A survey on mbse adoption challenges *The Systems Engineering Conference of the Europe, Middle-East and Africa (EMEA)* Berlin

IOP Conf. Series: Materials Science and Engineering 1097 (2021) 012004 doi:10.1088/1757-899X/1097/1/012004

Model-Based Design Workflows for Cyber-Physical Systems Applied to an Electric-Mechanical Coolant Pump

Gregor Höpfner[1,4], Georg Jacobs[1], Thilo Zerwas[1], Imke Drave[2], Joerg Berroth[1], Christian Guist[3], Bernhard Rumpe[2] and Jens Kohl[3]

[1]Institute for Machine Elements and Systems Engineering, RWTH Aachen University, Schinkelstraße 10, 52062 Aachen, Germany

[2]Chair of Software Engineering, RWTH Aachen University, Ahornstraße 55, 52074 Aachen, Germany

[3]BMW Group, Petuelring 130, 80788 München, Germany

[4]gregor.hoepfner@imse.rwth-aachen.de

Abstract. Cyber-Physical Systems (CPS) connect mechanics, electrics and electronics as well as software. Considering the interactions between these domains is a major challenge in product development. Model-Based Systems Engineering (MBSE) enables cross-domain system development, based on commonly understood system architectures. However, there is still a gap in using MBSE for development, especially in mechanics. Systematic evaluation of design decisions based on system architecture models is not done. The strong simulation methods predicting physical product behaviour are not well connected to the evolving system architecture. We propose a new approach for connecting simulation and design models in a System Modelling Language (SysML) system model using the example of a combustion engine's cooling circuit. The electric-mechanical coolant pump is chosen for design. The design processes of the hydrodynamic pump wheel and the volume flow controller are modelled as workflows in SysML. Here, we integrate design models for the pump wheel and the controller into the workflows. Thereby, parameters of the pump wheel and controller are calculated. Design results are automatically transferred to CAD. A workflow is created, testing the cooling circuit's behaviour against requirements, combining a Simulink control model and a Simcenter Amesim heat transfer model. The approach results in a framework for automated design and test processes.

1. Introduction

Cyber-Physical Systems (CPS) connect mechanics, electrics and electronics as well as software solutions. Considering the interactions between these different domains is one of the major challenges in modern product development. Model-Based Systems Engineering (MBSE) is a well-known approach, that enables cross-domain system development by modelling commonly understandable system architectures, e.g. for functional and product architectures, including interactions between sub functions of systems. Architecture models describe the system elements from requirement over functions to products. They may be used to link the artifacts that are created during system design. [1]

However, there is still a gap in using MBSE methods for product development, especially in the mechanical domain. MBSE often is only used up to abstract architecture definition. Design decisions are not evaluated systematically based on the system's functional architecture, especially in the mechanical domain. Instead, mechanical development is nowadays mainly driven by experience. Mechanical engineers know about their domain and how to design mechanical products. Strong simulation methods are used for product design and testing of component's physical behavior virtually using simulation models [2–6]. CPS are not component but function oriented. The mechanical domain has not yet developed a method to test the products with regard to functional requirements. A link between functional architecture and physical behavior models is not given.

Model-based design and test procedures, i.e. executable workflows allow for automated designing and testing of elements in the functional architecture at different stages of the design process. Such model-based workflows link the physical behavior models to the system's parameters in the functional architecture. By modelling the order of design and test steps in executable workflows, the parameters describing the system's geometrical, physical, material, and logical properties can be calculated automatically and the system can be tested at any system state. This may close the gap between functional architecture modeling and physical behavior models and lead to functional testing in mechanical engineering. In order to reach for such a modeling approach, we state the following research hypotheses:

- Structured, hierarchical workflows can link simulation models efficiently and independent of a simulation tool.
- Workflows efficiently parametrize the (principle) solutions in a solution architecture.
- Workflows help to redesign and test system solutions when requirements change (engineering change requests)

We derive a new modeling approach, connecting simulation and design models in a SysML system model, modelled in Cameo Systems Modeler (CSM) [7]. Based on the hypotheses, we analyze the functional architecture of a mechatronic system. To create design and behavior prediction models we use external tools and integrate the models within executable workflows using SysML activity diagrams and CSM. We propose two kinds of Activity Diagrams for modeling design workflows and test workflows, respectively. In order to show efficiency of the modeling technique, we chose the example of an engineering change request for a combustion engine's cooling circuit.

The article is structured as follows: section 2 presents the state of research regarding architecture and workflow modelling. Section 3 describes the system of interest and its functional architecture focusing on the solutions and principle solutions relevant for this paper. Section 4 derives a modelling technique for design workflows used to find parameter values for principle solutions using SysML Activity Diagrams. Section 5 integrates these detailed design workflows into top level design processes. Section 6 adds testing workflows, testing the system behavior. Section 7 gives the application case of the workflows, an artificial engineering change request. Section 8 gives a discussion on the presented work and section 9 concludes.

2. State of Research

In MBSE, the modelling language family SysML has become widely used. SysML is a general-purpose modeling language family for systems engineering [8]. It extends the Unified Modeling Language (UML) 2.5 [9] to represent aspects of both, software and hardware elements. Currently, SysML provides modeling languages for behavior, structure, and requirement modeling. The structure modeling languages are often used for architecture description. Internal Block Diagrams (IBDs) therein can be used to describe the internal structure of a block.

There are various methods for CPS architecture development in SysML. Their aim is to link requirements, functions and product components. Some methods coming from informatics model this link focusing on software system elements and software behavior [10–15]. However, they do not provide a link between function and product for mechanical system elements. For modeling the functional architecture of mechanical systems, the methods have been extended, e.g. in [16,17]. These techniques do not integrate the product describing parameters on the detailed levels of function and product description. In [18], the authors propose a technique that uses informal sketches which describe the

IOP Conf. Series: Materials Science and Engineering 1097 (2021) 012004 doi:10.1088/1757-899X/1097/1/012004

mechanical solution but are not usable for executable design workflows integrating simulations, as they do not contain parameters for mathematical calculations.

The seamless approach provided in [19] allows for product description up to detailed parameter level. Modeling languages that support this methodology in MBSE are specified in [20], which also provides a SysML profile as a specific language. These languages do not yet integrate simulation workflows for product design or testing.

Activity Diagrams are a commonly used language for modeling sequential behaviors of a system [21,22], functional requirements of CPS [23–25], as well as processes and workflows [26]. In [27] the process of developing a system model in SysML is described in an Activity Diagram. In [28] the authors use SysML Activity Diagrams to execute work processes in context of grid computing in software. To the best of our knowledge, Activity Diagrams have not yet been used to describe design and test workflows and link the activities to simulation models for mechanical engineering in an executable manner.

Workflows for simulations, however, are used in mechanical engineering to model and predict the physical behavior of a product. They combine simulation models to provide reusable and automated test procedures. Examples for tools providing such workflows are given in [29,30]. Workflows are then executable but not linked to the parameters in functional architecture and often limited to specific tools.

In total, there are methods to describe the functional architecture of CPSs that include the mechanical domain in SysML. However, these do not integrate simulation workflows and use these workflows to parametrize and test system architecture.

3. System of Interest: Cooling Circuit and Electric Mechanical Coolant Pump

To illustrate modelling concepts, we use the cooling circuit of a combustion engine from automotive engineering as a running example throughout the paper. The system describes the following physical behavior: Combustion drives convert the chemical energy held by fuel into mechanical energy. Due to the physical effect of combustion to serve this function, a portion of the chemical energy is converted into thermal energy, i.e. heat, during a chemical reaction, which causes the pressure in the combustion chamber to increase. The increasing pressure acts on the surface of the engine's piston causing the piston to move. The heat is partly removed from the engine by exhaust gas. However, some amount of heat causes the engine's temperature to increase by being transferred to the engine structure. Thus, the engines temperature increases. Due to material limitations, the engine stops functioning once a maximum temperature is reached. To prevent the engine from overheating, the thermal energy has to be removed from the engine. Often, water-cooling circuits take on this task. In our example, we use a water-cooling circuit with an electric mechanical cooling medium pump.

The overall function of a water-cooling circuit is to manage the engines heat flows (*manage heat flows*) in order to set a specific temperature of the engine. This function can be achieved by combining subfunctions: the heat flows from the two engine components crank case (CC) and cylinder head (CH) are considered as inputs for the function. The heat flows are removed from the engine by two volume flows (*distribute heat*) using the physical effect of convection, the surrounding air and an internally circulating cooling medium. The air is removed to the environment. The internal cooling medium is transferred to a cooler, where the heat is also released to the air using convection (*release heat*). The cooling medium has to keep circulating in order to keep heat transfer active, i.e. a volume flow is necessary, which is generated from electrical energy (*generate volume flow*). For setting a specific engine temperature, the amount of removed heat has to be controlled. In our system, this is done by controlling the volume flow of the cooling medium. Hence, the function *generate volume flow* can get a control signal, setting the required volume flow, from a controller (*control heat flows*), which evaluates the engine components temperature signals. Figure 1, top, shows the described internal structure of *manage heat flows* in an IBD.

In order to find a solution for the function *generate volume flow*, we have to continue the functional decomposition, Figure 1, bottom. To generate a volume flow, we apply mechanical energy to the fluid (*apply fluid with mechanical energy*). Hence, we have to get mechanical energy. In an electric mechanical coolant pump, the mechanical energy is converted from electrical energy (*convert el to mech energy*). In order to set specific volume flows based on a control signal, the electrical energy is increased or decreased before being converted (*increase/decrease electrical energy*).

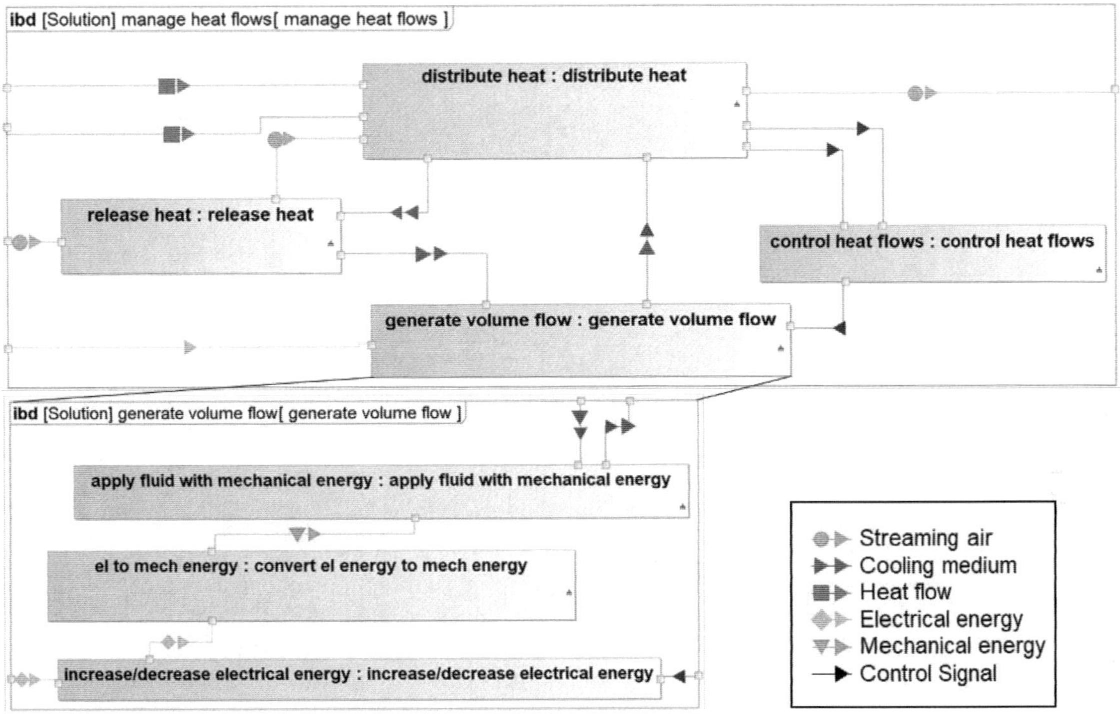

Figure 1. Functional architecture for *manage heat flows* and *generate volume flow*

The functional architecture of the cooling circuit has been modelled following the ideas of Koller, who described a system by elementary functions realized by principle solutions [19]. The described modeling follows the language profile SysML4FMArch, proposed in [20]. The functional architecture of the system has been developed in [31]. The functional decompositions of *generateVolumeFlow* and *distributeHeat* are discussed in [20] and [31], respectively. For each elementary function, a principle solution has been derived and modelled. Principle solutions realize elementary functions using physical effects or logics. Physical effects act between active surfaces from specific materials. The principle solution for *apply fluid with mechanical energy* is a hydrodynamic pump wheel. The physical effect is hydrodynamics, which accelerates an input fluid due to mechanical energy. The active surface required for this is a pump wheel. This active surface can be described by characteristic parameters, its inner and outer diameter, width, inlet and outlet angle and the number of wings in the hydrodynamic pump wheel. Based on these geometric parameters, the physical effect hydrodynamics is realized in a quantitative manner. The SysML4FMArch representation of the principle solution is shown in Figure 2. Parameters are modeled as values in SysML. From external requirements, there is a limitation for the pump wheels rotational velocity. The connection between the external requirement and parameter *rotational velocity* is modeled using a SysML-satisfy-connection.

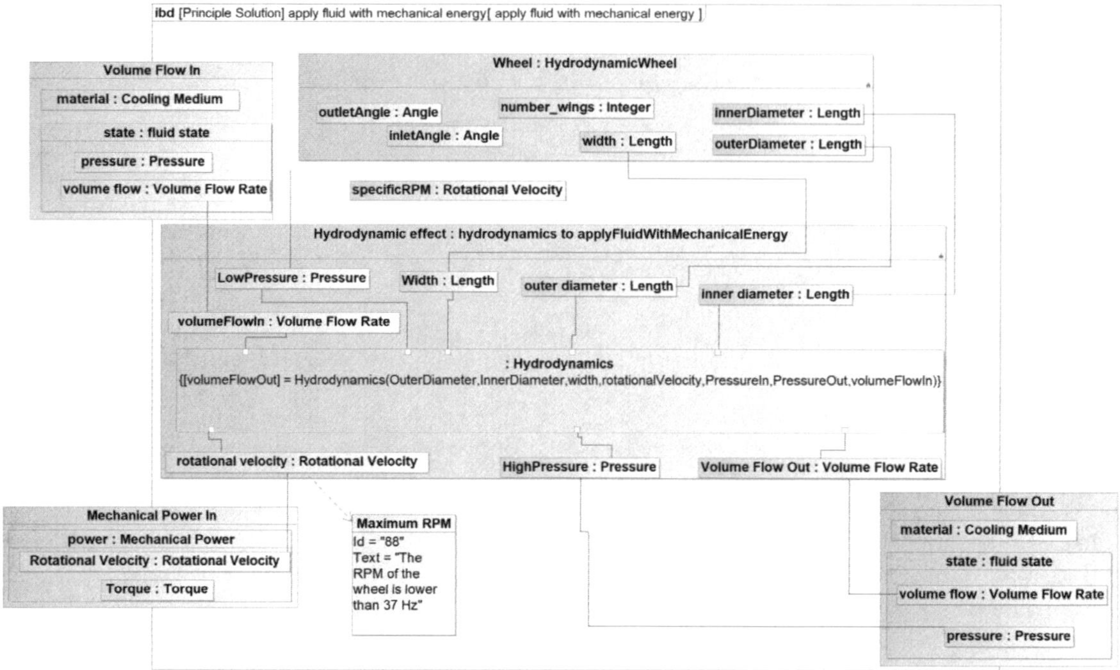

Figure 2. IBD for the principle solutions apply fluid with mechanical energy

Table 1 gives an overview about the further principle solutions, the realizing physical effects, geometric elements and relevant parameters with a description. The hydrodynamic pump wheel is driven by an electric drive providing rotational speed and torque. The volume flow is transferred to the engine and takes up an amount of heat by convection. The heat is released at the cooler by convection, reducing the pressure. To set volume flow rate based on heat, a control element is introduced, changing rotational speed of the pump.

Table 1. Selected functions, their physical effects, important parameters and parameter descriptions

Function	Solution/Principle Solution	Physical effect	Parameters	Parameter Description
distribute heat	Solution	Convection	maxVolFlow	Maximum volume flow required to satisfy the tempe-rature requirement
			minVolFlow	Minimum volume flow required to satisfy the tempe-rature requirement
release heat	Principle Solution	convection	deltaPOpt	Pressure difference of cooling medium between input and output port
control heat flows	Principle Solution	-	controlParameter	Control parameter of the P control
			controllVolumeFlow	Set point of volume flow in control
convert electrical to mechanical energy	Principle Solution	biot-savart law	RPM	Optimum rotational velocity
manage heat flows	Solution	-	RhoCooling	Density of the chosen cooling medium

The parameters describing the principle solution are given in the architecture. When investigating Figure 2, it becomes clear, that the values of the principle solution parameters, e.g. geometric parameters inner and outer diameter, have influence on how the physical effect fulfills the function, e.g. what quantity the volume flow rate is. The mechanical design process puts a lot of effort in finding the suitable parameter value set, in order to get the required behavior. In the following chapter, we describe a way of modeling the parameter estimation in the mechanical design process in design workflows and test the functional architecture with regard to requirement satisfaction.

4. Design workflow for the hydrodynamic pump wheel

Parametrization of principle solutions means finding a suitable set of values for the geometric, material and physical parameters in order to satisfy the requirements. Mechanical engineering has developed design procedures for standardized principle solutions in the past decades, which are given in norms, standards and design guidelines. Here, critical use cases are considered as input, standard parameters are declared, and their values are calculated so that the principle solution satisfies the use case. In order to find suitable parameter values for the geometric and physical parameters for the hydrodynamic pump wheel, we set up an automatized, executable workflow based on a given design guideline. In the following section, the design process of a hydrodynamic pump wheel is explained and a procedure to integrate it into the system model in an executable manner is proposed.

4.1. Design process of the hydrodynamic pump wheel

The design process of the hydrodynamic pump wheel aims to find a suitable value set for the parameters describing the hydrodynamic pump wheel's active surfaces, which are modelled in Figure 2. By fixing these geometric parameters, the hydrodynamic characteristic function is specified, i.e. the physical behaviour of the principle solution. We set up a parametrizable CAD model, which is adaptable based on the modelled geometric parameters.**Fehler! Verweisquelle konnte nicht gefunden werden.**

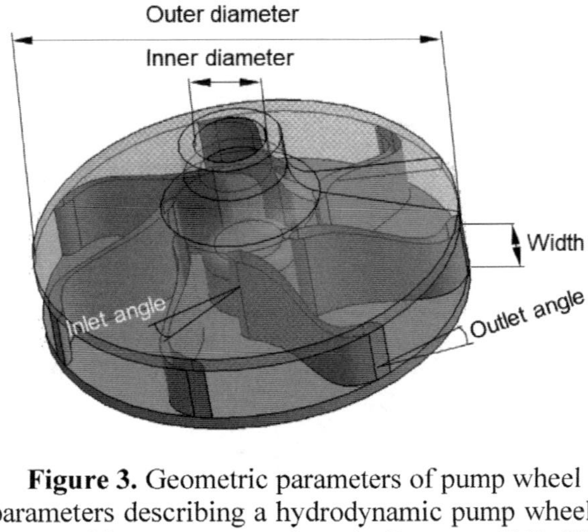

Figure 3. Geometric parameters of pump wheel

The aforementioned parameters describing a hydrodynamic pump wheel and a design guideline to calculate their values for a specific system are given in literature [32]. The guideline calculates the geometry of the pump wheel for a required operating point. The parameter values are calculated based on empirical equations and diagrams. The required inputs for the guideline are:

- Pressure delta, which the pump has to provide between input volume flow and output volume flow at optimum working point
- Rotational speed of the pump at optimum working point
- Density of the transported cooling medium at optimum working point
- Required volume flow at optimum working point

We implement the calculation steps of the guideline in an external software. Multiple tools may be chosen. As those guidelines in companies are often implemented in MS Excel, we choose this tool.

4.2. Concept for connecting functional architecture, workflow and simulation model

The functional architecture described in section 3 provides a structure for principle solutions and their describing parameters. The design guidelines calculate the values of principle solution parameters. They are not modelled directly in the functional architecture, as they do not serve a functional mean during product lifetime, but are only used during design. We model guidelines and other workflows due to their sequential character as SysML Activity Diagrams. A connection between the principle solution architecture and the Activity Diagram is needed. The parameters are the obvious connection, as they occur in both, IBD and Activity Diagram. While the IBD models the parameter structure and interdependencies, the Activity Diagram uses specific parameter values and calculates other parameter values from them, e.g. outer diameter of the hydrodynamic pump wheel from required volume flow. As storing parameter values is not the purpose of a systems modeller, especially for large systems, we create a separate parameter database and link IBD and Activity Diagram via this database, Figure 4. The structure of the database (2) is given by the principle solution IBD (1). In the IBD, parameters are modelled and organized. When a new parameter is added to the principle solution IBD, a database entry is created. Therefore, the database contains all product describing parameters from IBD and their system specific values.

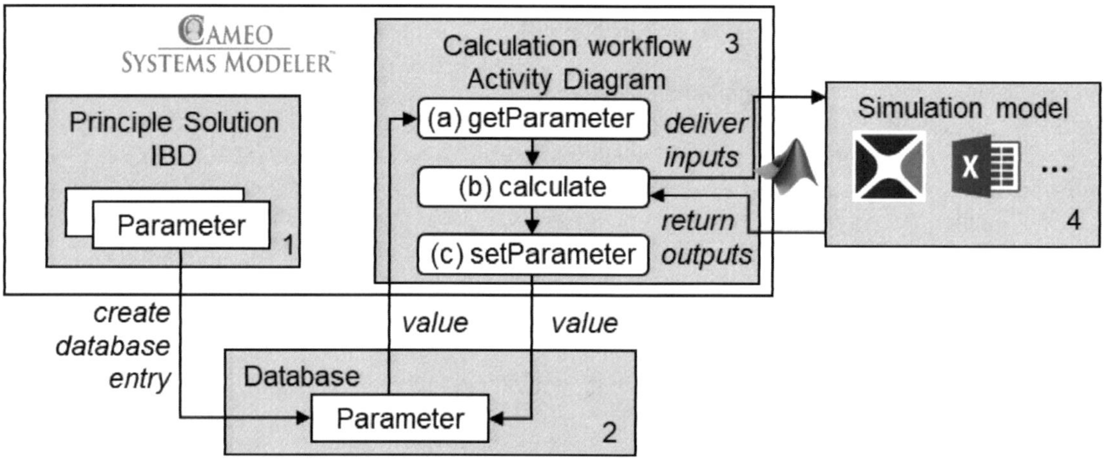

Figure 4. Concept of connecting Activity Diagram and architecture in CSM and integrating an external database and simulation models

Design workflows described in Activity Diagrams (3) need the current set of specified parameters in order to run calculations. Hence, the Activity Diagram needs to be able to load parameters from the database. For loading parameters, a standardized activity in the Activity Diagram is defined, named getParameter (a). The activity takes the value of a parameter from the external database and provides it for further calculations in the Activity Diagram.

After calculating parameter values in external tools, these parameter values have to be transferred to the database again, so that they are usable for further design and test steps. The parameter values are then saved in the database and can be loaded and used by further workflows. For saving a parameter value in a database, another standardized activity is defined, named setParameter (c). This activity takes the value, which has been calculated and stores it in the database at the position of its corresponding parameter.

Calculating values for principle solution parameters needs external calculation models, which are usually implemented in domain specific tools. External calculations or simulations are integrated in Activity Diagrams using a standardized interface activity calculate (b). The interface activity defines standard inputs for the calculation model and delivers them to the model in an external tool (4), starts the model and takes the result parameters from it. In the Activity Diagram, the calculate activity provides various input ports, to which the values from getParameter activities are delivered, and various output ports, from which further calculations or setParameter activities may take values in order to process the results.

In a prototypical implementation for the given use case, we used MATLAB data files (.mat) instead of a full database. The concepts are transferable to other database types. All external interfaces to both, database and simulation models where implemented in Matlab functions using CSM's Matlab interface.

4.3. Design workflow for the hydrodynamic pump wheel

The concept described in 4.2 is applied to the design workflow of the hydrodynamic pump wheel from 4.1. We model an Activity Diagram, which loads the required parameter inputs from the external database, calculates the geometric and physical parameter values and stores the calculation results in the database. Since the calculation has impact on geometric and physical behaviour of the principle solution, in a last step, further describing models as CAD models and physical characteristic curves are updated based on the new parameter value set, Figure 5.

The required values *deltaPOpt*, *RhoCooling*, *RPM* and *maxVolFlow* are defined by further principle solutions, which interact with the hydrodynamic pump wheel according to the functional architecture, Figure 1. The listed parameters are hence modelled in their respective principle solutions, Table 1:

- *deltaPOpt*, the pressure delta is defined by the principle solution *release heat*
- *RhoCooling* is the desnsity of the chosen cooling medium for *manage heat flows*
- *RPM* is the optimal rotational velocity for the principle solution of *convert electrical to mechanical energy*
- *maxVolFlow*, the required volume flow for the optimal working point is defined by the necessary volume flow of *distributeHeat*

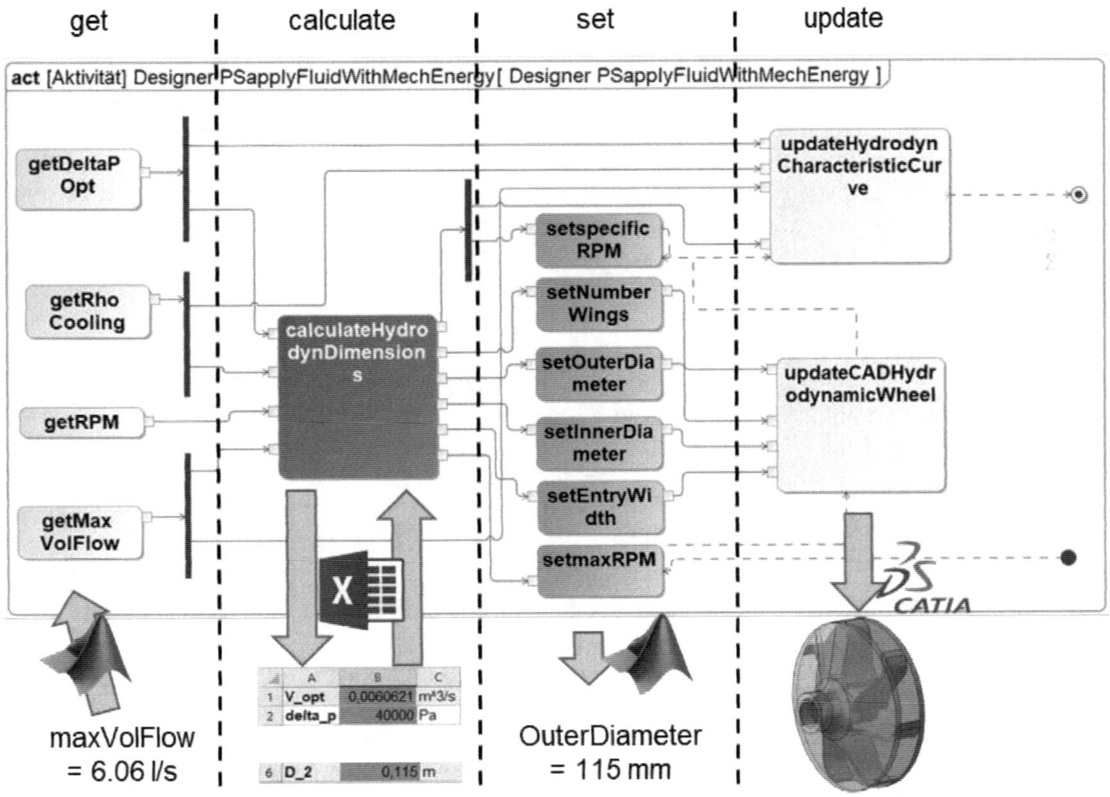

Figure 5. Design workflow for the hydrodynamic pump wheel in SysML

The calculation model implementing the design guideline is integrated in the activity *calculateHydrodynamicDimensions*. It estimates the previously described values of the hydrodynamic pump wheel geometric element. Additionally, two physical parameters are calculated, which result from the geometric parameter set: *specificRPM* and *maxRPM* are performance indicators describing the principle solution hydrodynamic pump wheel and its limitations and are used for estimating the characteristic curve of the principle solution. The estimated values are saved using individual setParameter-activities.

After parametrizing the principle solution, describing models have to be updated, so that they contain the new parameter set. In the case of the hydrodynamic pump wheel, describing models are the CAD model of the hydrodynamic wheel's active surfaces and the characteristic curve describing the behavior between pressure, volume flow and rotational velocity. The CAD model is prepared in a parametrized way: The parameters outer and inner diameter, outlet and inlet angle, and width are set as parameters in the CAD software Catia and exported to a parameter table. Changing the value in this parameter table automatically changes the value in the CAD model. The activity *updateCADHydrodynamicWheel* takes the calculated values and writes them into the aforementioned table. The activity *updateHydrodynCharacteristicCurve* recalculates the characteristic curve in MATLAB and stores it.

We have modelled an Activity Diagram which parametrizes the principle solution based on given requirements using design guidelines. As this describes the step to design the principle solution, we call this kind of Activity Diagram a *designer*. Designer Activity Diagrams are permitted to change geometric values of their principle solution in the database based on defined calculation models. They ensure that design decisions are made based on calculation models and document the calculations. As they are executable, design becomes automated.

5. Top level design workflow for the cooling circuit

In section 4, the principle solution of the hydrodynamic pump wheel is parametrized using a designer Activity Diagram. As described in 4.3, this designer requires input from other principle solutions. For example, the physical behaviour and parameter values from *distribute heat* define the required volume flow, as *distribute heat* describes the heat transfer functionally. Hence, the design of the principle solution hydrodynamic pump wheel cannot be done independently of the surrounding solutions. Further designers for other principle solutions are required. The designers have to be executed in an order, so that values can be estimated based on already calculated values from other principle solutions. Exemplary, we set up a second designer for *distribute heat*.

In distribute heat, the incoming heat from the engine is transferred to surrounding air and cooling medium and thereby removed from the system. Depending on the quantity of volume flow of the cooling medium, a different amount of heat is removed, resulting in different engine temperatures. We assume the geometric parameter values in distribute heat and the engine components as fixed. To estimate the required volume flow of the solution *distribute heat* we set up a physical model, describing the heat transfer behaviour in a physical simulation in Simcenter Amesim, Figure 6, a. The engine consists of two parts, the cylinder head and the crank case. Both may have different temperatures and are modelled as thermal capacities interacting with each other. The heat exchange with air is modelled. Heat introduced by engine and removed from cooling medium are thermal sources and sinks, introducing and removing an amount of heat from the capacities. As the engine components and the active surfaces are fixed, the heat transfer from engine and to air are fully described. The engine components temperatures are external requirements. The model can be used to estimate the required heat released to cooling medium in order to set the engine temperature. The heat released to cooling medium is varied until the temperatures fit the external requirements. Then from the required heat, the required volume flow is estimated. This is done for both, minimum and maximum required temperature, resulting in minimum and maximum required volume flow. The described calculation is implemented in the activity *calculateMinMaxVolumeFlow*. This activity takes the fixed values from engine geometry and *distribute heat* and the required temperatures and gives minimum and maximum volume flows from it. The maximum volume flow can then be used in the designer of hydrodynamic pump wheel. The full Activity Diagram for *distribute heat* is modelled in Figure 6 b).

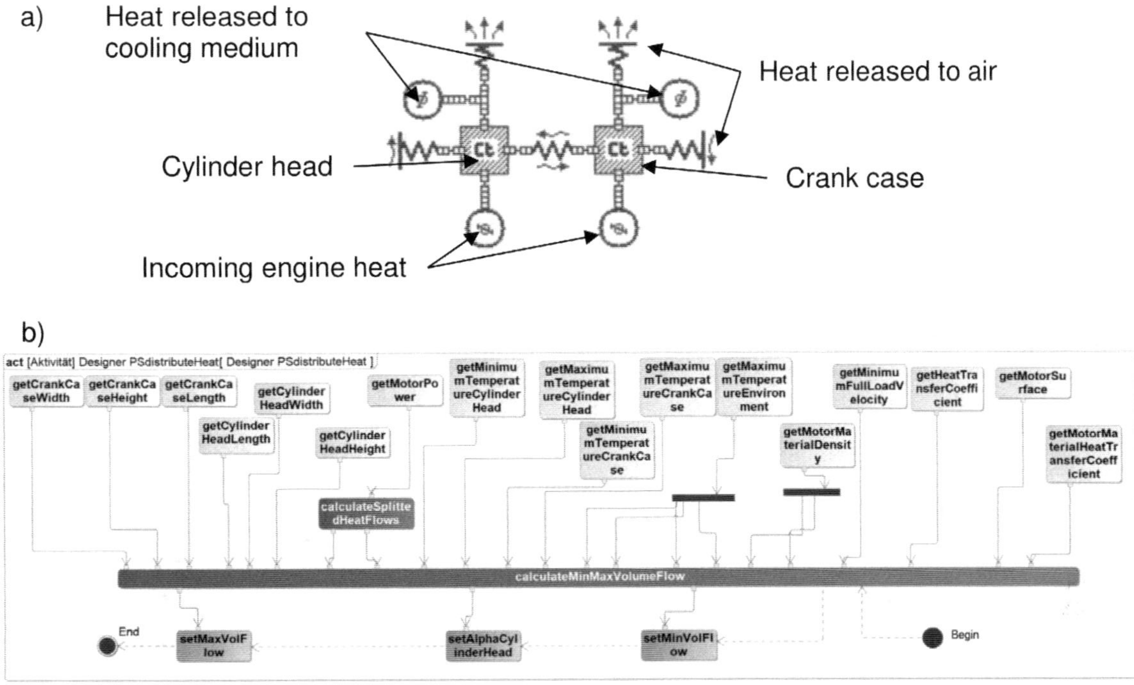

Figure 6. a) Heat transfer model b) workflow for calculating minimal and maximal necessary volume flow using the model from a)

As the designer applyFluidWithMechanicalEnergy, i.e. the hydrodynamic pump wheel from section 4.3 is only usable, when the value for *maxVolFlow* is set, an order for executing designers is necessary. This order describes the design activities necessary for the top-level solution *manage heat flows*. The top-level solution designer models the order in which principle solution designers are executed. The single designers are considered as one activity each in the overall designer for the full solution *manageHeatFlows*, Figure 7. This executable workflow can be used to redesign every principle solution one after another based on the formalized design process.

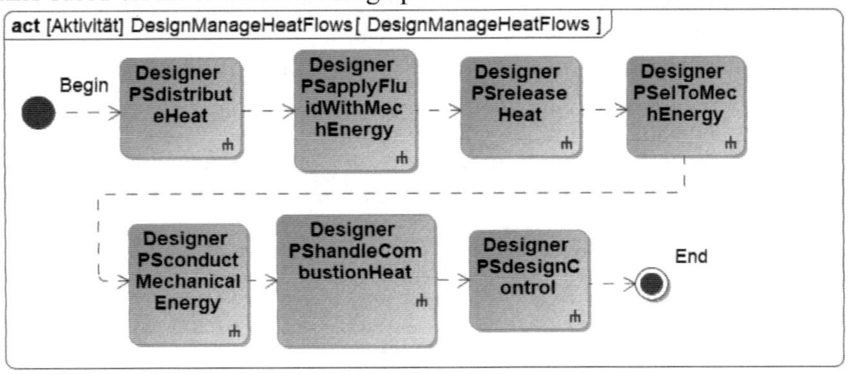

Figure 7. Overall design process of the solution *manageHeatFlows*, including the previously presented designers: Designer PSdistributeHeat from Figure 6 b) and Designer PSapplyFluidWithMechEnergy from Figure 5

6. Solutions Testing

The described design process uses explicit design cases to calculate principle solution parameters, e.g. the geometric values of the hydrodynamic pump wheel have been calculated for maximum volume flow.

Besides, not all requirements or use cases attached to the system are considered. Testing, if all requirements are satisfied for all use cases of the system, has to be done in a separate way. This is essential to deliver a fully working product and in case of requirement changes, in order to answer the question if a redesign is necessary.

Our system has three requirements to satisfy: temperature value ranges, the two engine components have to keep, and a maximum rotational speed for the hydrodynamic pump wheel. Testing the system behaviour has to regard the overall requirement fulfilment. We use two different simulation models to test the requirements. A Simulink model calculates the rotational speed of the hydrodynamic pump wheel and the volume flow created due to this rotational speed based on the hydrodynamic characteristic curve of the hydrodynamic pump wheel and the parameter values from *control heat flows* and *convert electrical to mechanical energy* (Table 1). The Simcenter Amesim model from Figure 6 a) is used to calculate the engine component temperatures, which set due to the calculated volume flow.

The previously named designers are used to set solution parameters, e.g. geometric parameters. However, they do not test further requirements and system interactions. To integrate the described tests into our system modeling approach, we introduce another workflow model. Tests are also modelled as Activity Diagrams; however, their purpose is different. A test does not manipulate values of system parameters, but uses the values to calculate physical states by executing physical models, e.g. the engine temperatures using the heat transfer model. The resulting physical states are then compared to values given as requirements. Test Activity Diagrams load solution parameters using getParameter activities as defined for designers before. They execute external simulations and calculations using calculation activities. The calculated parameter values are stored and compared to the requirements, which are linked by a *satisfy*-connection between the parameter and a design constrained.
The described testing procedure including both simulation models is modelled as a test Activity Diagram. The tests load values from all principle solutions which are contained in *manageHeatFlows*. The test then triggers the two simulation models: At first the Simulink model is started (*testRPMAndVolFlow*), which calculates the rotational speed due to the control parameter values and the volume flow rate resulting from rotational speed. The volume flow rate is then transferred to the Amesim model describing heat flows and temperatures in the engine components (*calculateTemperature*). Results of rotational speed and temperatures are compared to the requirements. The workflow including both models is shown in Figure 8.

7. Use case: Changing the engine power

The described workflow modelling allows for straightforward system testing and redesign when dealing with changed requirements. In an artificial scenario from automotive engineering, we show the usage of the functional architecture modelled in SysML4FMArch using the executable Activity Diagrams linked to principle solutions. The cooling circuit has been developed for a 135kW engine. The temperature requirements are set for temperature of cylinder head between 120°C and 150 °C and for crank case between 100°C and 130°C. The maximum speed of the pump is set to 2500 rpm. The system setup satisfies these requirements.

In a first change, the cooling circuit is tested for a higher engine class. The new engine has a power of 185 kW. Due to higher power, the heat production of the engine increases, which implies the need to retest whether the cooling system still fulfils all requirements. The test workflow from section 6 fails, showing, that the temperature of both engine components is now much to high (270°C for crank case and 347°C for cylinder head). Executing the Activity Diagram that designs the cooling circuit function *manage heat flows* yields two possibilities to change the system: Manipulating only the rotational speed in order to create a higher volume flow rate or to change geometric values of the hydrodynamic pump wheel. It is assumed, that changing a control parameter value to increase rotational speed is less effort than changing the geometry.

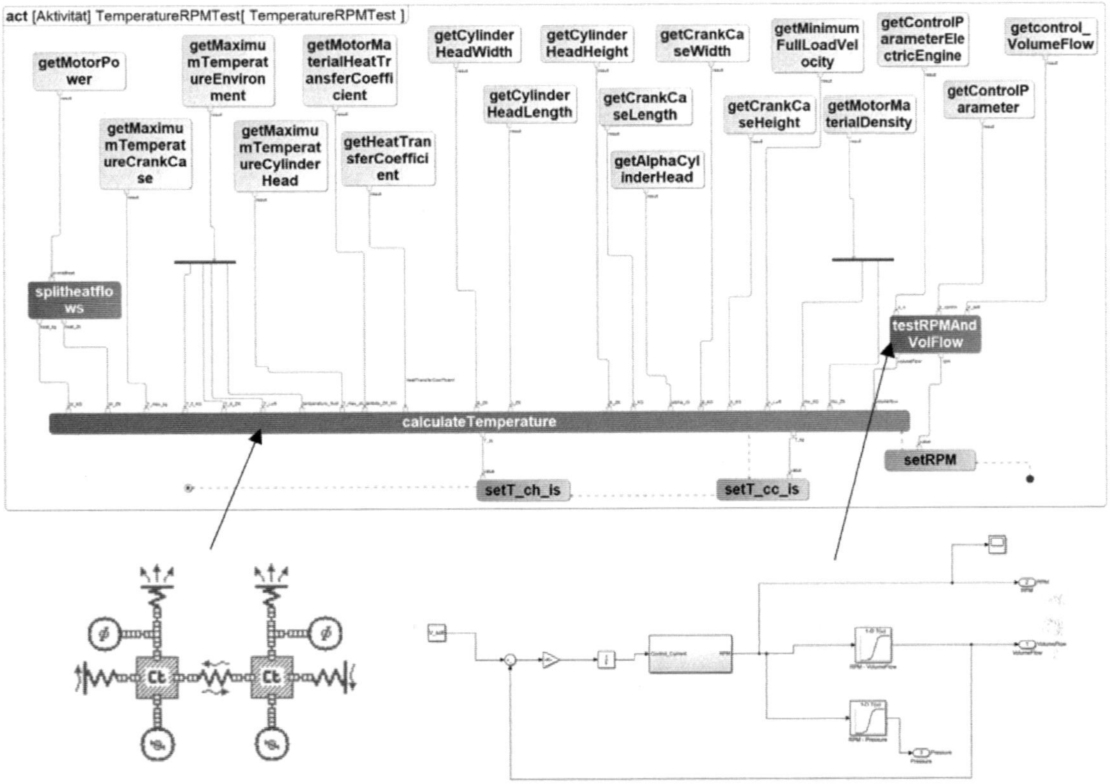

Figure 8. Test workflow to verify the engine components temperatures and the rotational speed of the pump wheel

Using the design Activity Diagram for control heat flows gives the new parameter value. Re-executing the functional tests reveals, that the temperature requirements are now satisfied. Rotational speed has increased to 2011 rpm.

In a second change, the engine power is raised to 250 kW. Controller redesign increases rotational speed once again, so temperatures are satisfied again. This time, rotational speed of the pump wheel becomes too high (2700 rpm). Therefore, the full redesign process is repeated. The pump wheel geometry is redesigned and the control parameter is adapted. The final repetition of the tests shows, that both, temperatures and rpm are now in range of the requirements (120°C crank case, 136 °C cylinder head and 1500 rpm). To achieve this, the pump wheel width has been increased by 5 mm. Modelling the design process in reusable design workflows and development of tests for all system requirements allows for automatic redesign and the information whether system requirements are fulfilled at any time.

8. Discussion

Integrating simulation models into architecture development using executable workflows provides the possibility to reproducible design the system and test it regarding requirement satisfaction. We have shown, that it is possible to model mechanical design processes in executable Activity Diagrams. Especially guidelines for single principle solutions, e.g. the pump wheel, can be modelled that way. Such guidelines are often given in mechanical engineering (i.e. norms and standards). Our approach makes such norms reusable in an easy way. Designers handle external requirement values and other boundary conditions as well as requirements coming from other internal system solutions. Design changes can be seamlessly proceeded and corresponding CAD- and other describing models can be updated automatically.

Executable models of test workflows enable agile development and allow to use strong simulations for virtual testing. Following the thoughts of continuous integration in software engineering, it is possible to run all test workflows overnight in order to prove system validity regarding requirements.

However, tool integration of such procedures is not optimal. Currently the get and set functions have to be written manually for every parameter. Storing parameter values in a mat-file does not scale, if the number of parameters exceeds a certain number and does not allow for role definitions. However, there are various data management techniques, which could be integrated in our approach to provide efficient, safe and distributed data handling.

A bigger disadvantage of the graphical Activity Diagrams is, that they get crowded from get and set functions, when many parameters are given for a model. Here a textual alternative for connecting get- and set-elements with models would be a solution.

9. Summary and Outlook

In the present study, we proposed an approach for modeling mechanical design and test processes in executable workflows using SysML Activity Diagrams and external tools. We developed two kinds of Activity Diagram standards for dimensioning (designers) and testing (testers), respectively. Designers load values of solution parameters, run simulations and calculations based on these values, and calculate values of other solutions' parameters. The parameter values are stored and describing models as CAD are updated automatically. Tests run simulation models based on the current values of solution parameters and calculate the values of physical states that way. Afterwards, the values of physical states are compared against the requirements to prove requirement satisfactions.

The concepts were explained by the design of an electric mechanical coolant pump from automotive industry. In the use case, the power of the combustion engine was adapted, and the cooling system was able to automatically adapt to the new requirements. The use case reveals the benefits of automated redesign and testing as a reaction to requirement changes.

Future work aims to standardize and classify design and test workflows. Simulation models have varying detail level. Higher detail levels usually require more parameters to be specified. This also has influence on the workflow detail level. Classification is here a step to have suitable libraries for both simulation models and design and test workflows using the models. Engineers can then use the best suitable workflow for their current system status.

Another step for automatization is ontology development. Developing an ontology for principle solutions, models and workflows allows for a stronger automatization and closes the gap between physical modelling and architecture. Hereby, strong design and model catalogues can be created, making agile development in mechanical engineering easier.

References

[1] International Council on Systems Engineering 2007 *Systems Engineering Vision 2020* (Seattle: International Council on Systems Engineering)

[2] Andary F, Berroth J and Jacobs G 2019 An Energy-Based Load Distribution Approach for the Application of Gear Mesh Stiffness on Elastic Bodies *Journal of Mechanical Design* 2019 **141**

[3] Berroth J, Jacobs G, Kroll T and Schelenz R 2016 Investigation on pitch system loads by means of an integral multi body simulation approach *Journal of Physics: Conference Series* **753**

[4] Golafshan R, Jacobs G, Wegerhoff M, Drichel P and Berroth J 2018 Investigation on the Effects of Structural Dynamics on Rolling Bearing Fault Diagnosis by Means of Multibody Simulation *International journal of rotating machinery* **2018**

[5] Pasch G, Jacobs G, Höpfner G and Berroth J 2019 Multi-Domain Simulation for the Assessment of the NVH Behaviour of a Tractor with Hydrostatic-Mechanical Power Split Transmission *Land.Technik AgEng* **2019**

[6] Wegerhoff M, Jacobs G and Drichel P 2019 Noise, vibration and harshness validation methodology for complex elastic multibody simulation models: With application to an electrified drive train *Journal of Vibration and Control* **25**

[7] No Magic Inc 2020 *MagicDraw 19.0 LTR SP4 User Manual*
[8] Object Management Group 2019 *OMG System Modeling Language (OMG SysML) Version 1.6*
[9] Object Management Group 2015 *OMG Unified Modeling Language (OMG UML) Version 2.5*
[10] Dänzer M, Kleiner S, Lamm J G, Moeser G, Morant F, Munker F and Weilkiens T 2014 Funktionale Systemmodellierung nach der FAS-Methode: Auswertung von vier Industrieprojekten *Tag des Systems Engineering* **2014**
[11] Pearce P and Hause M 2012 ISO-15288, Oosem and Model-Based Submarine Design
[12] Weilkiens T 2016 *SYSMOD-The Systems Modeling Toolbox-Pragmatic MBSE with SysML* (Fredesdorf: MBSE4U)
[13] Pohl K, Broy M, Daembkes H and Hönninger H 2016 *Advanced Model-Based Engineering of Embedded Systems* (Berlin: Springer)
[14] Alt O 2012 *Modell-basierte Systementwicklung mit SysML* (München: Hanser)
[15] Schlecht S and Alt O 2007 Strategien zur Testfallgenerierung aus SysML-Modellen *Software Engineering*
[16] Gausemeier J, Dorociak R and Nyßen A 2010 The Mechatronic Modeller: A Software Tool for Computer-Aided Modeling of the Principle Solution of an Advanced Mechatronic System *11th International Workshop on Research and Education in Mechatronics*
[17] Eigner M, Koch W and Muggeo C 2017 *Modellbasierter Entwicklungsprozess cybertronischer Systeme* (Berlin: Springer)
[18] Moeser G, Kramer C, Grundel M, Neubert M, Kümpel S, Scheithauer A, Kleiner S and Albers A 2015 Fortschrittsbericht zur modellbasierten Unterstützung der Konstrukteurstätigkeit durch FAS4M *Tag des Systems Engineering* **2015**
[19] Koller R and Kastrup N 1994 *Prinziplösungen zur Konstruktion technischer Produkte* (Berlin, Heidelberg: Springer)
[20] Drave I, Rumpe B, Wortmann A, Berroth J, Hoepfner G, Jacobs G, Spuetz K, Zerwas T, Guist C and Kohl J 2020 Modeling Mechanical Functional Architectures in SysML *Proceedings of the 23rd ACM/IEEE International Conference on Model Driven Engineering Languages and Systems*
[21] Hernandez C, Rodriguez M, Diaz I and Sanz R 2016 Model Based Engineering of Process Plants using SysML *26th European Symposium on Computer Aided Process Engineering*
[22] Biswas N, Chattopadhyay S, Mahapatra G, Chatterjee S and Mondal K C 2017 SysML Based Conceptual ETL Process Modeling *Communications in Computer and Information Science* **776**
[23] Markthaler M, Kriebel S, Salman K S, Greifenberg T, Hillemacher S, Rumpe B, Schulze C, Wortmann A, Orth P and Richenhagen J 2018 Improving model-based testing in automotive software engineering *IEEE/ACM 40th International Conference on Software Engineering: Software Engineering in Practice Track (ICSE-SEIP)*
[24] Drave I, Hillemacher S, Greifenberg T, Kriebel S, Kusmenko E, Markthaler M, Orth P, Salman K S, Richenhagen J, Rumpe B, et al. 2019 SMArDT modeling for automotive software testing *Software: Practice and Experience* **49** 2
[25] Drave I, Hillemacher S, Greifenberg T, Rumpe B, Wortmann A, Markthaler M and Kriebel S 2018 Model-Based Testing of Software-Based System Functions *44th Euromicro Conference on Software Engineering and Advanced Applications (SEAA)*
[26] Reiß D and Rumpe B 2012 Using Lightweight Activity Diagrams for Modeling and Generation of Web Information Systems *International United Information Systems Conference*
[27] Schonherr O and Rose O 2009 First steps towards a general SysML model for discrete processes in production systems *Proceedings of the 2009 Winter Simulation Conference (WSC)*
[28] Patel V V, McGregor J D, Goasguen S 2010 SysML-Based Domain-Specific Executable Workflows *IEEE International Systems Conference*
[29] MathWorks 2020. *Matlab 2020b - Manual*
[30] Siemens PLM Software 2018 *Teamcenter Rapid Start 11.6 - Setting Up Workflows for Product*

Development

[31] Zerwas T, Jacobs G, Höpfner G, Drave I, Berroth J, Guist C, Spuetz K, Konrad C, Rumpe B and Kohl J 2021 Mechanical Concept Development using Principle Solution Models *Antriebstechnisches Kolloquium 2021*

[32] Wesche W 2016 Radiale Kreiselpumpen: *Berechnung und Konstruktion der Hydrodynamischen Komponenten* (Berlin: Springer Vieweg)

IOP Conf. Series: Materials Science and Engineering 1097 (2021) 012005 doi:10.1088/1757-899X/1097/1/012005

Investigations on the Micro-Pitting and Wear Behavior of Nitrided Internal Gears

M Geitner[1], B Zornek[2], T Tobie[1] and K Stahl[1]

[1] Gear Research Centre (FZG), Technical University of Munich, 85748 Garching, Germany

[2] Magna PT B.V. & Co. KG, 74199 Untergruppenbach, Germany (former member of FZG)

Abstract. Nitriding processes can significantly increase the load carrying capacity of gears, compared e.g. to through hardened gears. Nitrided gears are generally characterized by a thin and hard compound layer on the surface and an adjacent diffusion layer that extends deeper into the material. Structure and properties of the compound layer significantly determine especially the tribologically dominated load carrying capacity characteristics of the tooth flank surface such as micro-pitting and wear. Previous investigations on nitrided gears were mainly focused on fatigue limits for the surface durability against pitting and tooth root bending strength. Furthermore, mostly external nitrided gears were investigated. Results of systematic investigations on the wear and micro-pitting behavior of nitrided internal gears are scarcely available. Based on the investigations in the research project FVA 482 IV, experimental test results confirm that nitrided internal gears in the material pairing "case hardened/nitrided" provide a significantly better performance regarding wear compared to through hardened or case hardened internal gears. Furthermore, it was shown that nitriding can have a positive influence on the micro-pitting behavior. A detailed analysis of the microstructure shows that the properties of the compound layer are primarily responsible for both, the micro-pitting and wear behavior.

1. Introduction

As the requirements considering load carrying capacity and manufacturing costs of transmission applications have increased within the last decades, alternative heat treatment processes compared to the established case hardening process became applicable. The shape stability of nitrided components during the hardening procedure is one of the main advantages of the nitriding treatment. Consequently, the effort regarding the finishing process can be reduced. Furthermore, nitriding leads to an increased corrosion and heat resistance. Within several investigations on nitrided components, an improved wear behavior was shown. [1, 2]

In numerous transmission applications (e.g. in automotive gear boxes such as automatic transmissions but also in industrial and wind energy transmission systems), planetary gear stages are used due to their high power density and compact design with coaxial in- and output. Nitriding can be a beneficial heat treatment process especially for internal gears.

Previous investigations on nitrided gears were mainly focused on fatigue limits for the surface durability against pitting and tooth root bending strength. Furthermore, mostly external nitrided gears were investigated. [3] However, results of systematic investigations on the wear and micro-pitting behavior of nitrided internal gears are scarcely available even though an increasing number of nitrided internal gears is used in different applications, e.g. in wind turbine gearboxes, where challenging tribological conditions can often be found.

Systematic theoretical and experimental investigations on the tribological behavior of nitrided internal gears were performed within the research project FVA 482 IV [4]. To identify load carrying capacity characteristics and to verify the applicability of the existing calculation methods for internal gears, an extensive program of gear running tests regarding micro-pitting and slow speed wear resistance have

IOP Conf. Series: Materials Science and Engineering 1097 (2021) 012005 doi:10.1088/1757-899X/1097/1/012005

been carried out. This paper is focused on the influence of different nitriding layer properties of internal gears resulting from gas nitriding processes on the micro-pitting and wear behavior.

2. Gear Failure Description – Micro-Pitting and Wear

2.1. Micro-Pitting

Niemann and Winter [5] as well as Linke [6] describe micro-pitting as a lubricant-related fatigue damage, which can occur on surface hardened gears even at low loads, primarily in the negative specific sliding area below the pitch circle. Due to a removal of the roughness peaks and a development of microscopic pores, the tooth flank is characterized by a matt and grey appearance. Micro-pitting can lead to profile deviations of the tooth flank and consequently to a changed dynamical behavior. Furthermore, micro-pitting can have an influence on the load carrying capacity considering consecutive failure types, such as pitting. The main influencing factors on the micro-pitting load capacity are the hardness of the material, the calculated lubricant film thickness, the roughness of the tooth flank as well as the used lubricant and its additives [7].

Extensive studies on the micro-pitting behavior of case hardened gears are available. An overview of the current state of knowledge on the micro-pitting resistance can be found in [8].

ISO/TS 6336-22 [9] represents a standardized calculation method of the micro-pitting load capacity of cylindrical gears. This technical specifiaction is mainly based on tests with case hardened gears, but its application is basically not limited. Acc. to ISO/TS 6336-22, gas nitrided gears with a surface hardness > 850 HV show a reduction of the micro-pitting resistance of approx. 33 % compared to case hardened gears.

2.2. Wear

According to Niemann and Winter [5], wear occurs primarily at low EHL film thicknesses, i.e. at low circumferential speeds. Wear is initially characterized by material removal in the flank areas with high sliding speeds above and below the pitch circle. The resulting scouring increases continuously with increasing running time and can later also lead to material removal in the area of the pitch point. Since wear is a continuously progressing damage mechanism, it is not counted among the classical fatigue damages.

Plewe [2] investigated the wear behavior of slow running spur gears systematically. The experimental investigations carried out at circumferential speeds of $v_t \leq 0.5$ m/s show that wear is highly dependent on the used lubricant, the paired materials and the surface hardness. According to Plewe [2], a calculation approach was developed to determine the amount of wear as a function of the lubricant film thickness. However, this approach only provides an approximation on the wear load capacity since the input data for the calculation approach refers to the pitch point. An equidistant abrasion over the tooth flank is assumed. On the basis of this calculation approach, Schudy [10] developed a local, iterative procedure for calculating the wear resistance.

3. Microstructure of Gas Nitrided Components

The thermochemical process of gas nitriding is characterized by interfacial reactions or diffusion processes in the formed boundary layers and in the base material. The sub-processes of the gas-solid reactions taking place in this process are shown schematically in Figure 1. Nitrogen is present on the surface, mostly within process-specific nitrogen compounds during nitriding. Depending on the process, it is released as atomic nitrogen by various mechanisms and can diffuse into the ferrous material of the component. A α-Fe-N mixed crystal with dissolved nitrogen is initially formed in the surface layer. If the nitrogen concentration in the α-Fe-N mixed crystal exceeds 0.11 mass %, cubic face-centered γ' nitride nuclei (Fe_4N) grow from the surface into the interior of the component. If the nitrogen solubility in γ' nitride (5.7 – 6.1 mass %) is exceeded, a phase transformation from cubic face-centered γ' nitride to hexagonal ε nitride (Fe2-3N) with a nitrogen solubility between 7.8 and 11.3 mass % occurs. By growing together, the γ' and ε iron nitride nuclei from a continous compound layer (CL). Below the compound layer, a diffusion layer of ferritic α-Fe with dissolved nitrogen follows [11, 12]. The compound layer thickness (CLT) and the phase structure of the compound layer as well as the depth and hardness distribution can be influenced by varying the nitriding conditions. To quantify the nitriding effects, the nitriding parameter K_N is defined including the ratio of ammoniak within the atmosphere during the nitriding process.

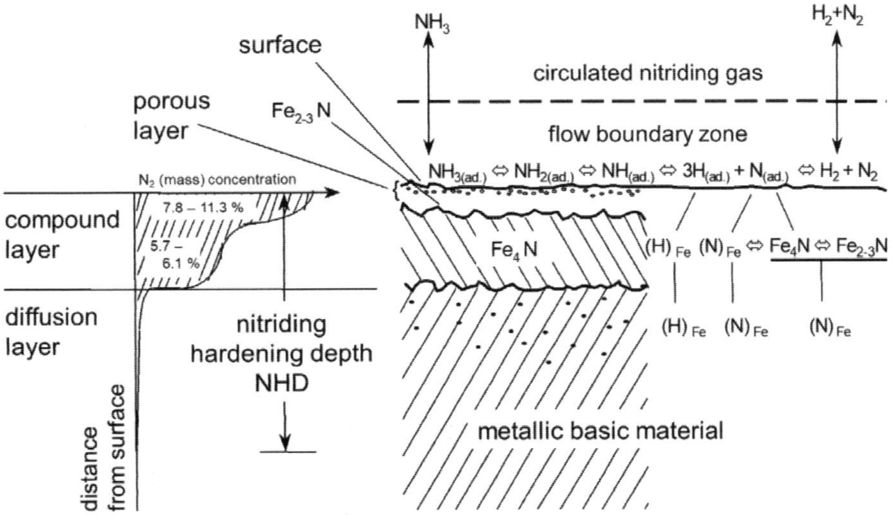

Figure 1. Microstructure of gas nitrided components, acc. to [13].

Depending on the nitriding conditions and the material composition, the compound layer is typically several microns thick and consists of iron and alloying element nitrides. In the near-surface area of the compound layer, pores form during nitriding depending on the material and the process. A porous layer thicknesses (CLT_p) of 30 – 40 % (relatively to the total thickness of the compound layer) is considered as characteristic. In contrast to the overlying compound layer, the diffusion layer extends to a depth of several tenths of a millimeter.

The nitriding hardening depth (NHD) is used as a measurable indicator for the thickness of the diffusion layer. It is determined from a hardness curve. In accordance with DIN 50190-3 [14], the NHD is the vertical distance from the surface to the point where the hardness corresponds to an appropriately defined limit value. It depends mainly on the nitriding temperature, the nitriding duration and the type and quantity of present alloying elements. For cylindrical gears, ISO 6336-5 [15] specifies a fixed limit value of 400 HV for determining the NHD. Only if the measured core hardness of the nitrided component exceeds a value of 380 HV, a limit hardness value of core hardness + 50 HV is considered. Consequently, the NHD is affected by the resulting strength within the previous through hardening process (quenching and tempering).

4. Experimental Investigations on Internal Gears

4.1. FZG Test Rig for Internal Gears

For investigations on the load carrying capacity of internal spur and helical gears, a special test rig was designed and constructed at the gear research centre (FZG) (see Figure 2). The FZG test rig for internal gears is based on the standardized FZG back-to-back concept [16]. Load carrying capacity tests for different gear pairings and materials can be performed. Within the test gearbox, both test gear and drive gear stage are included. The three planets positioned uniformly across the circumference are in mesh with the test and the drive ring gear. Within this concept, a sun gear is not enclosed and necessary since the focus of testing considers the convex-concave contact condition of external and internal gear. The planet carrier is driven by a continuously variable electric motor. The circumferential speeds can be adjusted variably by an additional shiftable countershaft transmission in a range of v_t = 0.05 – 10 m/s. Hence, the test rig is suitable especially for investigations at low and intermediate circumferential speeds. By a hydrostatic bracing unit, realistic and application-oriented stresses (maximum torque at the test ring gear $T_{2,max}$ = 6000 Nm) can be applied. [17, 18]

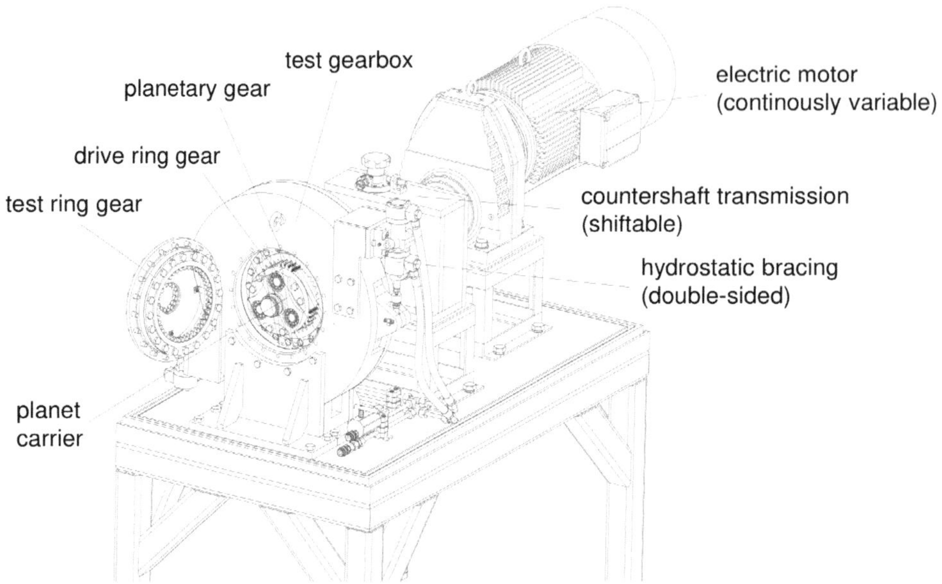

Figure 2. FZG test rig for internal gears [17].

4.2. Test Gears

For the experimental investigations, the established FZG internal reference spur gear design [8] was used. An overview of the main gear geometry data of the paired case hardened (Eh) planets and the nitrided (N) ring gears is given in Table 1.

Table 1. Data of the investigated test gear geometry, acc. to [4, 19].

Parameter	Unit	FZG internal spur gear	
		Planet (case carburized – Eh, material: 18CrNiMo7-6	Ring gear (gas nitrided – N, material: 31CrMoV9)
Normal module m_n	mm	4.5	
Number of teeth z	-	16	-42
Face width b	mm	16.0	10.0 [a)] (14.0)
Normal pressure angle α_n	°	20.0	
Working pressure angle α_{wt}	°	21.29	
Helix angle β	°	0	
Transverse contact ratio ε_α	-	1.437	
Addendum modification coefficient x	-	0.1817	-0.2962
Reference diameter d	mm	72.0	-189.0
Tip circle diameter d_a	mm	82.5	-185.0
Center distance a	mm	-59.0	

[a)] Considering an end relief with a length of 2 mm on each side.

The case hardened planets were conventionally ground after the heat treatment process. The nitrided ring gears were ground after hobbing to the targeted micro geometry before the gas nitriding process. A short involute tip relief (Ca = 35 ± 5 μm) was applied to all test gears. Additionally, the ring gears were ground with an adequate end relief (Ce = 12.5 ± 2.5 μm) with a length of 2 mm on each side to avoid bulges in the edge area after nitriding. The planets were characterized by an average flank surface roughness of Ra = 0.23 μm (min: 0.19, max: 0.31 μm) after grinding, while the ring gears showed an average of Ra = 0.46 μm (min: 0.34, max: 0.58 μm) after nitriding. With regard to tooth profile and flank shape as well as the pitch properties, almost all used planet gears fulfilled the quality requirement $Q \leq 5$ according to DIN 3962 [20] in new condition. Before nitriding, only some of the test ring gears showed qualities up to $Q \leq 7$, especially with regard to the flank shape. Nitriding additionally reduced the quality of the test ring gears, especially considering the profile angle deviation. Consequently, more frequent toothing qualities of up to $Q \leq 7$ were found. Figure 3 shows exemplary measurements on the ring gears of the tooth profile and flank shape after the gas nitriding process. In average, the nitriding process led to a reduction of the toothing quality by 1 – 2 stages.

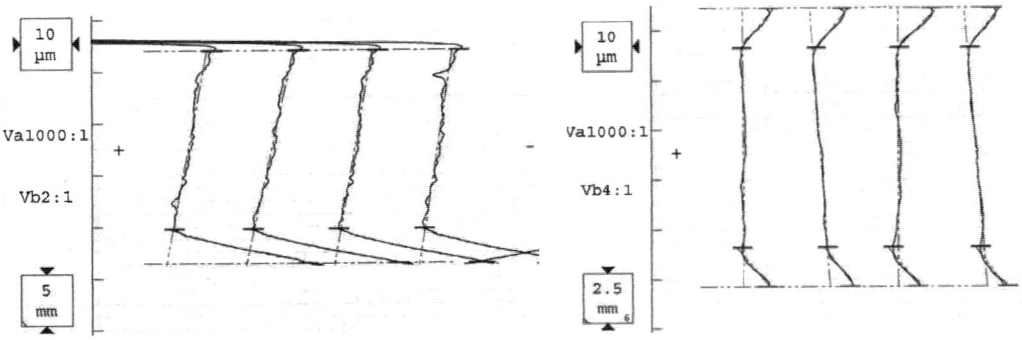

Figure 3. Exemplary measurements (nitrided ring gears) of the tooth profile (left) and flank (right) shape.

Within Table 2, the different investigated nitrided ring gears are specified. A reference variant, which was gas nitrided according to the recommendations of ISO 6336-5 [15] (NHD ≈ 0.3 – 0.4 mm for m_n = 4.5 mm), allows the comparison of the wear and micro-pitting behavior to variants with further variation of the nitriding parameters. To analyze the influence of the NHD, investigations on two variants (NHD1 and NHD2) with reduced values of the NHD were performed. The reduction of the NHD was mainly realized by a reduced nitriding time. Furthermore, two different ring gear variants (CL1 and CL2) with alternative compound layers (variation of CLT and phase composition) were included into the investigations.

Table 2 shows a comparison of the measured parameters of the nitriding hardening depth NHD, the compound layer thickness CLT, the porous layer thickness CLT_p, the core hardness and strength as well as the phase composition (determined by radiographic phase analysis and supplemented by Glow Discharge Optical Emission Spectroscopy) of the compound layers for the different investigated nitrided variants.

NHD1 and NHD2 showed a significantly reduced CLT compared to the reference, a porous layer could not be detected. CL1 is characterized by a reduced CLT and CLT_p, while the NHD is comparable to the reference variant R. For CL2, an increased NHD with comparable values of CLT and CLT_p to the reference is shown. Whereas mostly ε nitride phases can be identified for the reference, mainly γ‘ nitride phases are identified in the compound layers of the other variants.

Table 2. Overview of the nitriding layer properties of the investigated gas nitrided internal test ring gear variants.

Variant Nitriding treatment	NHD_{400HV} in mm	CLT in µm	CLT_p in µm	Core hardness in HV10	Core strength (acc. to ISO 18265) in N/mm²	CL phases
R (Reference) 510 °C, 30 h	0.38	11.4	3.7	276	861	ε nitride (mostly)
NHD1 500 °C, 4 h, $K_N = 3$	0.12	0.7	0	326	1020	γ' nitride
NHD2 500 °C, 12 h, $K_N = 3$	0.22	2.6	0	322	1003	γ' nitride
CL1 520 °C, 42 h, $K_N = 1$	0.43	4.1	0.5	326	1020	γ' nitride
CL2 520 °C, 42 h, $K_N = 3$	0.53	12.6	4.5	327	1020	γ' nitride (mostly)

In Figure 4, the hardness curve measurement (considering the tooth flank area near the pitch point C) as well as the microscopic overview of the nital etched specimen are illustrated exemplary for the reference.

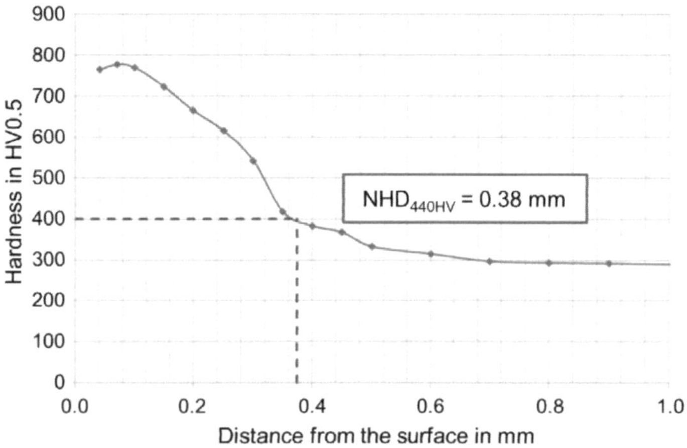

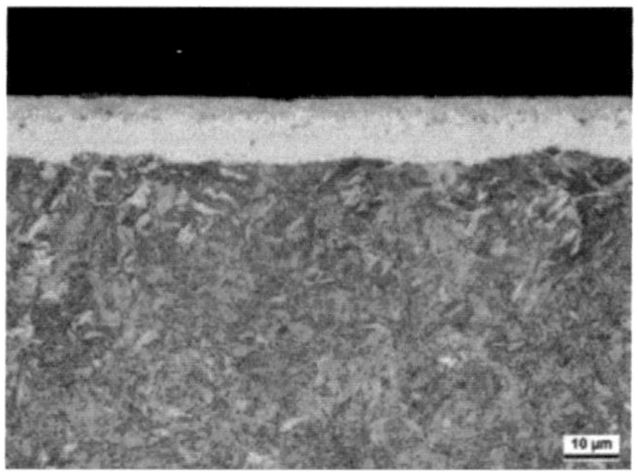

Figure 4. Hardness curve (left) and microscopic overview (right, nital etched, 1000:1) of the reference variant R.

4.3. Calculation of Representative Contact Pressure

The end relief on the investigated nitrided ring gears leads to a local load reduction in the edge area and to increased pressure levels in the center of the tooth width. The influences of this modified pressure distribution compared to uncorrected gear geometries were taken into account by means of the face load factor $K_{H\beta}$ (contact stress) according to ISO 6336-1 [21]. The factor $K_{H\beta}$ was therefore calculated depending on the applied loading on the internal gears using the FVA/FZG calculation program RIKOR [22]. All tests were performed under consideration of the adjusted Hertzian pressure p_C* in the pitch point C, which allows a comparison with results on test gears without modifications on the flank line. In the following, the relationship according to equation (1) is used for the considered load:

$$p_C{}^* = p_C \cdot \sqrt{K_{H\beta}} \tag{1}$$

4.4. Test Methods for Micro-Pitting and Wear

The investigations of the micro-pitting behavior include several load stage tests according to [8] with the internal gears. The test method follows the standardized micro-pitting test acc. to FVA 54/7 [23]. The tests were carried out at a circumferential speed of v_t = 2 m/s. Starting at p_C* = 800 N/mm² (BS6), the load was increased by 150 N/mm² in each load stage (BS). At a load of $p_C^* = 1550$ N/mm² (BS11), the test runs were completed. All load stage tests were performed using the reference oil FVA 3A (FVA 3 with an additive content of 4 % by volume Anglamol 99) at an oil injection temperature of $\vartheta_{in} = 60$ °C. Running time for each load stage was 16 hours.

For investigations on the wear behavior of the different variants, speed stage tests according to [8] (following the wear test acc. to DGMK 377 [24]) were carried out with the internal gears. The circumferential speed was reduced stepwise in each speed stage (GS) between $v_t = 2$ m/s (GS1) and 0.05 m/s (GS7). All speed stage tests were carried out with an oil injection lubrication using the non-additivated mineral oil FVA 3 (ISO VG 100) and at an injection temperature of $\vartheta_{in} = 60$ °C. Running time per speed stage was 16 hours.
The speed stage tests were initially performed at a constant load of $p_C^* = 1550$ N/mm² similar to [8]. After the occurrence of several damages on the nitrided ring gears, the load was reduced to $p_C^* = 1300$ N/mm² for all variants not yet completed at this time. Despite this adjustment of the applied load, the speed stage tests resulted in comparable lubricant film thicknesses, especially at the investigated low circumferential speeds as shown in Table 3.

Table 3. Comparison of the minimum lubricant film thickness h_{min} in the pitch point within the speed stage test at $p_C^* = 1550$ N/mm² and $p_C^* = 1300$ N/mm² (FVA 3, $\vartheta_{in} = 60$ °C) [4, 19].

v_t	2.0 m/s	1.5 m/s	1.0 m/s	0.5 m/s	0.25 m/s	0.10 m/s
h_{min} in µm ($p_C^* = 1550$ N/mm²)	0.093	0.081	0.066	0.045	0.030	0.017
h_{min} in µm ($p_C^* = 1300$ N/mm²)	0.109	0.093	0.075	0.050	0.032	0.018

In the following, the experimental results of the load and speed stage tests are presented. Unless otherwise marked, the data represent average values from two test runs. Due to an increased number of further damages to the test ring gears at high loads (see Figure 8), not all load stage tests were evaluated up to the highest load of $p_C^* = 1550$ N/mm².
For all tests on internal gears, the driving direction considers the start of contact in the tooth root flank area of the ring gear.

5. Micro-Pitting: Results of the Load Stage Tests on Nitrided Internal Gears

In the load stage test, the reference showed only limited micro-pitting in the root flank area of the case hardened planet gears from the first load stage (BS6, $p_C^* = 800$ N/mm²). This micro-pitting developed slowly further during the test up to the highest load level (BS11, $p_C^* = 1550$ N/mm²). However, no damage on the tooth flank could be detected for the mating nitrided test ring gear until the end of the test. In Figure 5, exemplary tooth flank photographs of the planet and ring gear after the end of the load stage test are illustrated.

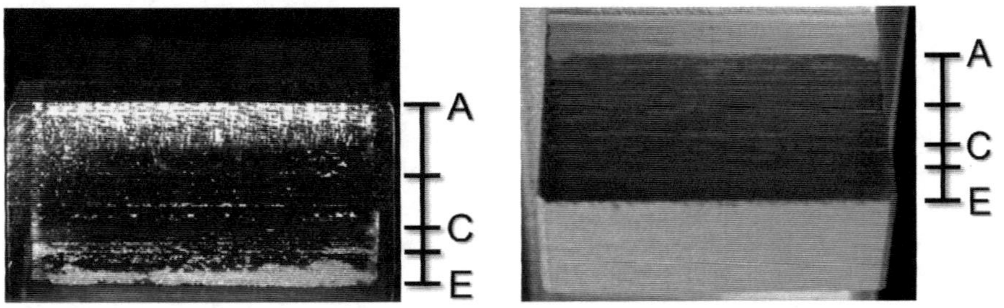

Figure 5. Exemplary tooth flank photographs of planet (left) and ring gear (right) for the reference after the end of the load stage test, after BS11 ($p_C^* = 1550$ N/mm²).

Figure 6 shows the development of the micro-pitted area GF_{Planet} and the average profile deviation $f_{fm,Planet}$ for the reference R (blue) determined on the planets after each load stage (BS). Although micro-pitting appeared to a limited area already in BS6 ($p_C^* = 800$ N/mm²), measurable profile deviations due to micro-pitting could only be observed from BS8 ($p_C^* = 1100$ N/mm²) on. In the last load

stage (BS11, $p_C{}^* = 1550\ \text{N/mm}^2$), a noticeable increase of the scouring depth in the root flank area of the case hardened planets was observed.
The measurements of the nitrided test ring gears showed no obvious changes in the profile shape. However, a slight material removal was observed in the area of the tip relief (especially at the tip edge) with increasing load and running time. By the end of the load stage test, an increase of the amount of the tip relief by approx. 5 µm could be determined. Consequently, the porous layer ($CLT_p \approx 3.7$ µm in new state) was completely removed near the tip edge.

5.1. Influence of the Nitriding Hardening Depth on the Micro-Pitting Behavior

To investigate the influence of the nitriding hardness depth on the micro-pitting resistance of nitrided internal gears, two variants with reduced NHD compared to the reference were tested. Both variants are characterized by comparatively thin compound layers without a porous layer. Variant NHD1 ($NHD_{400HV} \approx 0.1$ mm) did not show a continuous compound layer on the tooth flanks of the test ring gear at all. During the load stage test, the observed micro-pitting on the case hardened planetary gears clearly increased with decreasing nitriding depth and compound layer thickness. On the planets of the NHD1 and NHD2 variants, higher values of the micro-pitted area and average profile deviation were found compared to the reference (see Figure 6). For the variants NHD1 and NHD2 with reduced nitriding depth, no micro-pitting was observed on the tooth flanks of the nitrided test ring gears up to BS9 but further damages occurred in all test runs at loads of BS10 and higher ($p_C{}^* \geq 1400\ \text{N/mm}^2$) (see Figure 8).

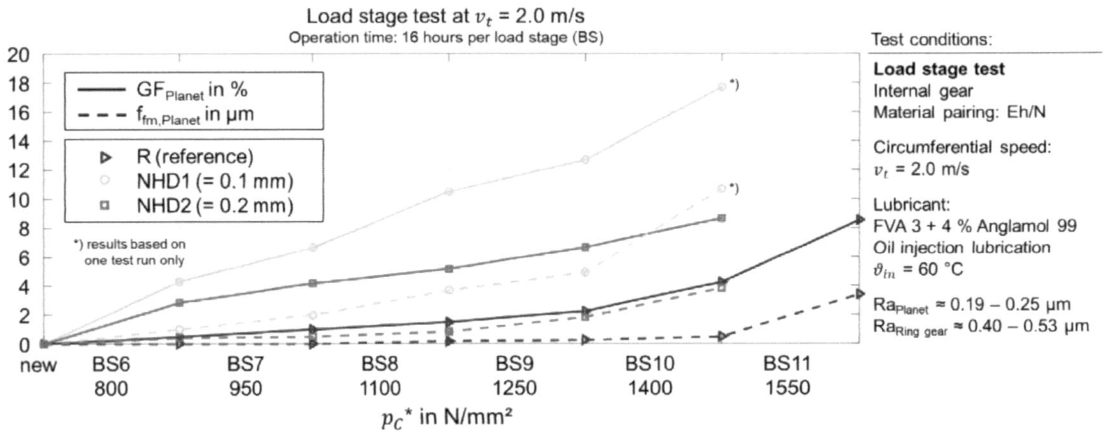

Figure 6. Micro-pitted area GF_{Planet} and average profile deviation $f_{fm,Planet}$ of the planet (variants NHD1, NHD2 and R) within the load stage test, evaluated after each load stage (BS), acc. to [4, 19].

5.2. Influence of the Compound Layer on the Micro-Pitting Behavior

The influence of thickness and composition of the compound layer on the micro-pitting behavior of nitrided internal gears was investigated using the variants CL1 ($K_N = 1$) and CL2 ($K_N = 3$). Figure 7 shows the development of the micro-pitted area GF_{Planet} and the average profile deviation $f_{fm,Planet}$ for these variants in comparison to the reference. In the pairing with the nitrided test ring gear of variant CL1, stronger micro-pitting was observed on the tooth flanks of the case hardened planets from BS6 ($p_C{}^* = 800\ \text{N/mm}^2$) on compared to the reference, both with regard to the micro-pitting area and the average profile deviation. On the tooth flanks of the nitrided test ring gears of CL1, a slightly higher material removal was observed in larger areas. It can be assumed that the thin porous layer ($CLT_p \approx 0.5$ µm) was already completely removed after the first intervals of the load stage test. In contrast, no micro-pitting on the planets could be identified during the entire load stage test for the variant CL2. Thereby, slight changes of approx. 3 µm in the profile shape of the nitrided test ring gear were found. Due to the increased porous layer thickness of CL2 ($CLT_p \approx 4.5$ µm), it is probable that even after completion of the load stage test there was still a porous layer existing on the tooth flanks of the ring gears of variant CL2.

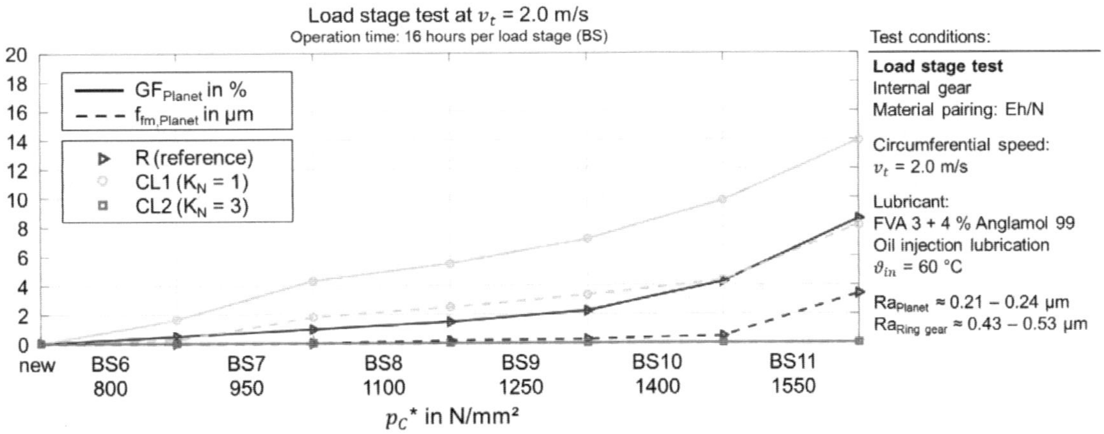

Figure 7. Micro-pitted area GF_{Planet} and average profile deviation $f_{fm,Planet}$ of the planet (variants CL1, CL2 and R) within the load stage test, evaluated after each load stage (BS), acc. to [4, 19].

5.3. Further Damages on the Nitrided Ring Gears

Except for the reference ring gear (31CrMoV9, industrially gas nitrided), the tooth flanks of the nitrided internal gears were damaged in at least one of the two test runs for each variant tested in the load stage test. Figure 8 shows an exemplary photograph of a damaged tooth of a nitrided test ring gear of the variant NHD1. The damage is located mainly in the tip area of the internal gear. It is assumed that the failure is initiated near the tip edge.

No comparable damage was found within further investigations on nitrided external gear teeth (not part of this paper) even at nominally higher values of the Hertzian pressure at the pitch point C (up to p_C* = 1850 N/mm²). Therefore, a local pressure distribution was calculated and analyzed using the FVA/FZG calculation program RIKOR [22] (see also Figure 8).

Due to the concave/convex pairing, internal gears show favorable radii of curvature over a wide range of the contact path (compared to external gears). Therefore, the planets of an internal gearing are supposed to transmit a significantly higher torque than the pinion of an external gearing in order to set comparable Hertzian pressures at the pitch point C. However, the radius of relative curvature ρ_{red} at the tip edge of the test ring gear shows comparable values to a related external gear pairing (see also [8]). Consequently, the tip edge area close to the end of contact (point E) of the test ring gear is exposed to unfavorable high loads, which likely lead to the observed damage.

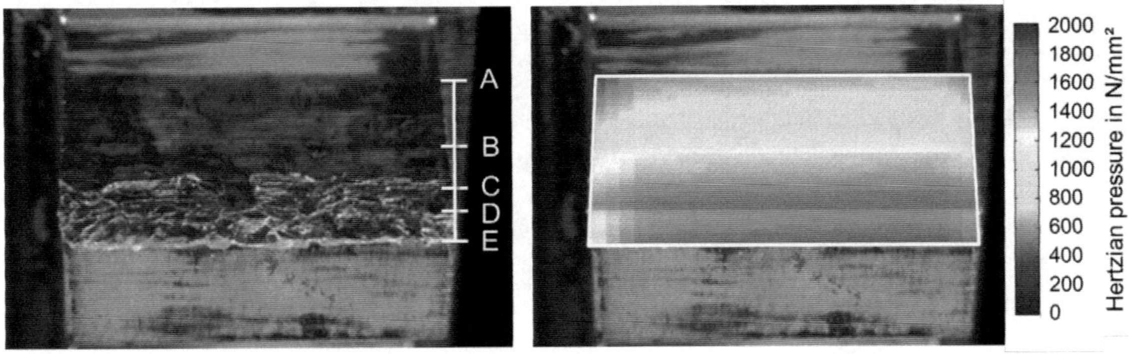

Figure 8. Exemplary photograph of the damaged tooth flank of a nitrided ring gear of the variant NHD1 after BS10 (left), local pressure distribution analysis for p_C* = 1550 N/mm² using RIKOR (right).

6. Wear: Results of the Speed Stage Tests on Nitrided Internal Gears

Figure 9 shows exemplary tooth flank photographs of the reference variant R for planet and ring gear from the speed stage test.

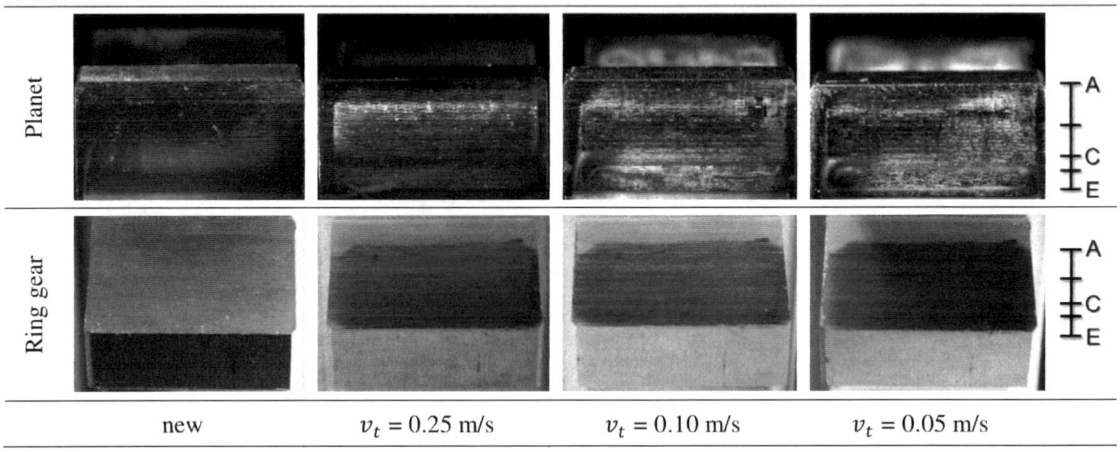

Figure 9. Exemplary tooth flank photographs of the planets and ring gears (variant R) in new state and for damage relevant speed stages from $v_t \leq 0.25$ m/s.

On the tooth flanks of the case hardened planet gears, no micro-pitting could be detected. From circumferential speeds of $v_t = 0.25$ m/s on, a wear related change in the flank surface of the case hardened planet gears was observed. In contrast, the nitrided ring gear showed no wear related damage on the tooth flanks until the end of the speed stage test.

In accordance with the occurring wear on the flank surface, strongly increasing material losses of the reference planets (see Figure 10, blue) could be observed at circumferential speeds of $v_t < 0.25$ m/s. The nitrided test ring gear showed hardly a mass decrease at low circumferential speeds. The shown values of the mass reduction in Figure 10 were converted to 100,000 load cycles per speed stage assuming a linear wear behavior. The development of the material losses of the ring gears within comparable speed stage tests (average of two test runs, [18]) for the paring Eh/V (performed at $p_C = 600$ N/mm², lubricant FVA 3A) and Eh/Eh ($p_C = 600$ N/mm², FVA 3A) is shown.

6.1. Influence of the Nitriding Hardening Depth on the Wear Behavior

In order to investigate the influence of the nitriding hardening depth on the wear behavior of nitrided internal gears, the variants NHD1 and NHD2 (with a smaller nitriding depth compared to the reference) were tested. Due to several damages concerning the nitrided test ring gears of these variants that occurred in the load stage tests, the speed stage tests were carried out at a load of $p_C^* = 1300$ N/mm². Despite the reduced load, damage to the tooth flanks of the nitrided test ring gears occurred in one of the test runs for NHD1 and in both test runs for NHD2. The mass reduction of the ring gears was therefore not evaluated.

Figure 10 shows the amounts of the material losses for the planets compared to the reference. For both variants NHD1 and NHD2, increased material losses of the planets compared to the reference can be identified. For NHD1 with a nitriding hardness depth of $NHD_{400HV} \approx 0.1$ mm, the thin and not completely continuous compound layer (CLT ≈ 0.7 µm) on the tooth flanks of the ring gears appears to affect the increased wear on the case hardened planets. The planets of the NHD2 variant showed a further increased wear than the NHD1 variant. It is assumed that the occurred damage on the tooth flanks of the ring gear has caused an increased release of especially hard particles out of the compound layer (CLT ≈ 2.6 µm), which has reinforced the wear on the planets. It can be assumed that the continuous removal of the porous layer produces particles with a maximum size characterized by the distance between the pores. Larger particles can be produced when the compound layer breaks out.

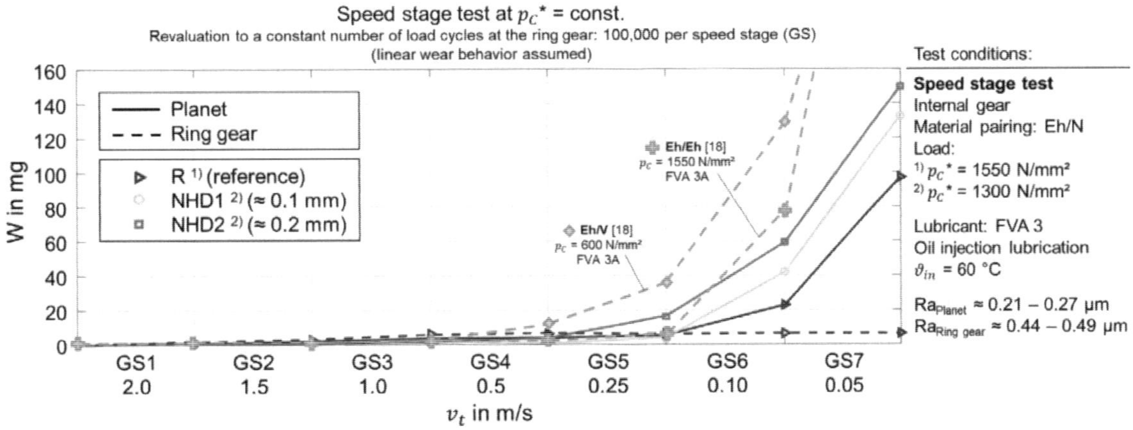

Figure 10. Cumulated material losses of the planets and ring gears (variants NHD1, NHD2 and R) within the speed stage test, revaluated to a constant number of load cycles per speed stage (GS), acc. to [4, 18, 19].

6.2. Influence of the Compound Layer on the Wear Behavior

In Figure 11, the amounts of the material losses are shown for the planets and ring gears of the variants CL1 (K_N = 1) and CL2 (K_N = 3) in comparison to the reference. In contrast to the reference, an increase in mass losses was determined on the test ring gears of both variants. For CL2, the highest values of mass reduction were identified. Due to the composition of the compound layer (15 % ε nitride) for the CL2 ring gear, the wear particles in the lubricant were characterized by an increased hardness compared to variant CL1. Correspondingly, a higher mass reduction was observed on the planets of variant CL2 compared to the reference. Due to the reduced hardness of the wear particles from the compound layer (consisting of γ' nitride) of the CL1 ring gear, this variant showed less negative effects on the wear behavior of the case hardened planets. Thereby, a more favorable wear behavior was observed on the planets in comparison to the reference.

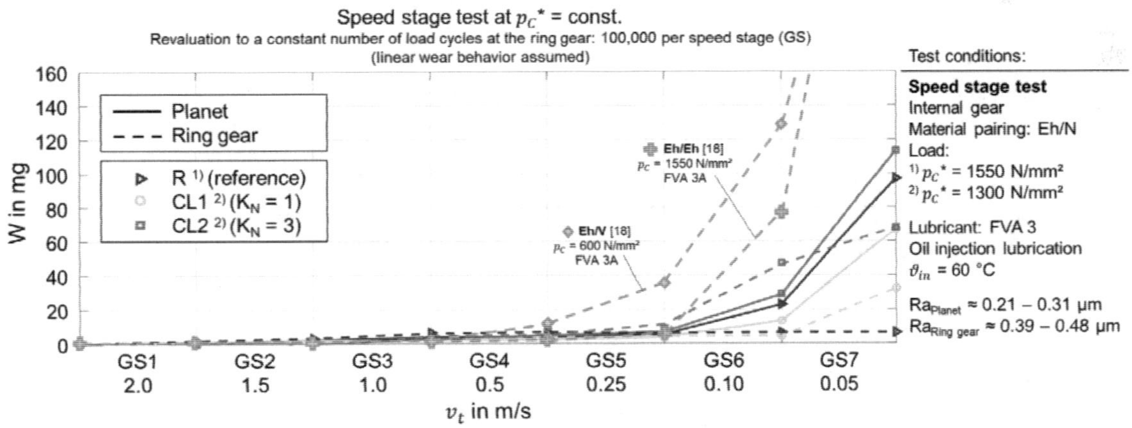

Figure 11. Cumulated material losses of the planets and ring gears (variants CL1, CL2 and R) within the speed stage test, revaluated to a constant number of load cycles per speed stage (GS), acc. to [4, 18, 19].

7. Summary and Outlook

Based on the results of the load stage tests (following the standardized micro-pitting test according to FVA 54/7 [23]) on nitrided internal gears in combination with case hardened planets, it was determined that the porous layer on the surface of the nitrided ring gears in particular has a significant influence on the micro-pitting behavior for the pairing Eh/N. As long as a continuous porous layer was available on the surface of the nitrided ring gear tooth flanks, no micro-pitting was observed. As soon as the porous layer of the tooth flank is removed (locally), micro-pitting can occur in the corresponding areas of the tooth flanks of the case hardened planets. The development of micro-pitting progresses particularly quickly with compound layers containing predominantly (less hard) γ' nitride whereas compound layers including a certain percentage of ε nitride are characterized by an increased (micro) crack growth restistance due to a raised hardness. To achieve a high micro-pitting resistance, a mixed γ'+ε nitride compound layer with a noticeable porous layer is recommended. It was shown that appropriate nitriding can have a positive influence on the micro-pitting behavior. [19]

For the classificiation of the wear resistance for nitrided internal gears, several speed stage tests (following the standardized wear test acc. to DGMK 377 [24]) were carried out. For a high wear resistance, the presence of a continuous compound layer and a suitable composition of this compound layer apperantly is required. A removal of the compound layer leads to a negative effect on the wear behavior of the pairing Eh/N, especially compound layers with only a small percentage (10 – 20 %) of ε nitride. Significant effects of thick porous layers on the wear resistance of the gears could not be determined within the experimental investigations.
All in all, the test results confirm that nitrided internal gears in the material pairing Eh/N provide a significantly improved performance regarding wear compared to through hardened or case hardened internal gears paired with case hardened planets. [18, 19]

Investigations on the pitting load capacity of nitrided internal gears were scarcely available so far. Consequently, systematic experimental tests comparably to [25] could complete the knowledge about the effects of nitriding properties of internal gears on the behavior considering the main flank damage types.

References

[1] Reese U 2000 Vergleich von thermochemischen Behandlungsverfahren mit und ohne Gefügeumwandlung *Stahl* vol 6 pp 62–64
[2] Plewe H-J 1980 Untersuchung über den Abriebsverschleiß von geschmierten, langsam laufenden Zahnrädern *PhD Thesis* (Munich: Technical University of Munich)
[3] Sitzmann A, Tobie T, Stahl K and Schurer S 2019 Influence of the Case Properties after Nitriding on the Load Carrying Capacity of Highly Loaded Gears *ASME Proceedings 2019* (Anaheim) IDETC2019-97405
[4] Zornek B, Tobie T and Stahl K 2017 Nitrierte Innen-Außenverzahnung: Tribologische Tragfähigkeit von nitrierten Innen- und Außenverzahnungen bei geringen Umfangsgeschwindigkeiten *Final report FVA 482 IV* vol 1206 (Frankfurt: Forschungsvereinigung Antriebstechnik e.V.)
[5] Niemann G and Winter H 2003 *Maschinenelemente: Getriebe allgemein, Zahnradgetriebe-Grundlagen, Stirnradgetriebe* vol 2 (Berlin Heidelberg New York: Springer) 2nd ed
[6] Linke H 2010 Stirnradverzahnung (München Wien: Hanser) 2nd ed
[7] Schönnenbeck G 1984 Einfluß der Schmierstoffe auf die Zahnflankenermüdung im Umfangsgeschwindigkeitsbereich 1 … 9 m/s *PhD Thesis* (Munich: Technical University of Munich)
[8] Schudy J 2008 Flankentragfähigkeit Innen-/Außenverzahnung: Flankentragfähigkeit von innen- und außenverzahnten Stirnrädern bei geringen Umfangsgeschwindigkeiten *Final report FVA 482 I* vol 867 (Frankfurt: Forschungsvereinigung Antriebstechnik e.V.)
[9] International Standard Organisation 2018 Calculation of load capacity of spur and helical gears: Calculation of micro-pitting load capacity *ISO/TS 6336-22*
[10] Schudy J 2010 Untersuchungen zur Flankentragfähigkeit von Außen- und Innenverzahnungen: Einflüsse auf das Grübchen-, Grauflecken- und Verschleißverhalten, insbesondere bei langsam laufenden Getriebestufen *PhD Thesis* (Munich: Technical University of Munich)
[11] Somers M A J and Mittemeijer E J 1992 Verbindungsschichtbildung während des Gasnitrierens und des Gas- und Salzbadnitrocarburierens *Härterei-Technische Mitteilungen – HTM* vol 47 pp 5–13

[12] Spies H-J and Bergner D 1992 Innere Nitrierung von Eisenwerkstoffen *Härterei-Technische Mitteilungen – HTM* vol 47 pp 346–356
[13] Klümper-Westkamp H 1989 Entwicklung und Anwendung eines Nitriersensors zur in-situ-Erfassung des Nitrierprozesses *VDI-Fortschritts-Berichte* vol 5
[14] Deutsches Institut für Normung e.V. 1979 Härtetiefe wärmebehandelter Teile: Ermittlung der Nitrierhärtetiefe [eng.: Hardness depth of heat-treated parts: Determination of the effective depth of hardening after nitriding] *DIN 50190-3*
[15] International Standard Organisation 2016 Calculation of load capacity of spur and helical gears: Strength and quality of materials *ISO 6336-5*
[16] International Standard Organisation 2000 Gears - FZG test procedures: FZG test method A/8,3/90 for relative scuffing load-carrying capacity of oils *DIN ISO 14635-1*
[17] Höhn B-R, Stahl K, Schudy J, Tobie T and Zornek B 2011 Investigations on the Flank Load Carrying Capacity in the Newly Developed FZG Back-to-Back Test Rig for Internal Gears *Proceedings Fall Technical Meeting 2011* (Alexandria: AGMA)
[18] Stahl K, Höhn B-R, Schudy J, Tobie T and Zornek B 2013 Investigations on the Flank Load Carrying Capacity of Internal Gears *Proceedings International Conference on Gears 2013, VDI-Berichte* vol 2199
[19] Zornek B 2018 Untersuchungen zur Flankentragfähigkeit vergüteter und nitrierter Innen- und Außenverzahnungen *PhD Thesis* (Munich: Technical University of Munich)
[20] Deutsches Institut für Normung e.V. 1978 Toleranzen für Stirnradverzahnungen Toleranzen für Abweichungen einzelner Bestimmungsgrößen *DIN 3962*
[21] International Standard Organisation 2019 Calculation of load capacity of spur and helical gears: Basic principles, introduction and general influence factors *ISO 6336-1*
[22] Hein M, Fingerle A, Tobie T, Otto M and Stahl K 2018 RIKOR Lastkollektiv: Auslegung von Verzahnungskorrekturen bei Lastkollektiven *Final report FVA 722 I* vol 1309 (Frankfurt: Forschungsvereinigung Antriebstechnik e.V.)
[23] Forschungsvereinigung Antriebstechnik e.V. 1993 Micropitting: Test procedure for the investigation of the micro-pitting capacity of gear lubricants *Information Sheet 54/7*
[24] Höhn B-R, Bayerdörfer I and Michaelis K 1997 Method to Assess the Wear Characteristics of Lubricants - FZG Test Method C/0,05/90:120/12 *Information Sheet DGMK 377-1* (Hamburg: Deutsche Wissenschaftliche Gesellschaft für Erdöl, Erdgas und Kohle e.V.)
[25] Geitner M, Zornek B, Tobie T and Stahl K 2019 Investigations on the pitting load capacity of internal spur and helical gears *Forschung im Ingenieurwesen* vol 83 pp 553–561 URL: https://link.springer.com/article/10.1007/s10010-019-00327-8

Acknowledgments
The presented results are based on the research project IGF no. 17730-N/1 (FVA 482/IV) undertaken by the Research Association for Drive Technology e.V. (FVA); supported partly by the FVA and through the German Federation of Industrial Research Associations e.V. (AiF) in the framework of the Industrial Collective Research Programme (IGF) by the Federal Ministry for Economic Affairs and Energy (BMWi) based on a decision taken by the German Bundestag. The authors would like to thank for the sponsorship and support received from the FVA, AiF and the members of the project committee. The research project was carried out in collaboration with Leibnitz-IWT Bremen where the nitriding of the test gears and several analyses for the characterization of the nitring layers were performed. The authors would like to thank Leibnitz-IWT and its associates for the support and the close cooperation.

IOP Conf. Series: Materials Science and Engineering 1097 (2021) 012006 doi:10.1088/1757-899X/1097/1/012006

Extended Evaluation of Pitting Degradation Tests to Increase the Remaining Useful Life of Gear Wheels

Yvonne Gretzinger[1], Andreas Kroner[1,2], Mark Henß[1], Martin Dazer[1], Bernd Bertsche[1]

[1] Institute of Machine Components, Pfaffenwaldring 9, 70569 Stuttgart, Germany
[2] Mercedes-Benz AG, 70546 Stuttgart, Germany

yvonne.gretzinger@ima.uni-stuttgart.de

Abstract. Within this work, the potential of an operating strategy is analysed based on test runs in laboratory environment, field tests as well as simulations. The operating strategy is used to increase the remaining useful life of gear wheel pitting by reducing the local stress at the weakened tooth. The local stress reduction is achieved by applying an adapted periodic input torque to the transmission and shift the minimum in the area of the weakened tooth. All the other teeth undergo a higher stress and therefore compensate the reduction. To analyse the potential of this strategy, the evaluation of pitting degradation tests has to be extended to all teeth using data from test runs as well as field tests. Both analyses show great potential as many teeth are totally intact. Finally, a simulative approach was carried out to quantify the potential and to show a reference value for the accuracy of the measuring system. As a result, for an increase in lifetime of 10 %, the measurement system has to locate pittings with a size smaller than 1 % based on the total tooth flank. If the pitting is located even earlier, a maximum of 38 % increase in lifetime is possible.

1. Introduction

Many transmissions fail due to pitting at the gear wheels. This flank spalling is caused by material fatigue and is normally located below the pitch circle [1]. If the gear wheel is case hardened, usually one tooth is responsible for the failure of the whole system of the gear wheel only and thus for the whole transmission. Sometimes there are few other small pittings at the circumference, but most of the teeth are still intact at the standardized end of life criterion of 4 % pitting area based on the total tooth flank [2]. The possible service life of these teeth is therefore not used which is why the system gear wheel is well suited for a Prognostics and Health Management (PHM) approach.

There are five basic areas within a PHM approach, the system, data, diagnosis, prognostics and optimization. First of all, the system has to be analysed and different data has to be collected, e.g. with sensors. This data can then be used to analyse the actual state of the product within the diagnosis. In the prognostics, the current state of the system is forecasted to the future under assumed usage scenarios. With these scenarios, the end of life (EOL) can be calculated and following this the remaining useful life (RUL) which is defined as difference between EOL and the current time. In the last step, this information can be used to optimize the service life of the system. Exemplarily, the stress of the system can be reduced and therefore an increase in service life is possible. This is why this last step is also called Health Management. [3]

In the field of automotive transmission, such a PHM approach is presented by Foulard [4]. The input torque and speed are measured and an online simulation performs a linear damage accumulation of the gear wheels. If a defined sum of damage is reached, the total input torque of the transmission and therefore the stress at the weakest tooth is reduced. With this torque reduction an increase in service life is possible. Unfortunately, the transmission loses the maximum transferable power due to the reduced torque. Moreover, the defect within the gearing is not measured directly, the reduction of the input torque is only based on a simulation.

Within this work, an operating strategy is presented that reduces the torque specifically at the weakest and therefore previously damaged tooth. The total power remains the same as the other intact teeth undergo a higher torque. The operating strategy is consequently based on the principle that the strength of the individual tooth is scattering in a certain range because of the material and production. So, the main target of this paper is the examination of the potential of the operating strategy. This potential is divided into two parts. In a first step, the statistical potential is analysed by means of extended evaluations of pitting tests. Here, the scattering in tooth strength at the circumference is examined in two test environments with different conditions and requirements, a laboratory test bench (chapter 3) and a field test (chapter 4). In a second step, based on the findings, the potential of the operating strategy is quantified within a simulative approach (chapter 5).

2. Operating Strategy

In practice, gears are usually loaded by a constant torque, which can change due to a specific load spectrum. But concerning one revolution of the gear wheel, the torque remains nearly constant. The operation of the gear with the specific load spectrum continues until system failure that is defined as 4 % pitting area related to the active tooth flank. Shortly before the failure, the damage can be detected by accelerometers at the transmission housing. This information is used to take the transmission out of service but not to influence the input torque, as seen in figure 1.

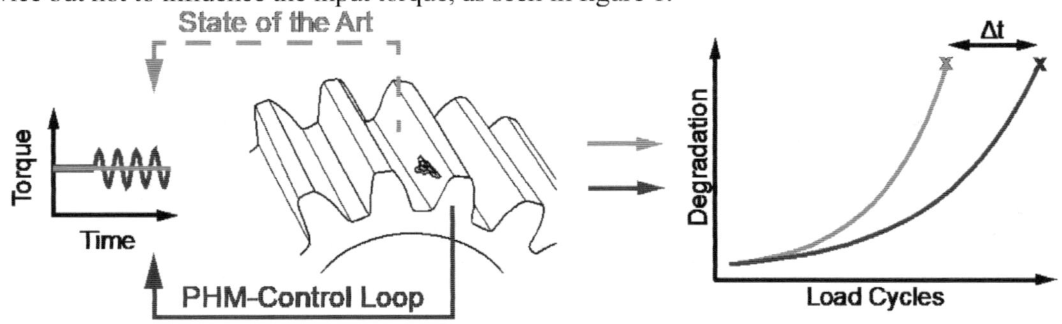

Figure 1. Schematic diagram of the operating strategy.

In contrast, the information about an upcoming failure is used in the presented operating strategy to gain a local stress reduction at the weakened tooth. Here, a small pitting (< 4 %) is detected and localized at the circumference. Within a PHM-Control Loop, this information is used to influence the input torque, see figure 1. An adapted periodic input torque is applied and the minimum is shifted in the area of the pre-damaged and therefore weakened tooth. For example, the input torque can follow a sine signal or a jump function. As an advantage of this operating strategy it should be mentioned, that the overall power of the transmission remains the same. This is possible due to a higher torque/stress at the intact tooth at the circumference.

Pitting degradation is a soft failure that begins with a very small area. This small pitting grows with operating time and the speed of growth depends on the applied torque, as shown in figure 1. This leads to an increase in service life, if the operating strategy is used.

3. Evaluation of pitting degradation tests in laboratory environment

The increase in service life due to the operating strategy is only possible if the strength of each individual tooth is varying within a certain range. For this reason, a phenomenological investigation of the failure mechanism pitting is necessary and the evaluation of pitting tests must be extended to all teeth.

3.1. Electrical Stress Test Bench and Specimen

In previous work [5], the degradation of gear wheel pitting was investigated on a test bench. Here, the focus of the analysis was based on the weakest tooth of the gear wheel only. The following test setup was used:

- Electrical stress test bench with two electrical machines
- Single stage test transmission with serial gear wheels with 21 teeth (pinion) and 41 teeth (gear)
- Helical gearing
- Constant stress level: 200 Nm
- Constant output speed: 1300 rpm
- Oil temperature: 90 °C
- Number of specimens: 8

As the aim of the prior work was the analysis of the degradation path of pitting, the test was interrupted at a constant interval of 0.25 million cycles. The flanks of the pinion were visually analysed and a negative imprint of some of the biggest pittings were made. The pitting areas were measured afterwards and the test was terminated as soon as one pitting achieved the standardized end of life criterion of 4 % pitting area. Only this pitting was used for further investigations. Smaller pittings were not considered because they are not relevant for the system failure.

3.2. Results concerning all teeth in laboratory environment

The available data of the mentioned test runs are now used to analyse all teeth at the circumference. Therefore, the investigations are expanded concerning the location of the pittings at the circumference and the degradation path of all pittings. In a downstream step, the degradation paths are used to extrapolate all pittings to the end of life (EOL) criterion of 4 % pitting area and thus, the failure distribution can be extended.

3.2.1. Localization of the pittings at the circumference

A total of eight pinions are examined concerning pitting failure. For each pinion, the location and the size of the pittings are determined in a radar chart, see figure 2. The radii are the 21 tooth and the drawn black line represents the pitting size at the individual tooth. In the example of pinion I, there are four pittings at the circumference with a range of size between 1.7 % and 6.0 %. Moreover, it turns out that many pittings occur at neighbouring tooth flanks, as it can be seen in figure 2 at tooth number three and four.

This observation was further investigated, see figure 3. As the number of teeth is 21, the total sum of teeth is 168 for the eight pinions. First, the probability of the appearance of a pitting at the individual tooth is 31 %. At 116 teeth and therefore 69 % no pitting occur and the teeth are totally intact. Second, all teeth with pitting (in total 52) are analysed. Here, the probability of a direct neighbouring pitting is very high with 88 %. So, there are 17 pittings without neighbour, eleven pairs of pitting, three trios and even one quartet. For more information about the depending and competing failure modes of pittings see [6].

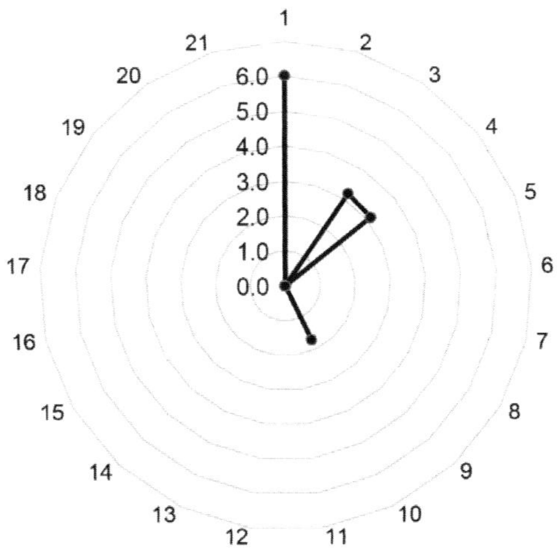

Figure 2. Radar chart of pinion I.

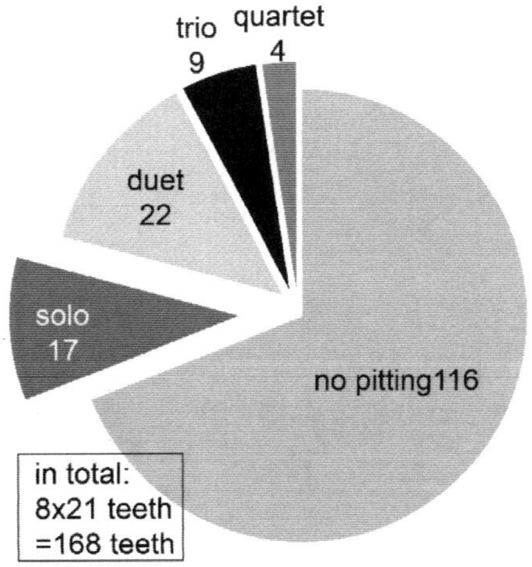

Figure 3. Pie chart of all investigated teeth.

It can be concluded that the analysis of the location of the pittings at the circumference of the pinions show a great potential of the presented operating strategy. Most of the teeth do not have a pitting or preliminary stage of pitting (see figure 3) at the termination of the test run although the gear wheel counts as failure. Therefore, these teeth can undergo a higher stress and compensate the power of the transmission. As these teeth did not show any sign of a pitting, their strength must be very high. The initiation of the pitting lies in the future and even then, a certain time of pitting growth is available until these teeth fail. The potential of these teeth is therefore enormous. Furthermore, the neighbouring pittings are advantageous for the operating strategy. If a sinusoidal input torque is implemented, these neighbouring teeth also undergo a reduced stress and therefore the degradation speed is automatically reduced due to the lower loads. This advantage also applies to the jump function. Neither the measurement nor the control of the torque is ideal. Consequently, the reduced torque is always at a certain angle of the gear wheel so that the neighbouring teeth are in the reduced area as well.

3.2.2. Degradation of gear wheel pitting and failure distribution

Pitting damage usually develops progressively after an incubation period [7]. This was also observed in the available measurements of the eight pinions. As an example, figure 4 shows the degradation path of a pitting that did not lead to failure. The black circles mark the measured pitting size at different load cycles and the grey line represents the progressive curve fit. The initial pitting occurs at 6.25 million load cycles and at the termination of the test (11 million load cycles), its size is 2.15 %. With this degradation path an extrapolation to the EOL criterion 4 % pitting area is possible. In total, the initial pittings were only visible after more than 70 % of the total test run. The initial pitting size is round about 0.45 %.

The eight pinions have 52 pittings at the end of the test run, but the focus was on the weakest tooth only. Therefore, the smaller pittings were not considered in the test runs and the degradation path is only available for 23 pittings. Equation 1 was used for the curve fitting of all degradation paths:

$$f(x) = a \cdot e^{bx} + c \tag{1}$$

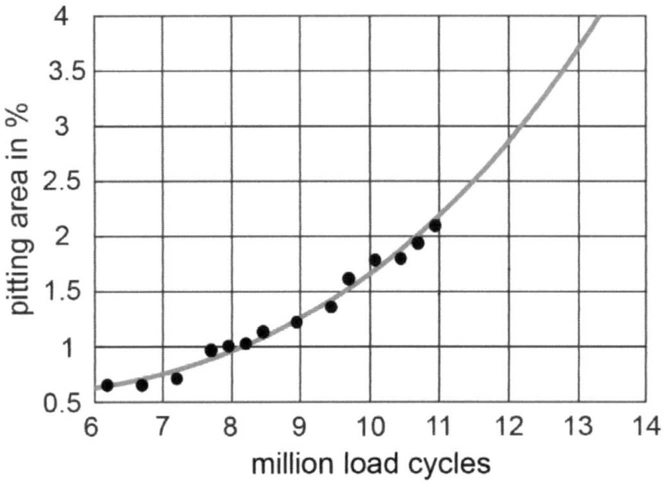

Figure 4. Measurement points and regression of one exemplary degradation path.

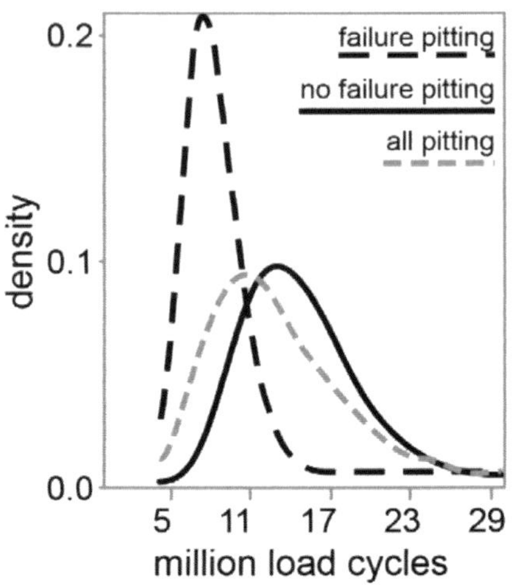

Figure 5. Failure density curves concerning all pittings

After the regressions, an extrapolation to the EOL criterion was carried out and the pittings were classified in three groups:

- group 1: pittings that lead to failure,
- group 2: pittings that did not lead to failure,
- group 3: all pittings together.

The failure distributions of those groups were analysed, which is shown in figure 5 in the form of the density function. An adaption test shows a lognormal distribution function for all groups. The dashed black line represents the state of the art with group 1, the pittings that lead to failure. The solid black curve are all the other pittings that did not lead to failure. The median values of the two groups show a quite large offset. For group 1 the median value is 8.59 million load cycles and for group 2 14.18. Hence, the difference between the two groups is 5.59 million load cycles. Furthermore, the density function of group 1 is narrower and therefore higher than the density function of group 2. That means the scattering of group 1 is smaller than the scattering of group 2 and from this, the very different strength of the individual teeth can be read off directly. The scattering of the strength is very advantageous for the operating strategy as the stress at the pre-damaged tooth is local reduced and all the other teeth undergo a higher stress. It is obvious, that the strategy only works if those teeth with higher load do not fail earlier than the pre-damaged tooth with the reduced stress. At this point it must be emphasized once again that in group 2 only the smaller pittings with degradation paths are collected. The many teeth that are totally intact and do not have any pitting are not considered here, compare figure 3. There is no information about the initiation of pitting or the pitting growth available from these teeth. Therefore, they cannot be included in the evaluation. But nevertheless, it can be concluded that the distribution of all other teeth at the circumference is much broader and the offset of the median values is therefore much higher.

Up to this point, the two groups have been considered separately to show the difference between failure pitting and other pittings at the circumference. But the density functions of group 1 and 2 are overlapping to a large extent and moreover, the failure mechanism is the same which indicates that the pittings follow one distribution. Therefore, a hypothesis test was done to clarify if there is a significance between the two distributions. As expected, no significance was observed and the pittings were analysed all together in the third group (dashed grey line). Compared to group 1, there is a higher scatter and therefore a broader distribution. The median value is 11.8 million load cycles and thus the high potential of the operating strategy is obvious. Therefore, the feasibility of the operating strategy could be demonstrated by means of the laboratory test as a high difference in the strength of the individual teeth was shown.

4. Evaluation of field tests

The previously described investigations were performed in a laboratory environment and at a constant load. For this reason, it was investigated whether the phenomena can also be observed in the field. Only then, the statistical potential of the operating strategy can be confirmed.

First, the transmissions were described and the method to determine the real field loads of a single use case of the individual transmissions. Afterwards, the diagnosis of the gear wheels is presented and compared to the results of the laboratory tests.

4.1. Transmissions of the field tests

For the investigation of the field data an automotive transmission from a large series was selected. An automatic transmission has been chosen because it minimizes the operator's influence on the shifting process and thus, on possible load peaks that can occur during fast clutch engagements. The automotive sector has extremely high quantities and the production process as well as the geometry of the gearing is optimized. Therefore, only a marginal influence of the manufacturing on the pitting degradation is expected. Finally, a 9-speed planetary gearbox was selected, see figure 6. This transmission is used for vehicles with Standard drive and their all-wheel drive derivatives.

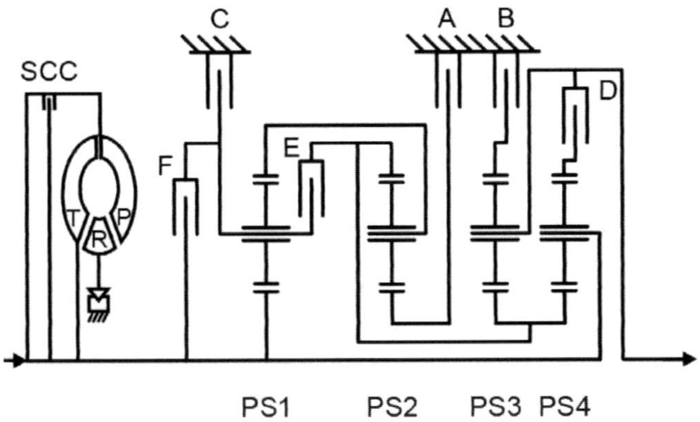

S	Brake			Clutch			i	Gear step
	A	B	C	D	E	F		
1	•	•			•		5,35	1,65
2		•			•	•	3,24	1,44
3	•	•				•	2,25	1,37
4	•	•		•			1,64	1,36
5	•			•		•	1,21	1,21
6				•	•	•	1,00	1,15
7	•			•	•		0,87	1,21
8			•	•	•		0,72	1,20
9	•		•	•			0,60	Total
R1	•	•	•				-4,80	8,91

Figure 6. Gearbox diagram and gearshift pattern of the 9-speed automatic transmission. SCC: slip-controlled torque converter lock-up clutch, T: turbine, R: reactor, P: pump, PS: planetary gear set, S: speed, i: ratio [1]

Table 1. Number of teeth of the four planetary gear sets

	planet gear set 1	planet gear set 2	planet gear set 3	planet gear set 4
sun	46	44	37	34
planet	26	27	23	26
ring gear	98	100	83	86

The chosen transmission is made up of four planetary gear sets, three brakes and three clutches. Advantage of the planetary transmission is that in general all planetary gear sets are used for the nine forward and one reverse speed of the transmission. Therefore, they are all under load and the amount of specimen is very high although only some transmissions have to be analysed. Each planetary gear set has three or four planets, depending on the maximum input torque. For this reason, the planets are chosen for further investigations as the number of specimens is the highest. Moreover, the number of teeth of the different planets is in the same range as the pinion of the laboratory tests, which makes a comparison very easy. All numbers of teeth of the four planetary gear sets are listed in table 1.

4.2. Determination of the real field load on the transmissions

Automatic car transmissions like the 9-speed automatic transmission have an electronic control unit (ECU) which is integrated into the vehicle network. Information and requirements of the complete powertrain can be exchanged for example by the controller network (CAN) signals like the engine torque, the engine output speed and the current gear-speed, which is achieved by an interconnection of the four planetary gears sets. Based on the damage accumulation approach, the transmission ECU is able to determine an operating point dependent damage value for each gear-speed. This value is calculated based on the information on the drivetrain-CAN like the engine output torque and speed, compare [4]. At the time of the analysis of the transmission, the obtained gear-speed dependent damage values of the real field loads can be used to do a reverse calculation to the loads of each planetary gear set. For this purpose, the geometric data and the gear ratio of the transmission is used to calculate a damage value depending on the different planetary gear sets. Further information such as the deposited mileage and ECU operating time are used to estimate and assign the field load spectra. In a last step it is necessary to normalize the calculated damage values to known loads. For this purpose, damage values generated from success run tests are defined as reference value. From this, percentage damage values can be derived. The back calculation and scaling of this approach is presented in detail in [8].

4.3. Diagnosis of the gear wheels

The examined transmissions are mainly early failures and the failure cause is varying, for example the mechatronic control unit with the integrated ECU and sensors are failed. Therefore, the mileage is relatively low concerning the design point and the gearing is not failed in the examined transmissions. Consequently, the tooth flanks do not show any pitting, but there is a pre-stage visible, namely micropitting. This observation corresponds with the results from the test bench. The first pitting outbreak was visually observed at about 6 million load cycles. The termination of the test was at about 8.5 million load cycles. Therefore, the pitting grows in the last third of the lifetime. It is obvious, that there are no pittings visible in the determined field tests.

Micropitting is indicated by grey zones on the tooth flank as fine surface pittings break out due to mixed friction as a result of the oil viscosity and the surface structure [1]. The effect of micropitting on macropitting was investigated in literature, for example [9-11]. Micropitting affects the location of the beginning pitting on the tooth flank. As the micropittings mostly emerge in the area of the tooth root, pittings occur preferably at the border between micropitting and the micropitting free area. However, the direction of growth is determined exclusively by the sliding conditions. [10] The greater the micropitting area, the greater the probability of pitting [9].

In the following, micropitting is regarded as preliminary stage to pittings and is therefore evaluated in the same way as in chapter 3.2.1. In total, four transmissions were analysed concerning micropitting. As the planetary gear set 4 has the highest damage value, only those planets were determined concerning all teeth. Therefore, four planets per transmission with 26 teeth each were examined. The pie charts for the four transmissions are shown in figure 7.

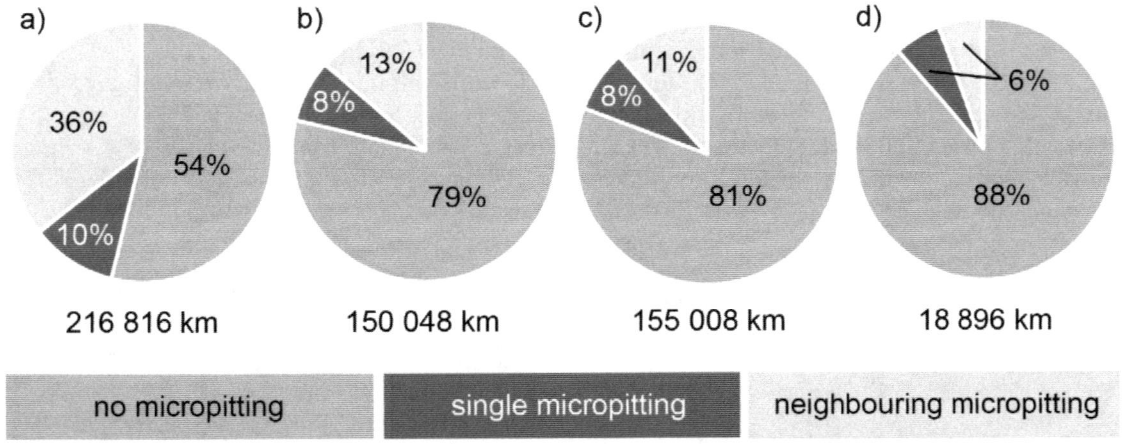

Figure 7. Pie chart of the four planetary gear sets. a) transmission A (damage value 38%) b) transmission B (damage value 53%) c) transmission C (damage value 59%) and d) transmission D (damage value 63%)

In all pie charts of figure 7 it is visible that most of the teeth do not have any damage. Furthermore, the amount of damage at neighbouring teeth is higher than the amount of single damage. The observations within the laboratory tests could therefore be confirmed.

Within this chapter, the four transmissions were evaluated separately as the mileage and the damage value vary in a wide range. The total number of teeth examined in each transmission is 104, four planets with a number of teeth of 26. Transmission A has a high mileage with above 200 000 km and nevertheless the percentage of the intact teeth is 54 % and therefore very high. With a medium mileage of about 150 000 km at transmission B and C the proportion of intact teeth increases to 80 %. The very small mileage of transmission D with less than 19 000 km shows an extremely high number of intact teeth with 88 %. But despite the low mileage of only 6 % of the design goal, 12 % of the teeth already show micropitting, which was not expected. For the analysis, however, the damage value must also be considered since this allows a statement about the field load of the individual transmission. Transmission D has with 63 % the highest damage value of the four transmissions although the mileage is the lowest. Therefore, the customized stress of transmission D was very high and far above the average. With this point of view, the visible micropittings are obvious.

Transmission A shows a contrary user profile. The mileage is very high with above 70 % of the expected mileage whereas the damage value is very low with 37 %. Although the mileage suggests that pitting is beginning to occur, only preliminary stages can be seen. The damage value indicates a very weak load of the transmission during the customized load spectra. Therefore, it is obvious that there are no pittings visible. Those two transmissions (A and D) show the difficulty in designing new transmissions as the field load spectrum varies in a wide range with passenger cars. More information about the stress and strength of different transmissions can be found in previous work [8].

The transmissions B and C are also analysed further. The mileage as well as the damage value are nearly the same, above 150 000 km and 50 %. Those two transmissions can therefore be treated together and as expected, the pie charts in figure 7 look nearly the same. Slight differences always occur due to scattering in the material, the manufacturing process of the transmission and the customized load spectra.

Last but not least, the percentage of the micropitting at neighbouring teeth was analysed. The four pie charts in figure 7 show that the percentage increases with the mileage. At the same time, the total number of micropitting rises so that the probability of neighbouring is statistically increased. In addition to the statistical approach, the neighbouring can be explained phenomenologically. In practice, more than one tooth is in contact within meshing. For example, the contact ratio of the spur gear pair of the laboratory investigations is 2.7. Thus, two or three teeth are simultaneously in mesh. If one tooth has a

IOP Conf. Series: Materials Science and Engineering 1097 (2021) 012006 doi:10.1088/1757-899X/1097/1/012006

defect, the load capacity is reduced and therefore the two neighbouring teeth have to compensate the load. Consequently, the load rises and the probability of a defect increases [6].

During the analysis of the individual tooth flanks, it was also noticed that the zone of micropitting moves along the tooth flank. Transmission D has a relatively high damage value, but the mileage is very small with almost 19 000 km which indicates a high torque. The photo from the microscope in figure 8 shows a micropitting area at the tooth root. In contrast, the transmission B has a relatively low damage value, but the mileage is with about 150000 km significantly higher. Here, the photo in figure 9 shows micropitting above the pitch circle and the input torque is expected to be lower.

Figure 8. Microscope photo of one tooth from transmission D. Micropitting zone in the area of the tooth root.

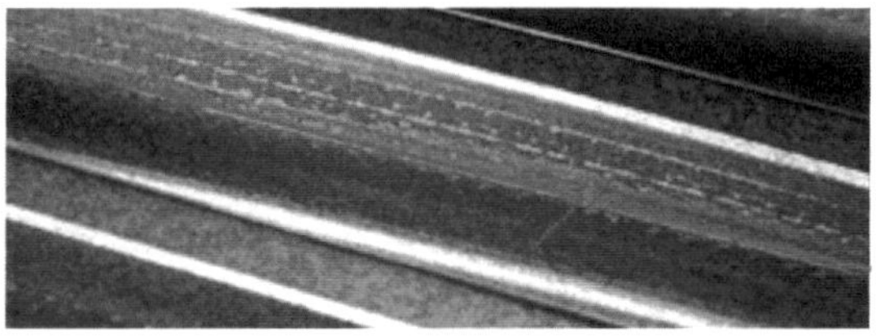

Figure 9. Microscope photo of one tooth from transmission B. Micropitting zone in the area above the pitch circle.

These images suggest that the highest load point on the flank moves with different input torque. It is assumed that this effect occurs due to the tooth stiffness. At the addendum, the stiffness is lower than at the dedendum. Therefore, the tooth is bending with higher load and the highest load point moves to the tooth root. With lower input torque the bending stress is lower and the highest load point is higher at the tooth flank. This effect shows additional potential for the operating strategy as the highest load point is moved away with the adapted input torque. With regard to a local damage accumulation on the tooth flank this effect has advantages on both the pre-damaged flank and the intact flank.

In summary, it can be said that the qualitative potential of the operating strategy has been demonstrated within the extended evaluations. The strength of the individual teeth is scattering in a wide range as most of the teeth do not have any pitting. Futhermore, it is advantageous for the operating strategy that the pittings occur preferably at neighbouring teeth and therefore the local reduction at these teeth is done automatically. Within the analysis of the field tests, an additional potential concerning the tooth flank was detected. All these results show the feasibility of the operating strategy. Therefore, the quantitative potential of the strategy is analysed in the next chapter by means of a simulation.

5. Simulative results

Within the simulation study the influence of a load shift on the percentage lifetime extension is investigated. For this purpose, the failure behaviour is simulated by means of a model, which is then run through in several iterations according to the Monte Carlo principle. The initiated failure behaviour is defined by the initial values (pitting size a_0, load cycle n_0, damage d=0), the EOL values (pitting size a_{EOL}=4 %, load cycle n_{EOL}, damage d=1), as well as pitting growth a=f(n) and damage d=g(n). The initiated failure behaviour corresponds to the actual failure behaviour if all teeth are loaded with a constant moment without the strategy. If the moment is varied, the damage increment following from the moment is calculated by means of a linear damage accumulation.

Furthermore, the Wöhler Curve of the spur gear pair from chapter 3 is used to get the new failure time with the adapted input torque of the operating strategy. This is done by a parallel shift of the Wöhler Curve in the failure time without the operating strategy. Linear damage accumulation and a load-independent relationship $a = f(g^{-1}(d))$ are assumed, whereby the corresponding pitting size can be calculated for each damage.

Finally, a two-stage load spectrum is calculated for each tooth and the difference of the failure time with and without operating strategy is considered. The boundary conditions of the simulation are the following:

- Failure distribution (nEOL): Normal with mean 11.39 million load cycles and standard deviation 1.053e6
- Distribution of pitting initiation time (n0): Normal with mean 6.875 million load cycles and standard deviation 1.053e6
- Distribution of pitting initiation size (a0): Normal with mean 0.45 % and standard deviation 0.205
- Wöhler exponent: 10.5
- Adapted torque function: ideal jump function with reduction only at the weakest tooth
- Standard torque: 200 Nm
- Reduced torque at the weakest tooth: 190 Nm
- Increased torque at the other teeth: 200,5 Nm
- Only weakest tooth is considered
- Iterations: 100
- Variation: startpoint (direct at pitting initiation, 7, 8, 9, 10 million load cycles)

The starting point of the operating strategy is varied within the simulation and the results are collected in table 2. In the first simulation, the starting point was directly the initiation of the pitting that is drawn from the distribution. This initiation is simulated as a normal distribution. After that, the starting point was varied between 7 and 10 million load cycles, which corresponds to an average pitting size of 0.49 % and 2.28 %.

Table 2. Simulative results: potential of the operating strategy with different starting points

Starting point	With pitting initiation	7 million load cycles	8 million load cycles	9 million load cycles	10 million load cycles
Medium pitting size	0.45 %	0.49 %	0.87 %	1.44 %	2.28 %
Median of lifetime extension	20.46 %	19.61 %	11.83 %	4.21 %	1.14 %
Maximum of lifetime extension	38.11 %	25.99 %	17.88 %	11.50 %	5.68 %

The greatest potential is presented when the operating strategy is started directly at the pitting initiation. Here, the reduced torque is applied to the weakest tooth for the longest and a maximum extension of 38 % lifetime could be achieved. Moreover, the median of the lifetime extension is reduced to only 1.1 % if the operating strategy is started at 10 million load cycles. From table 2 it can be concluded that the detection of the pitting has to be very sensitive. If a lifetime extension of 10 % should be reached, a pitting size smaller than 1 % has to be detected. The later the operating strategy starts, the less extension can be achieved. But the potential of the operating strategy could be proven definitively by simulation and further consideration is recommended.

Within this simulation, the weakest tooth only is considered. This is sufficient for this simulation, because the input torque at all the other teeth is only increased by 0.5 Nm and those teeth do have a higher strength (compare to chapter 3 and 4). Additionally, for all starting points 100 iterations were calculated with a percentage error in the median value lower than 5 % in the Monte Carlo simulation. In further work, the number of iterations should be increased, to get a more accurate result. Moreover, the simulation should be extended to more teeth at the circumference compared to chapter 3 and 4 and it should be validated within laboratory test runs.

6. Conclusions and outlook

Based on the results of the extended pitting evaluation of laboratory and field tests and the simulative approach, several conclusions can be drawn:

- There are some pittings at the circumference, but only one pitting causes the failure of the system
- In 88 % of pitting failure, a neighbouring flank has a pitting as well
- 70 % of the tooth flanks of the test runs do not have any pitting
- The pitting degradation paths show an exponential growth and therefore a failure distribution of all teeth could be extrapolated
- The evaluation of all pittings results in a significantly higher scattering whereas the totally intact tooth should be included in further work
- The field tests also show a great potential as the same observations as in the laboratory test runs could be examined although only a pre-stage of pitting was visible, namely micropitting.
- The field tests show an additional potential for the operating strategy as the micropitting moves on the flank with different input torque. It is therefore expected that the highest load point on the tooth flank moves as well.
- The simulative approach shows a maximum increase of service life of 38 %
- The later the strategy starts the lesser the potential is. Therefore, the measurement of the pitting location has to be very accurate and should detect pitting with a size smaller than 1 % pitting area.

In further work, the simulation has to be extended to more pittings at the circumference. Additionally, the adapted input torque should be analysed as it is an ideal jump function in this simulation. A verification in an experimental test run is also still pending.

References

[1] Naunheimer H, Bertsche B, Ryborz J and Novak W 2011 *Automotive Transmissions – Fundamentals, Selection, Design and Application* 2nd edition Springer-Verlag Berlin Heidelberg
[2] ISO 6336-2 2006 *Calculation of load capacity of spur and helicalgears - Calculation of surface durability (pitting)*. International Standard
[3] Henß M and Bertsche B 2019 *AutoEncoder basierte automatisierte Zustandsdiagnose von Wälzlagern* Tagung Technische Zuverlässigkeit, Nürtingen, Germany
[4] Foulard S, Rinderknecht S and Ichchou M 2015 *Real-time and online lifetime monitoring system for automotive transmissions* The 14th IFToMM Word Congress, Taipei, Taiwan (Oct.)
[5] Beslic Z, Mueller P, Yan S and Bertsche B 2017 *Planning optimal degradation tests in consideration of budget and statistic accuracy on different stress levels* Proc. ISSAT 2017 Conference, 3rd – 5th August 2017, Chicago, USA, ISBN 978-0-9910576-4-1, pp. 108-115
[6] Gretzinger Y, Dazer M and Bertsche B 2020 *Using life data of competing failure modes to increase the remaining useful life of gear wheels* Proc. RAMS 2020-Conference, 27th – 30th January 2020, Palm Springs, USA.
[7] Sommer K, Heinz R and Schöfer J 2018 *Verschleiß metallischer Werkstoffe - Erscheinungsformen sicher beurteilen* 3rd edition Springer Vieweg
[8] Kroner A, Luo Z and Bertsche B 2020 *Prediction of the Remaining Useful Life of Transmission Components within a Remanufacturing Process to Ensure the Second Product Life* Proc. IRF2020: 7th International Conference Integrity-Reliability-Failure, 6th – 10th September 2020, Funchal, Portugal
[9] Felbermaier M 2016 *Untersuchungen zur Graufleckenbildung und deren Einfluss auf die Grübchentragfähigkeit einsatzgehärteter Stirnräder* Dissertation TU München
[10] Radev T 2005 *Einfluss des Schmierstoffes auf die Grübchentragfähigkeit einsatzgehärteter Zahnräder* Dissertation TU München
[11] Bartz W J 1999 *Schäden an geschmierten Maschinenelementen* 3rd edition expert verlag

IOP Conf. Series: Materials Science and Engineering 1097 (2021) 012007 doi:10.1088/1757-899X/1097/1/012007

Investigation of lubricant supply in rolling point contacts under starved conditions using CFD simulations

D Fischer[a]*, S von Goeldel[a], G Jacobs[a], A Stratmann[a] and F König[a]

[a] Institute for Machine Elements and Systems Engineering, RWTH Aachen University, Germany

* Corresponding author: dennis.fischer@imse.rwth-aachen.de

Abstract. High friction and wear in rolling bearings are prevented by a lubricating film, which separates the contacting surfaces. In rolling contacts with limited lubricant supply, the film thickness decreases at high rolling speeds, as more lubricant is displaced than replenished. This lubricant depletion is known as starvation and may induce mixed lubrication due to a low film thickness. When estimating the bearing lifetime, a fully flooded lubrication, i.e. an unlimited lubricant supply, is usually assumed. This leads to an overestimation of film thickness in operating ranges where starvation occurs. Thus, considering the onset of starvation is crucial to reduce the risk of premature bearing failures due to high friction and wear in the mixed lubrication regime. Therefore, this contribution presents a method, based on computational fluid dynamics (CFD) simulations, to calculate the onset of starvation in oil lubricated point contacts

Keywords. Starvation, Lubricant Supply, Wettability, CFD Simulation

Nomenclature

Latin characters

a_{Hertz}	Hertzian radius [m]
Ca	Capillary Number [-]
E	Elastic modulus [GPa]
F	Force [N]
G	Dimensionless material parameter [-]
h	Film thickness [m]
k	Ellipticity [-]
L	Length [m]
m	Inlet meniscus distance [m]
m^*	Critical inlet meniscus distance [m]
p	Pressure [Pa]
R	Radius [m]
u	Velocity [m s^{-1}]
U	Dimensionless speed parameter [-]
V	Volume [m^3]
W	Dimensionless Load parameter [-]

Greek characters

α_p	Pressure-viscosity coefficient [GPa^{-1}]
η	Dynamic viscosity [mPa s]
θ	Contact angle [°]
ν	Poisson's ratio [-]
σ	Surface tension coefficient [N m^{-1}]

Subscripts

c	Central
ff	Fully flooded
i	Initial
s	Starved
$red.$	Reduced
x	x-direction
y	y-direction

1. Introduction

In machine elements, such as rolling bearings, oil or grease lubrication is applied to reduce friction and wear. For an efficient operation with long service life, the contacting surfaces of rolling elements and raceways must be separated by a lubricating film. The surface separation for oil lubricated contacts is ensured in the hydrodynamic lubrication regime, if the viscosity and the rotational speed are sufficiently high to form a complete separating film. However, if the amount of lubricant is limited, a complete film formation and thus, an operation in the hydrodynamic lubrication regime cannot be guaranteed. Mostly, grease lubricated rolling bearings operate under limited lubricant supply, since the amount of base oil, bleeding from the grease bulk, is limited and the bearing is not re-lubricated [1–3]. This can lead to insufficient lubricant supply and to a film thickness decrease, known as starvation, which might induce boundary or mixed lubrication under operating conditions. Consequently, premature bearing failures due to high friction and wear are likely to occur.

As presented in the authors' previous work [4], the lubricant supply by replenishment in front of a rolling contact is significantly affected by the wettability, the available amount of oil as well as operating and lubricant properties, like rolling speed and viscosity. As mentioned in [5,6], the lubricant supply can be correlated to the inlet meniscus distance between contact centre and the boundary between oil and air in front of the contact. If the meniscus approaches the contact and the meniscus distance is reduced below a critical value, the film thickness decreases due to starvation [7–11]. The effects of speed and viscosity on the lubricant supply as well as on the onset of starvation were examined by several authors using experimental setups [9,12–21] and recently also by a few authors in numerical simulations [5,11,22,23]. HAMROCK and DOWSON derived a correlation (equation (1)to (4)) between inlet meniscus distance and the film thickness for oil lubricated contacts, which can be used to calculate the onset of starvation [24]. However, a major challenge in using their model is to determine the actual inlet meniscus distance in dependence of the operating conditions and the lubricant properties as well as the available oil amount [8].

Recently, NOGI et al. [11] presented an equation based on numerical and experimental investigations, to determine the meniscus distance (equation (5)). To derive the equation, a numerical model was used under the assumption of a single-phase oil flow and the available oil amount is represented by the height of the initial oil film in front of the contact. However, the oil film height is mostly unknown in an application and can hardly be determined [2,18,25]. Thus, following the idea by NOGI et al. [11] but using the two-phase CFD model presented in [26], this work demonstrates the derivation of an equation to calculate the meniscus distance considering the available oil volume. Moreover, on the basis of published work [7,12,20,27], a significant impact of the oil wettability on the meniscus distance is expected. The work by HUANG et al. [28] points out, that in particular for grease lubricated contacts, the contact angle between bled oil and a thickener layer on the contacting surfaces significantly changes in comparison to oil on a steel surface. Hence, the presented model in this contribution additionally considers the effect of wettability, i.e. contact angle, on the meniscus distance.

The following contribution describes the derivation of an equation to calculate the inlet meniscus distance. Therefore, the CFD model presented in [26] is used for a parametric study to identify the effect of capillary number (Ca), contact angle and oil volume on the inlet meniscus distance. In accordance to [11], a least square approximation is applied on the simulation results to derive a general equation for the meniscus distance. Applying this equation to the HAMROCK and DOWSON model (equation (1) to (4)), the effects of contact angle and available oil volume on the onset of starvation are examined and discussed.

2. Governing Equations

The equation used to calculate the central (c) film thickness for fully flooded (ff) point contacts is given by HAMROCK and DOWSON [29] as:

$$h_{c,ff} = 2.69 * U^{0.67} * G^{0.53} * W^{-0.067} * \left(1 - 0.61e^{-0.73k}\right) * R_x \tag{1}$$

with the dimensionless parameters for speed U, load W, material G and ellipticity k:

$$U = \frac{\eta * u_{mid}}{E_{red} * R_x}; \quad G = E_{red} * \alpha_p; \quad W = \frac{F}{E_{red} * R_x^2}; \quad k = 1.03 * \left(\frac{R_y}{R_x}\right)^{0.64} \tag{2}$$

with η being the dynamic viscosity, u_{mid} the rolling speed, E_{red} the reduced modulus of elasticity, $R_x/R_y = 1$ as $R_x = R_y$ is valid for point contacts, F the applied load and α_p the pressure-viscosity coefficient. However, under starved lubrication, equation (1) is extended by a film thickness reduction factor, which depends on the inlet meniscus distance m and the critical meniscus distance m^*. The critical inlet meniscus distance m^* describes the meniscus distance, which leads to the onset of starvation and is formulated in [24] as

$$m^* = \left[1 + 3.06 * \left(h_{c,ff} * \frac{R_x}{a_{Hertz}{}^2}\right)^{0.58}\right] * a_{Hertz} \tag{3}$$

with a_{Hertz} as the HERTZian contact radius. For $m \leq m^*$, the central (c) film thickness in the range of starvation *(s)* can be calculated using

$$h_{c,s} = h_{c,ff} \left(\frac{m - a_{Hertz}}{m^* - a_{Hertz}}\right)^{0.29} \tag{4}$$

Finally, the application of equation (4) requires the knowledge of the actual inlet meniscus distance m, which depends on lubricant and operating properties. As described in section 1, NOGI et al. [11] recently published an equation for the meniscus distance:

$$m_{Nogi} = 1.94 * \left(\frac{h_i * R_x}{a_{Hertz}{}^2}\right)^{0.59} * \left(\frac{h_{c,ff} * R_x}{a_{Hertz}{}^2}\right)^{-0.061} * e^{(-1.95 * Ca^{0.45})} * a_{Hertz} \tag{5}$$

where h_i is the initial film height, a_{Hertz} the HERTZian radius and Ca the capillary number

$$Ca = \left(\frac{\eta_0 * u}{\sigma_{oil,air}}\right) \tag{6}$$

with $\sigma_{oil,air}$ as the surface tension coefficient. However, in order to consider the wetting behaviour as well as the actual oil volume, another equation for the calculation of the inlet meniscus distance is presented in this work.

3. Numerical and Experimental Methods

In the following, the CFD model is described, which is used for the simulations of lubricant supply to determine the meniscus distance m. As mentioned before, the simulation results for m are applied to equation (4) to calculate the film thickness. Furthermore, the ball-on-disc tribometer as well as lubricant properties and operating conditions used for film thickness measurements are presented. In section 4.2 the measurement results are compared to the calculated film thickness in order to verify the application of the inlet meniscus equation.

3.1. CFD Model for Simulation

The geometry in the vicinity of a point contact of a ball over a plane was modelled using the open source software OPENFOAM [30]. The setup of this CFD model is described in detail in the authors' previous work [26] and will be briefly presented in the following. **Figure 1** shows the simulation domain with a flow field in characteristic butterfly shape in the vicinity of the contact. The surfaces of plane and ball are partially wetted with a thin oil film, while the ambient free volume is filled with air. Both, oil and air, were modelled as incompressible, isothermal and isoviscous NEWTONian fluids. The body surfaces were defined as rigid walls with no-slip boundary condition and constant velocity. The wetting behaviour was considered by the boundary condition of a static contact angle between oil and the solid surfaces. Pure rolling condition was achieved by setting the circumferential velocity equal to the plane velocity. To simulate the effect of oil displacement and replenishment due to over rolling of two balls in series, a periodic boundary condition was applied. Thus, the flow fields, including pressure, velocity and oil/air distribution, at the outlet of the simulation domain were projected onto the inlet. This method allows to consider the influence of lubricant displacement and replenishment on the meniscus distance between two contacts. The resulting meniscus distance was determined as the distance between the center of the contact and the boundary between oil and air in front of the contact, see **Figure 1**. For further information on the CFD model, the simulation process and the meniscus distance evaluation, the reader is referred to [26].

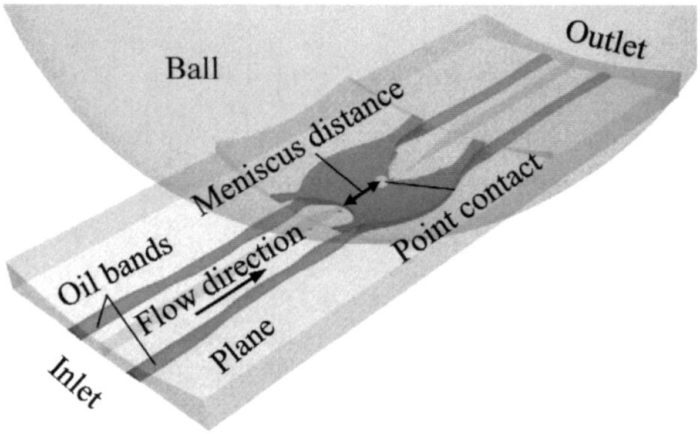

Figure 1. Simulation domain and flow field in the CFD model [26].

This CFD model is used in this work to simulate the fluid flow of oil around a point contact and thereby, determine the meniscus distance in dependence of *Ca*-number, contact angle and oil volume. These parameters were varied in 46 simulations, so that the effect of each parameter on the meniscus distance can be identified. The parameter combinations and results for the meniscus distance are listed in **Table 2** in the appendix.

3.2. *Experimental Setup*

In order to verify the results of film thickness calculation presented in this contribution, the film thickness of a point contact was measured on a ball-on-disc tribometer EHD2 from PCS Instruments by the principle of interferometry, which is described in detail in [31]. For the experiments the contact was lubricated with a limited amount of polyalphaolefine (PAO) oil containing no additives. Therefore, the weight of disc, ball and ball carriage were determined in dry condition before applying 10 µl oil to the disc. A running in phase of 24 disc revolutions at constant speed 50 mm/s was set before the actual measurement to distribute the oil droplets along the track, so that a continuous oil band was formed on the track. During measurement, the polished steel ball was loaded with 47 N against the glass disc at a track radius of 37.5 mm and the rolling speed was increased in the range from 20 mm/s to 1000 mm/s. Both, ball and disc, were driven to ensure pure rolling and the measurements were performed for two temperatures at 42 °C and 24 °C. **Table 1** gives material, lubricant and load parameters.

Table 1. Parameters of film thickness measurements.

Parameter	Value	Parameter	Value
F	47 N	α_p	15 GPa^{-1}
p	0.7 GPa	R_x	9.525 mm
u_{mid}	20 – 1000 mm/s	E_{steel}	207 GPa
V_{oil}	10 µl	ν_{steel}	0.29
$\eta_{1,(42\ °C)}$	73.30 mPas	E_{glass}	75 GPa
$\eta_{2,(24\ °C)}$	195.34 mPas	ν_{glass}	0.22

For each temperature two tests were performed and after each measurement the weight of disc, ball and ball carriage was determined. From the weight difference of the components before and after the measurement the amount of oil attached to the components was identified.

4. Results

As mentioned before, the CFD model presented in the authors' previous work [26] was used in this contribution to determine the inlet meniscus distance in dependence of the parameters *Ca*-number, contact angle and available oil volume. In a parametric study the lubricant supply is simulated for 46 parameter variations (see **Table 2** in appendix) to identify the impact of each parameter on the meniscus distance. Similar to [11,23], a least square approximation is applied on the results to derive an equation, which predicts the meniscus distance as a function of the mentioned parameters. In the following, the derived equation for the inlet meniscus distance is presented and compared to the equation for point contacts presented by NOGI et al. in [11] for verification. Furthermore, the equation is applied to the HAMROCK and DOWSON equation (4) to calculate the film thickness under starved lubrication and the results are compared to measurement results on the ball-on-disc tribometer.

4.1. Meniscus Distance Equation

To find a general formulation for the meniscus distance, which describes the combined effects of the *Ca*-number, the contact angle and the oil volume, a least square fit is applied on the results for the meniscus distance. The simulation results and the simulation parameters are listed in **Table 2** in the appendix. Thus, the meniscus distance equation, which gives m_{CFD} in metres, yields:

$$m_{CFD} = 4.533 * e^{(-2.571*Ca^{0.384})} * cos(\theta)^{-0.249} * \left(\frac{V_{Oil}}{L_{Track}} * 10^{-6}\right)^{0.446} \quad (7)$$

The oil volume factor V_{Oil}/L in µl/mm describes the available oil volume related to the track length in rolling direction. This factor is introduced since the available oil volume is distributed over the whole track length during operation. Hence, if the total amount of applied oil volume for a lubricated system is known, the oil volume factor allows to apply equation (7) to point contact systems with different track lengths. This will be shown exemplarily for a ball-on-disc tribometer in section 4.2. **Figure 2** shows exemplarily simulation results (see appendix, case #7 and #41), which emphasis the meniscus distance at different contact angles 11° and 66°.

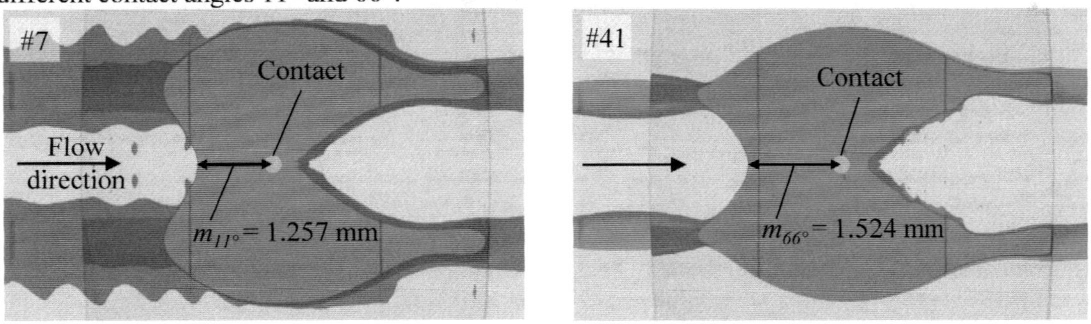

Figure 2. Simulation results of case #7 and #41.

To verify the results for the meniscus distance equation (7), it is compared to results using equation (5) presented by NOGI et al. [11] in **Figure 3**. For the comparison, *m* is presented as a function of the capillary number *Ca*, which is defined in equation (6). *Ca* is varied by increasing the rolling speed, but the surface tension and the viscosity are kept constant to $\sigma_{oil,air} = 0.028$ N/m and $\eta = 73.30$ mPas, respectively. The contact angle is set to $\theta = 0°$ in equation (7), since it is not considered in equation (5). Load, geometry, lubricant and material parameters, which are required for equation (5), are chosen exemplarily according to the experimental setup described in section 3.2. As already mentioned in section 1, the amount of oil volume in equation (5) is represented by the initial oil film height h_i in front of the contact. According to [11] an exemplarily value of $h_i = 20$ µm is chosen for the comparison of m. The oil volume for equation (7) is estimated according to [11,25] from the initial oil film height, the width and the track length. The oil film width is measured in the experiments, described in section 3.2, to a width of 2.0 to 2.5 mm, so an exemplarily width of 2.25 mm is chosen for the comparison in **Figure 3**. Since the oil volume in equation (7) is related to the track length, considered by the oil volume factor V_{oil}/L, a track length is not required to compare both equations. From $h_i = 20$ µm and a width of 2.25 mm, the oil volume factor is determined to $V_{oil}/L = 0.045$ µl/mm. In **Figure 3** the critical meniscus distance m^* is additionally presented exemplarily for the chosen parameters, so that the onset of starvation can be determined at the intersection between the curves for m and m^*.

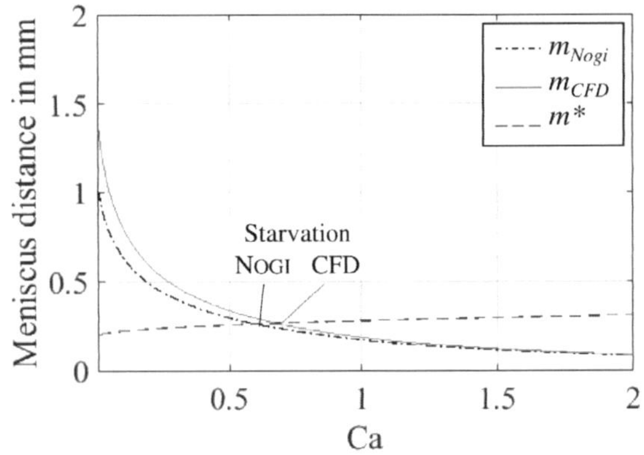

Figure 3. Comparison of derived equation (7) with equation (5) by NOGI et al. [11] for $V_{Oil}/L = 0.045$ µl/mm, $\eta = 73.30$ mPas, $\theta = 0°$.

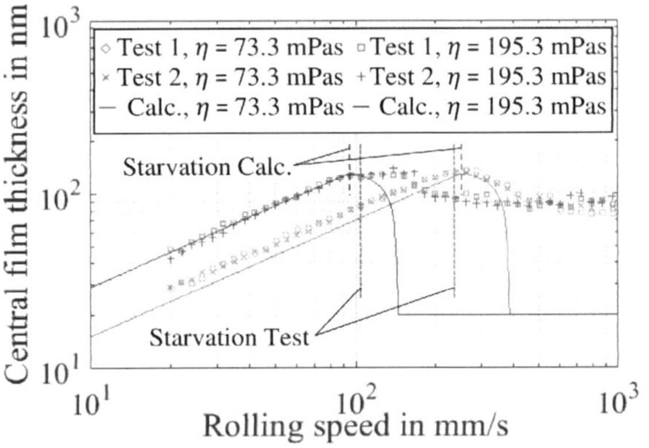

Figure 4. Comparison of calculated and measured film thickness at two different viscosities for $V_{Oil}/L = 0.042$ µl/mm, $\theta = 8°$.

The results for m and the onset of starvation at the intersection are in good agreement for both models. A slight deviation for Ca-numbers from 0 to 1 might be a result of an inaccurate estimation of V_{oil}/L from the initial oil film height and width. Moreover, both equations are based on different types of simulation methods with different boundary conditions and simplifications [11,26]. While equation (5) is dependent on the film thickness under fully flooded condition ($h_{c,ff}$), the film thickness is not considered in the derived equation (7). As explained in [26], the influence of this factor is assumed to be negligibly small. This is confirmed by the low exponent of -0.061 at the film thickness term in equation (5).

4.2. *Film Thickness Measurement and Calculation*

To calculate the film thickness under starved lubrication, the derived meniscus distance equation (7) is applied to the HAMROCK and DOWSON equation (4). To verify the calculated film thickness, the results are compared to film thickness measurement results under limited oil supply. The setup of the experiments on the ball-on-disc tribometer is described in section 3.2. For the measurements, an oil volume of $V_{Oil} = 10$ µl is distributed on the disc at a track radius of 37.5 mm, i.e. on a track length of $L_{Track} = 235.62$ mm. From the measured weight difference of the components before and after each measurement, no significant oil loss could be identified. Moreover, it is found, that about 90 % of the oil mass sticks to the disc. Therefore, it is assumed, that the entire supplied oil volume of 10 µl is present on disc and ball during the measurements. **Figure 4** shows the measured central film thickness in comparison to the calculated central film thickness as a function of rolling speed for two different viscosities of PAO oil. The static contact angle between PAO oil and glass disc is measured to approx. 8° at room temperature and is applied to equation (7). The pressure-viscosity coefficient $\alpha_p = 15$ GPa^{-1} is determined from high pressure measurements at $p = 0.176$ GPa and 40 °C. The presented results show a good agreement for the onset of starvation for both viscosities.

However, with increasing rolling speed, the calculation of film thickness in the range of starvation predicts a film thickness decrease with a high gradient. This behaviour differs significantly from the measurement results, which show a low gradient in the film thickness decrease in accordance to literature [32,33]. Moreover, the measured film thickness reaches a residual constant film thickness between 80 and 100 nm from approx. 600 mm/s. Both, the gradient in film thickness decrease as well as the residual film thickness cannot be predicted accurately by the HAMROCK and DOWSON equation (4). The minimum film thickness in the range of starvation is therefore artificially bounded for the calculation to 20 nm, as presented in [32,33]. This value is set constant for all calculations regardless of lubricant and operating properties, like viscosity and rolling speed.

To predict film thickness for oil lubricated contacts in the range of starvation, further work should focus on enhancing the HAMROCK and DOWSON equation (4) with a physics-based model, describing the gradient in film thickness decrease and formation of a residual constant film thickness. Nevertheless, the presented model can be applied to predict the onset of starvation in dependence of the *Ca*-number, contact angle and available amount of oil volume. In the following, the effects of the mentioned parameters on the film formation, with focus on the onset of starvation, are discussed more in detail.

5. Discussion

The presented measurement and simulation results in section 4 are focused on oil lubricated contacts. As mentioned in section 1, especially grease lubricated rolling bearings often operate under starved lubrication due to a limited amount of bled oil bleeding from the grease bulk. Moreover, as described in [28], the wettability of bled oil on the thickener material is significantly depended on the thickener type. In comparison to the measured contact angle of PAO oil on the glass disc with approx. 8°, the contact angle of oil on thickener material can reach values up to 57° [28]. As shown in the authors' previous work [26] the contact angle significantly affects the meniscus distance and as emphasised in [20], the wettability determines the film formation. Hence, in the following, the derived meniscus distance equation is applied to calculate the film thickness and the effects of contact angle and available oil amount on the onset of starvation are discussed in the context of grease lubrication.

5.1. Effect of Contact Angle

With the derived meniscus distance equation (7) applied to the HAMROCK and DOWSON equation (4) for starved lubricated point contacts, the effect of contact angle on the onset of starvation is examined for the three cases of $\theta_1 = 11°$, $\theta_2 = 55°$ and $\theta_3 = 66°$. The calculated film thickness for these contact angles is shown in **Figure 5** and compared to the fully flooded results. For the load, geometry and material parameters given in **Table 1**, the onset of starvation for a contact angle of 11° can be determined at approx. 250 mm/s rolling speed and 131 nm film thickness. Increasing the contact angle to 55°, leads to shift of the onset of starvation to 290 mm/s (+ 16 %) and 143 nm (+ 10 %). For 66° starvation is shifted to 315 mm/s (+ 26 %) and 151 nm (+ 15 %).

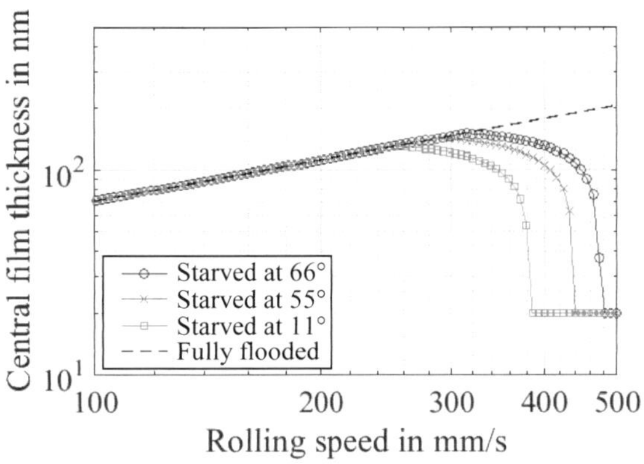

Figure 5. Onset of starvation for different contact angles for η = 73.3 mPas, V_{Oil}/L = 0.042 µl/mm.

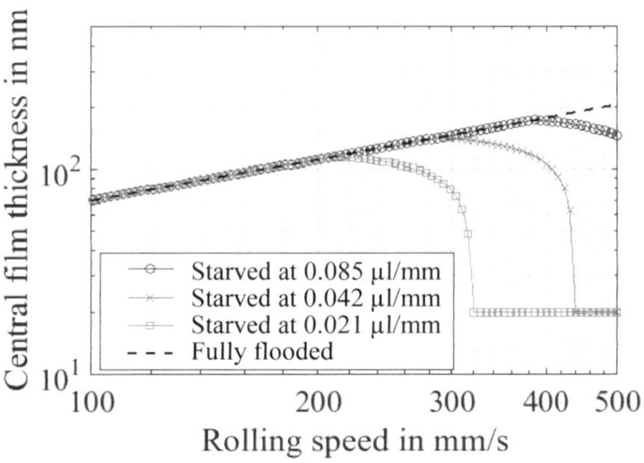

Figure 6. Onset of starvation for different oil volume factors for η = 73.3 mPas and θ = 55°.

This analysis shows, that the impact on the shift of the starvation speed and film thickness is more dominant at higher contact angles. Thus, greases with thickener/base oil combinations, that lead to high contact angles support the lubricant supply and reduce the risk of the onset of starvation.

5.2. Effect of Oil Volume

Another parameter, which significantly affects the lubricant supply in grease lubricated contacts is the available amount of bled oil, which is released from the grease bulk due to oil bleeding. As shown in [26], a higher available oil amount increases the meniscus distance at constant lubricant and operating properties. The effect of increasing the available oil volume on the onset of starvation is exemplarily shown in **Figure 6** for three oil volume factors $(V_{Oil}/L)_1$ = 0.021 µl/mm, $(V_{Oil}/L)_2$ = 0.042 µl/mm and $(V_{Oil}/L)_3$ = 0.085 µl/mm. For the lowest oil volume factor, the onset of starvation can be determined to 210 mm/s rolling speed and 115 nm film thickness. Increasing the available oil volume by factor 2 shifts the onset of starvation to 287 mm/s (+ 37 %) and 143 nm (+ 24 %). Increasing the oil volume by factor 4 leads to a shift to 386 mm/s (+ 47 %) and 174 nm (+ 51 %). These results demonstrate that the available amount of oil has a significant impact on the onset of starvation and the maximum film thickness, which can be reached with increasing rolling speed. In comparison to the effect of contact angle, a change of the oil volume has a more pronounced impact on the onset of starvation. Thus, greases with a low bleeding behaviour increase the risk of the onset of starvation, which is in accordance with literature [34–36].

Furthermore, using the derived meniscus distance equation (7), the minimal required oil amount to ensure an operation in the hydrodynamic lubrication regime at a lambda ratio $\lambda \geq 3$ can be determined. If a composite surface roughness of 80 nm is assumed, a minimum film thickness of h = 240 nm must be reached to fulfil $\lambda \geq 3$. This film thickness is built up under fully flooded conditions at approx. 625 mm/s for a viscosity of η = 73.30 mPas as well as for material, lubricant and load parameters given in **Table 1**. **Figure 7** shows exemplarily the minimal required oil volume factor for different contact angles to reach a film thickness of h = 240 nm in the range of starvation. With increasing speed > 625 mm/s, a higher oil amount is required to obtain a film thickness of h = 240 nm. Additionally, the results point out, that at higher contact angles less oil volume is needed to ensure an operation in the hydrodynamic lubrication regime.

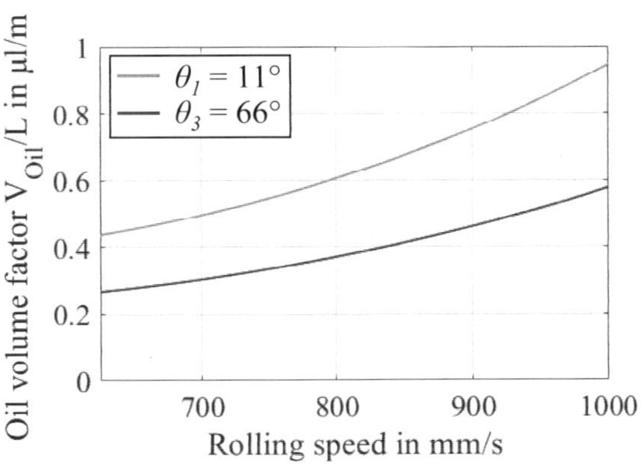

Figure 7. Minimal required oil amount to reach $h = 240$ nm.

5.3. *Validity of the derived Meniscus Distance Equation*

The presented meniscus distance equation is based on simulation results for a limited parameter range shown in Table 2 in the appendix. In accordance to the experiments, all simulations consider a ball radius of 9.525 mm. Applying the equation to extended parameter ranges, as demonstrated in the sections 4.1 and 4.2, the results show a good agreement with a similar model presented in [11] as well as experimental measurements. However, for an extensive validation, more experiments with a varying contact angle and varying oil volume should be performed. Moreover, for an accurate prediction of the film thickness in the range of starvation, both the HAMROCK and DOWSON equation (4) and the meniscus distance equation (7) should be enhanced with focus on an advanced description for the formation of the residual constant film thickness in the range of starvation.

6. Conclusion

To calculate the film thickness under starved lubrication and determine the onset of starvation, the model by HAMROCK and DOWSON, described in equations (1) to (4), can be used. However, to apply this model, the knowledge of the actual inlet meniscus distance in dependence on operating conditions is inevitable. Therefore, this contribution presents an equation (7) to calculate the inlet meniscus distance for point contacts. The derived equation is based on a least square approximation of CFD simulation results for the meniscus distance under variation of operating and lubricant properties. The results can be summarised as follows:

- A comparison between the derived equation (7) and equation (5) presented by NOGI et al. [11], shows good agreement of the calculated meniscus distance. However, in addition to equation (5), the derived equation (7) in this work considers the wettability of oil on the contacting surfaces and the available oil volume for lubrication.
- Applying the derived meniscus distance equation (7) to the HAMROCK and DOWSON equation (4) for film thickness calculation under starved lubrication, the onset of starvation agrees with film thickness measurement results. However, in the range of starvation the calculated film thickness results show significant deviations of the measured film thickness.
- The analysis of the effects by contact angle and amount of oil volume on the onset of starvation reveal, that with higher contact angle and oil volume, the onset of starvation is shifted to higher speeds. However, an increase of oil volume has a more pronounced impact on the onset of starvation than a higher contact angle.
- For a given film thickness, the meniscus distance equation (7) allows to determine the required oil amount depending on operating and lubricant properties.

To describe the film formation in the range of starvation more accurately, future work should focus on an enhancement of both, the HAMROCK and DOWSON model as well as the presented meniscus distance equation (7). In accordance to [23], extended geometries, like different ball radii and elliptical or line contacts should be examined to enhance presented model by effects of geometry on the meniscus distance. Nevertheless, the derived meniscus distance equation can be used to determine the onset of starvation depending on contact angle and available oil volume, which is of significance describing the film formation in grease lubricated contacts. Thus, the presented equation can be combined with existing models, taking e.g. the thickener effect on the film formation into consideration, like recently developed in [37].

Acknowledgements
The authors would like to thank the Research Association for Drive Technology FVA (Forschungsvereinigung Antriebstechnik e. V.) as well as the participating member companies for the support of the IGF project N/1 19027, which is funded by the German Federation of Industrial Research Associations AiF (Arbeitsgemeinschaft industrieller Forschungsvereinigungen) within the framework of the program for the promotion of the Industrial Collective Research IGF (Industrielle Gemeinschaftsforschung) by the Federal Ministry for Economic Affairs and Energy BMWi (Bundesministerium für Wirtschaft und Energie) based on a resolution of the German Bundestag. Simulations were performed with computing resources granted by RWTH Aachen University under project ID rwth0628.

Appendix

Table 2. CFD simulation parameters and results for m.

Case	θ (°)	*Ca* (-)	V_{Oil}/L (µl/mm)	*m* (10^{-3} m)	*Case*	θ (°)	*Ca* (-)	V_{Oil}/L (µl/mm)	*m* (10^{-3} m)
#1	11	0.090	0.173	1.592	#24	44	0.332	0.174	0.909
#2	11	0.098	0.174	1.559	#25	44	0.107	0.290	1.880
#3	11	0.100	0.173	1.545	#26	44	0.107	0.305	1.938
#4	11	0.111	0.173	1.480	#27	44	0.107	0.318	1.969
5	11	0.120	0.173	1.461	#28	44	0.127	0.274	1.716
#6	11	0.127	0.173	1.395	#29	44	0.127	0.332	1.913
#7	11	0.166	0.173	1.257	#30	44	0.127	0.452	2.181
#8	11	0.249	0.173	1.035	#31	55	0.120	0.174	1.656
#9	11	0.332	0.174	0.886	#32	55	0.127	0.174	1.560
#10	11	0.100	0.222	1.831	#33	55	0.166	0.174	1.260
#11	11	0.100	0.274	1.860	#34	55	0.249	0.174	0.999
#12	11	0.100	0.318	1.968	#35	55	0.332	0.174	0.896
#13	11	0.127	0.274	1.730	#36	55	0.127	0.274	1.836
#14	11	0.127	0.332	1.873	#37	55	0.127	0.332	2.025
#15	11	0.127	0.452	2.094	#38	55	0.127	0.452	2.479
#16	11	0.127	0.531	2.222	#39	55	0.127	0.498	2.502
#17	44	0.100	0.174	1.555	#40	66	0.136	0.174	2.048
#18	44	0.107	0.174	1.521	#41	66	0.166	0.174	1.524
#19	44	0.111	0.174	1.496	#42	66	0.249	0.174	1.311
#20	44	0.120	0.174	1.533	#43	66	0.332	0.174	1.038
#21	44	0.127	0.174	1.467	#44	8	0.096	0.126	1.389
#22	44	0.166	0.174	1.199	#45	8	0.184	0.126	1.056
#23	44	0.249	0.174	0.957	#46	8	0.356	0.126	0.677

References

[1] Franke E, Poll G. Service life and lubrication conditions of different grease types in high-speed rolling bearings. In: Dowson D, Priest M, Taylor CM, Ehret P, Childs THC, Dalmaz G et al., editors. Tribology Series Lubrication at the Frontier. Elsevier; 1999, p. 601–609.

[2] Cann PME, Lubrecht AA. Bearing performance limits with grease lubrication: the interaction of bearing design, operating conditions and grease properties. Journal of Physics D: Applied Physics 2007;40(18):5446–51. https://doi.org/10.1088/0022-3727/40/18/S05.

[3] Cen H, Lugt PM. Film thickness in a grease lubricated ball bearing. Tribology International 2019;134:26–35. https://doi.org/10.1016/j.triboint.2019.01.032.

[4] Fischer D, von Goeldel S, Jacobs G, Stratmann A. Numerical Investigation of Effects on Replenishment in Rolling Point Contacts using CFD Simulations. Tribology International 2021(accepted). https://doi.org/10.1016/j.triboint.2021.106858.

[5] Chevalier F, Lubrecht AA, Cann PME, Colin F, Dalmaz G. Starvation Phenomena in E.H.L. Point Contacts: Influence of Inlet Flow Distribution. In: The Third Body Concept Interpretation of Tribological Phenomena. Elsevier; 1996, p. 213–223.

[6] Damiens B, Venner CH, Cann PME, Lubrecht AA. Starved Lubrication of Elliptical EHD Contacts. Journal of Lubrication Technology 2004;126(1):105–11. https://doi.org/10.1115/1.1631020.

[7] Poll G, Li XM, Bader N, Guo F. Starved Lubrication in Rolling Contacts - A Review. Bearing World Journal 2020:69–81.

[8] Wedeven LD, Evans D, Cameron A. Optical analysis of ball bearing starvation. Journal of Lubrication Technology 1971;93(3):349–61. https://doi.org/10.1115/1.3451591.

[9] Chiu YP. An analysis and prediction of lubricant film starvation in rolling contact systems. ASLE Transactions 1974;17(1):22–35. https://doi.org/10.1080/05698197408981435.

[10] Wolveridge PE, Baglin KP, Archard JF. The Starved Lubrication of Cylinders in Line Contact. Proceedings of the Institution of Mechanical Engineers 1970;185(1):1159–69. https://doi.org/10.1243/PIME_PROC_1970_185_126_02.
[11] Nogi T. Film Thickness and Rolling Resistance in Starved Elastohydrodynamic Lubrication of Point Contacts With Reflow. J. Tribol 2015;137(4). https://doi.org/10.1115/1.4030203.
[12] Jacod B, Pubilier F, Cann PM, Lubrecht AA. An analysis of track replenishment mechanisms in the starved regime. Tribology Series 1999;36:483–92. https://doi.org/10.1016/S0167-8922(99)80069-8.
[13] Chevalier F, Lubrecht AA, Cann PME, Dalmaz G. The evolution of lubricant film defects in the starved regime. Tribology Series 1998;34:233–42. https://doi.org/10.1016/S0167-8922(98)80078-3.
[14] Gershuni L, Larson MG, Lugt PM. Lubricant Replenishment in Rolling Bearing Contacts. Tribology Transactions 2008;51(5):643–51. https://doi.org/10.1080/10402000802192529.
[15] Ali F, Křupka I, Hartl M. Enhancing the parameters of starved EHL point conjunctions by artificially induced replenishment. Tribology International 2013;66:134–42. https://doi.org/10.1016/j.triboint.2013.05.003.
[16] Kostal D, Sperka P, Svoboda P, Krupka I, Hartl M. Experimental observation of elastohydrodynamically lubricated contacts replenishment. MM Science Journal 2015;3:640–4. https://doi.org/10.17973/MMSJ.2015_10_201525.
[17] Kostal D, Sperka P, Svoboda P, Krupka I, Hartl M. Influence of lubricant inlet film thickness on elastohydrodynamically lubricated contact starvation. Journal of Tribology 2017;139(5):1–6. https://doi.org/10.1115/1.4035777.
[18] Nogi T. An Analysis of Starved EHL Point Contacts with Reflow. Tribology Online 2015;10(1):64–75. https://doi.org/10.2474/trol.10.64.
[19] van Emden E, Venner CH, Morales-Espejel GE. Aspects of flow and cavitation around an EHL contact. Tribology International 2016;95:435–48. https://doi.org/10.1016/j.triboint.2015.11.042.
[20] Li XM, Guo F, Wong PL, Zhao Y. Regulation of lubricant supply by wettability gradient in rolling EHL contacts. Tribology International 2018;120:565–74. https://doi.org/10.1016/j.triboint.2018.01.020.
[21] Liang H, Guo D, Ma L, Luo J. The film forming behavior at high speeds under oil–air lubrication. Tribology International 2015;91:6–13. https://doi.org/10.1016/j.triboint.2015.06.010.
[22] Chan RTP, Martinez-Botas RF, Gohar R. Isoviscous flow past a rigid sphere partially immersed in a thin oil film. Lubrication Science 2007;19(3):197–212. https://doi.org/10.1002/ls.42.
[23] Nogi T, Shiomi H, Matsuoka N. Starved Elastohydrodynamic Lubrication With Reflow in Elliptical Contacts. J. Tribol 2018;140(1). https://doi.org/10.1115/1.4037135.
[24] Hamrock BJ, Dowson D. Isothermal Elastohydrodynamic Lubrication of Point Contacts: Part IV - Starvation Results. Journal of Lubrication Technology 1977;99(1):15. https://doi.org/10.1115/1.3452973.
[25] Cann PM, Damiens B, Lubrecht AA. The transition between fully flooded and starved regimes in EHL. Tribology International 2004;37(10):859–64. https://doi.org/10.1016/j.triboint.2004.05.005.
[26] Fischer D, von Goeldel S, Jacobs G, Stratmann A. Numerical Investigation of Effects on Replenishment in Rolling Point Contacts using CFD Simulations. Tribology International 2021(accepted).
[27] Cann PME. Thin-film grease lubrication. Proceedings of the Institution of Mechanical Engineers Conference 1999;213:405–16.
[28] Huang L, Guo D, Liu X, Xie G, Wan GTY, Wen S. Effects of nano thickener deposited film on the behaviour of starvation and replenishment of lubricating greases. Friction 2016(4):313–23. https://doi.org/10.1007/s40544-016-0123-9.
[29] Hamrock BJ, Dowson D. Isothermal Elastohydrodynamic Lubrication of Point Contacts: Part III - Fully Flooded Results. Journal of Lubrication Technology 1977;99(2):264. https://doi.org/10.1115/1.3453074.

[30] Weller HG, Tabor G, Jasak H, Fureby C. A tensorial approach to computational continuum mechanics using object-oriented techniques. Comput. Phys. 1998;12(6):620. https://doi.org/10.1063/1.168744.
[31] Johnston GJ, Wayte R, Spikes HA. The measurement and study of very thin lubricant films in concentrated contacts. Tribology Transactions 1991;34(2):187–94. https://doi.org/10.1080/10402009108982026.
[32] Guangteng G, Cann PM, Spikes HA. A study of parched lubrication. Wear 1992;153(1):91–105. https://doi.org/10.1016/0043-1648(92)90263-8.
[33] Guangteng G, Spikes HA. The Role of Surface Tension and Disjoining Pressure in Starved and Parched Lubrication. Proceedings of the Institution of Mechanical Engineers, Part J: Journal of Engineering Tribology 1996;210(2):113–24. https://doi.org/10.1243/PIME_PROC_1996_210_487_02.
[34] Fischer D, Jacobs G, Stratmann A, Burghardt G. Effect of base oil type in grease composition on the lubricating film formation in EHD contacts. Lubricants 2018;6(2):32. https://doi.org/10.3390/lubricants6020032.
[35] Cann PM. Starved grease lubrication of rolling contacts. Tribology Transactions 1999;42(4):867–73. https://doi.org/10.1080/10402009908982294.
[36] Huang L, Guo D, Wen S. Starvation and reflow of point contact lubricated with greases of different chemical formulation. Tribology Letters 2014;55(3):483–92. https://doi.org/10.1007/s11249-014-0376-2.
[37] Nogi T, Soma M, Dong D. Numerical Analysis of Grease Film Thickness and Thickener Concentration in Elastohydrodynamic Lubrication of Point Contacts. Tribology Transactions 2020;63(5):924–34. https://doi.org/10.1080/10402004.2020.1778147.

IOP Conf. Series: Materials Science and Engineering 1097 (2021) 012008 doi:10.1088/1757-899X/1097/1/012008

Numerical Transfer Path Analysis for NVH System-Simulation Models

Gerwin Pasch[1,2], Georg Jacobs[1] and Joerg Berroth[1]

[1]Institute for Machine Elements and Systems Engineering, RWTH Aachen University, Schinkelstraße 10, 52062 Aachen, Germany

[2]gerwin.pasch@imse.rwth-aachen.de

Abstract For the assessment of the Noise, Vibration and Harshness (NVH) behavior of complex systems in early phases of the development process, validated modelling methods are available that allow the prediction of structure-borne and airborne noise at system level. However, due to large model sizes, the identification of weak spots in the vibroacoustic transfer behavior and the derivation of optimization measures are highly complex, time-consuming, and mostly not practical. In the field of experimental NVH analysis, transfer path analysis (TPA) has been established as target-oriented methods for identifying vibroacoustic anomalies at system level. In this contribution, TPA methods from the field of measurement technology are selected and applied to numerical NVH system models, aiming for high accuracy and low computational effort in post-processing. The applications of the methods are shown and discussed using an elastic multi-body simulation model of a tractor drivetrain with a transient run-up of the vehicle speed as an example. The classical direct-force and component-based blocked-force TPA methods selected and adapted for this study allow for efficient calculation of the sound contributions of numerical models. At the same time, they overcome typical challenges of experimental TPA, such as exact force determination or consideration of rotational degrees of freedom. In addition, the comparison of the two methods shows that the path contributions are in general different for classical and component-based TPA and only under specific conditions the same. Both numerical TPA methods allow for the identification of weak spots in the NVH behavior in an efficient and target-oriented way.

1. Introduction

Manufacturers and suppliers are increasingly aiming to shorten development times and reduce development costs in order to remain competitive in the market. Long development times and high costs are mainly caused by prototypes, measurements and product changes in late phases of the product development process (PDP). To reduce the number of prototypes and measurements to a minimum and to avoid changes in prototyping phase, methods of virtual product development are currently used in many industrial sectors.

In addition to the central goal of function fulfilment, comfort features such as the Noise, Vibration and Harshness (NVH) behavior are increasingly becoming the focus of attention for customers and consumers and are therefore an important requirement and development goal. The NVH behavior describes all audible and perceptible vibrations on machines and vehicles. According to the machine acoustic transfer chain, sound arriving at a receiver point, e.g. the driver's ear, is characterized by excitation, transmission and radiation properties. Sound is transmitted along airborne or structure-borne sound paths, the so-called transfer paths. The propagation of structure-borne sound depends on the structural dynamics of the individual subsystems as well as on their couplings and thus on the system topology. Due to the high dependencies and interactions, the prediction of the NVH behavior requires investigations at system level. In the field of virtual product development, numerous validated modelling methods are available today which provide reliable noise predictions for entire systems up to the medium and higher frequency range (<5 kHz), e.g. [1–4]. However, the interpretation of simulation results and the identification of root causes of acoustic anomalies at system level can be highly time consuming and challenging. The large number of components and couplings involved in a vibration event and the huge amounts of data often make an interpretation and target-oriented analysis impossible or only possible with great effort.

Conventional methods of evaluating NVH results from numerical models can be divided into punctual and distributed evaluation methods. Punctual evaluation methods consider single sensor positions analogous to experimental measurements. For analysis, the signals are typically transformed into the frequency domain and visualized in the form of spectrograms. The selected sensor points often show a high sensitivity with regard to the exact location for evaluation. In addition, individual sensor points usually provide only an insight into a specific section of the system behavior. Therefore, with evaluation of individual points the derivation of optimization measures influencing the global system behavior is difficult. Punctual evaluation methods are, nonetheless, suitable for the verification of requirements at specified interfaces.

Distributed evaluation methods typically consider the entire structure involved in the vibration event. For evaluation, quantities such as deformations, vibration velocities or accelerations are typically used, for example for analyzing mode shapes or operational deflection shapes. Sometimes also quantities such as energy distribution or structure intensity are used for analysis of structure-borne sound. The selected quantities are evaluated either for individual timesteps or individual frequencies. Typically, a large number of evaluation points are considered at the same time. The analysis is hardly manageable with complex three-dimensional structures or many interacting components. Due to the high complexity, the identification of weak spots is very costly and partly not manageable.

In the field of measurement technology, however, target-oriented evaluation methods of analyzing the NVH behavior of entire systems are known: the methods of transfer path analysis (TPA). The aim of these methods is to identify individual sound contributions to the total sound at defined receiver points. The receiver points are typically at positions in the system for which specific sound requirements are defined, such as maximum acceleration levels at specified interfaces. The sound contributions can be allocated to individual sound sources or individual transfer paths. With the knowledge of dominant sound sources or paths, optimization measures can be determined and derived in a target-oriented way. However, the experimental methods are applicable only in later phases of the PDP when hardware prototypes are available. Changes implemented in later phases of the PDP as well as the large number of prototype variants required generate high costs.

In this publication, methods known from experimental TPA are applied for the target-oriented evaluation of NVH simulation models. For this purpose, TPA methods are selected which deliver sound contributions with high accuracy and minimum calculation effort in post-processing. Section 2 shows the principles of TPA and gives an overview of the available TPA methods. In section 3, two suitable methods are selected and the process for determining of sound contributions with the use of numerical simulation models is presented. Section 4 shows the application and results of the methods for an exemplary system model of a tractor drivetrain. The results are discussed in section 5 and a comparison of the sound contributions of the classical and component-based methods is given.

2. Principles of Transfer Path Analysis

The basic idea of TPA is the decomposition of the total sound at an evaluation point into individual sound contributions, also known as paths. Complex vibration phenomena are reduced to a limited number of paths. The TPA considers all properties that are relevant for the NVH behavior. This includes the interaction between excitation and structural dynamics or transfer paths as well as interference effects caused by phase differences, that result in increasing or decreasing sound amplitudes. The sound contributions help to determine dominant sources or paths in a structured and target-oriented way. The sound contributions are calculated by multiplying forces or sound quantities from operational measurements with measured transmission properties. According to the paradigm of dynamic substructuring [5], the considered system AB is therefore divided, i.e. substructured, into the subsystems A and B, Figure 1.

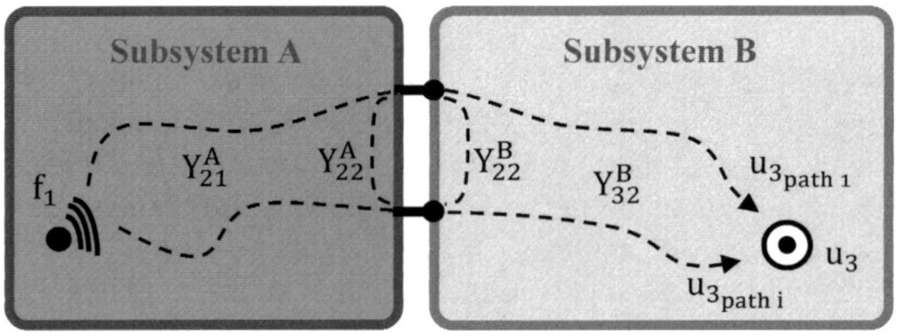

Figure 1. General procedure for TPA: substructuring of the System AB.

The interfaces and thus the paths are determined by the choice of the cutting plane. Subsystem A is usually the active side which contains the excitation sources with the excitation forces f_1. Subsystem B is the passive side which is excited by system A via several interfaces and which contains the receiver points. The analyzed sound quantity u can be displacements, velocities or accelerations. The complex-valued summation of the sound contributions $u_{3_{path\ i}}$ of all n paths results in the total sound u_3 at the receiver point:

$$u_3 = \sum_{i=1}^{n} u_{3_{path\ i}} \quad (1)$$

The transfer characteristics of the structure are described in the form of the frequency response functions (FRFs) $Y^A_{21}, Y^A_{22}, Y^B_{22}$ and Y^B_{32}. The superscript indices indicate the subsystem and the subscript indices indicate the points between the FRFs are obtained.
Over the last decades various TPA methods have been developed. In [6], an overview and a classification of the methods is given. The methods are divided into three TPA families:

- Classical TPA
- Component-based TPA
- Transmissibility-based TPA

The classical and the component-based TPA are physical methods, whereas the transmissibility-based TPA is based on correlations and statistics. The transmissibility-based TPA offers the advantage of a reduced measurement effort. For application to numerical models, the measurement effort is irrelevant. In this publication, only the two physical methods will be discussed.

The calculation of classical and component-based TPA can be derived using the general approach for substructuring. On the conditions of displacement compatibility $u^A_2 = u^B_2$ and force equilibrium $g^A_2 + g^B_2 = 0$ in the interfaces, the total sound u_3 at the receiver caused by the source force f_1 can be determined as follows:

$$u_3 = Y^B_{32} \cdot (Y^A_{22} + Y^B_{22})^{-1} \cdot Y^A_{21} \cdot f_1 \quad (2)$$

For both, classical and component-based method, the sound contributions are determined from time-dependent forces and frequency-dependent transmission properties in the form of FRFs. The methods differ in how the quantities of equation (2) are combined. The selected combination results in different subdivisions of the overall system for the measurements.

In classical TPA, the forces are determined in the assembled system AB. These forces correspond to cutting forces respectively interface forces. The FRFs are determined in the isolated passive structure B. The product of interface forces g_2^B and FRFs Y_{32}^B provides the path contributions.

$$u_3 = Y_{32}^B \cdot (Y_{22}^A + Y_{22}^B)^{-1} \cdot Y_{21}^A \cdot f_1 = Y_{32}^B \cdot g_2^B \tag{3}$$

with

$$g_2^B = (Y_{22}^A + Y_{22}^B)^{-1} \cdot Y_{21}^A \cdot f_1 \tag{4}$$

For the component-based TPA, the equivalent forces f_{eq}^A are determined which depend only on the source structure A. Together with the transmission properties Y_{32}^{AB}, determined in the assembled system AB, the path contributions are calculated. The derivation is shown in (5), where (2) is extended by $Y_{22}^A \cdot (Y_{22}^A)^{-1}$.

$$u_3 = Y_{32}^B \cdot (Y_{22}^A + Y_{22}^B)^{-1} \cdot Y_{22}^A \cdot (Y_{22}^A)^{-1} \cdot Y_{21}^A \cdot f_1 = Y_{32}^{AB} \cdot f_{eq}^A \tag{5}$$

with

$$Y_{32}^{AB} = Y_{32}^B \cdot (Y_{22}^A + Y_{22}^B)^{-1} \cdot Y_{22}^A \tag{6}$$

and

$$f_{eq}^A = (Y_{22}^A)^{-1} \cdot Y_{21}^A \cdot f_1 \tag{7}$$

As shown in equation (7), the equivalent forces are dependent only on the active system A.

In experimental TPA, the determination of forces is challenging. Since the integration of force sensors changes the structural dynamics of the system, methods for indirect force determination were developed. For this purpose, acceleration or displacement quantities and transmission properties are measured, from which the forces are subsequently determined, for example by matrix inversion. An overview of the methods is given in Table 1, that is an excerpt of [6].

Table 1. Overview and classification of TPA methods according to [6].

Family	**Method**	Operational measurement		Source characterization			Apply to
		Quantity	On system	Quantity		Using	FRFs
	Direct-force	g_2^B	AB	Interface force	g_2^B	–	Y_{32}^B
Classical	Mount-stiffness	u_2^A, u_2^B	AB	Interface force	g_2^B	Z^{mt}	Y_{32}^B
	Matrix-inverse	u_4	AB	Interface force	g_2^B	Y_{42}^B	Y_{32}^B
	Blocked-force	g_2^{bl}	A (blocked)	Equiv. force	f_2^{eq}	–	Y_{32}^{AB}
	Free-velocity	u_2^{free}	A (free)	Equiv. force	f_2^{eq}	Y_{22}^A	Y_{32}^{AB}
Component-based	Hybrid-interface	g_2^R, u_2	AR	Equiv. force	f_2^{eq}	Y_{22}^A	Y_{32}^{AB}
	In-situ	$u_2(u_4)$	AR/AB	Equiv. force	f_2^{eq}	$Y_{22\,(42)}^{AR/AB}$	Y_{32}^{AB}
	Pseudo-forces	u_4	AR/AB	Pseudo force	f_{ps}	$Y_{4ps}^{AR/AB}$	Y_{3ps}^{AB}

Only for the classical direct-force and the component-based blocked-force method, the forces are determined directly without further calculations. For all other methods, the forces are determined indirectly e.g. with mathematically expensive operations like matrix inversion. The blocked-force method uses the fact that the equivalent forces are equal to forces that fixate the active system rigidly. These forces are called blocked forces g_2^{bl} and can be derived with the boundary condition $u_A = 0$.

$$u_A = 0 = Y_{21}^A \cdot f_1 - Y_{22}^A \cdot g_2^{bl} \tag{8}$$

By rearranging equation (8) the following is obtained:

$$g_2^{bl} = (Y_{22}^A)^{-1} \cdot Y_{21}^A \cdot f_1 \tag{9}$$

This corresponds exactly to the equivalent forces f_{eq}^A shown in equation (7) that shows that both forces are equal.

The use of TPA for the evaluation of numerical models is known in the literature as numerical TPA (NTPA). In previous publications the application of two TPA methods to numerical models has been shown, although in each case only under consideration of a single operation point: [7] and [8] show the applications of the classical direct-force method. In [9], the application of the component-based in-situ method for a linearized system in frequency domain is presented. Due to the indirect force determination of the in-situ method, extensive post-processing is necessary. The advantage of numerical models of direct availability of forces without further complex calculations is not used.

In contrast, this publication extends the NTPA methods by the component-based blocked-force method to achieve high accuracy and low computational effort. The parallel use of the classical direct-force method allows for a direct comparison of the sound contributions between the classical and component-based TPA families.

3. Solution approach

In early phases of the PDP, elastic multi-body simulation (EMBS) is widely used for analyzing the NVH behavior of virtual prototypes. The EMBS enables efficient calculation of the structural dynamics of large, flexible structures considering non-linear transmission elements and transient excitations under variable operating conditions in frequency as well as in time domain. Coupling elements such as bushings or roller bearings are typically modeled using concentrated force elements. Variables like acceleration or forces can be determined at any desired location in the system. Typical weaknesses of experimental TPA, such as exact force determination, consideration of rotational degrees of freedom, installation space limitations and measurement noise, do not exist for numerical simulations.

For selecting of suitable TPA methods for the application to numerical models, two requirements in particular are considered: high accuracy and low computational effort especially in post-processing. The two requirements are met by selecting physical TPA methods with direct-force determination. As described above, the concentrated force elements of the EMBS allow for direct force signal output at interfaces without further calculations. The amount of data exchanged is also kept low by the direct-force determination, which enables efficient calculation in the post-processor. Therefore, the classical TPA with direct-force and the component-based TPA with blocked-force determination are selected.

For both methods the procedure and the calculation chain were created. Each method requires two separate simulations which can be derived based on an initial model covering the overall system AB. The methods differ mainly in the definition of the system boundaries of the simulation models.

Figure 2 shows the method proposed for determining path contributions according to the classical TPA method with direct-force determination for numerical models.

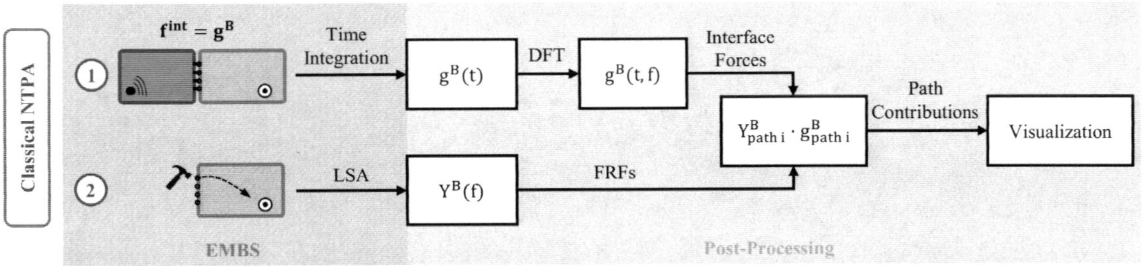

Figure 2. Proposed method for performing classical NTPA with direct-force determination.

As shown in the previous section, in classical TPA the interface forces are obtained in the assembled system AB. Using time integration, the time-dependent interface forces are calculated in time domain for example for a run-up of a drivetrain. Both, translational and rotational forces are directly available as output variables. Using discrete Fourier transform (DFT), the time signals are transferred into frequency domain. The FRFs for the classical NTPA are determined in frequency domain on the passive structure using linear system analysis (LSA) calculation. For this purpose, the active subsystem is deactivated in the simulation model. The FRFs are determined successively between all interface degrees of freedom (DOFs) and the receiver DOFs. The path contributions are obtained by multiplying the complex-valued forces and FRFs.

The proposed method for determining path contributions using the component-based TPA method with blocked-force determination for numerical models is presented Figure 3.

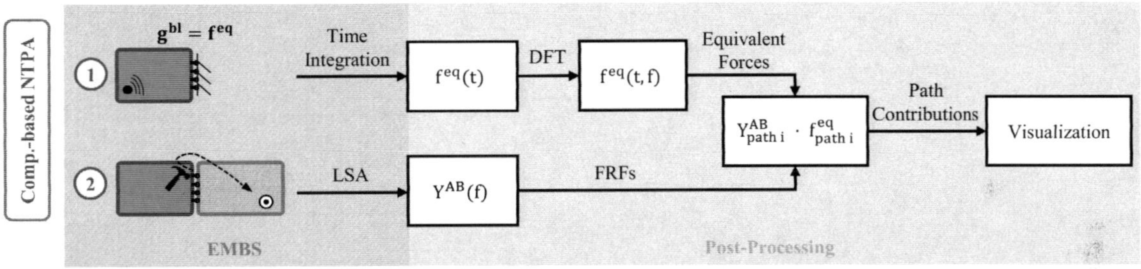

Figure 3. Proposed method for performing component-based NTPA with blocked-force determination.

The equivalent forces required in the component-based TPA equals to the so-called blocked forces. The blocked forces are determined by fixing the source structure to an infinitely stiff environment. In the model this is achieved by deactivating the passive structure and connecting the interfaces to the ground. The LSA is performed on the assembled system. The operations for post-processing are the same as in the classical method.

The number of paths results from the number of interface DOFs and receiver DOFs. In general, all force-transmitting interface DOFs should be considered. Receiver DOFs can be freely selected by the user according to his interests. Translational as well as rotational DOFs can be considered. For clarity, individual paths can be summarized in the visualization tool. For example, the interface DOFs of individual interfaces can be summarized and the receiver DOFs can be combined to one receiver signal.

4. Application of NTPA to a tractor drivetrain

The NTPA methods presented above are applied to a demonstrator EMBS model for illustration. A validated model of a tractor drivetrain with a hydrostatic-mechanical power split transmission is used, which is described in detail in [3]. The model used in this study is shown in Figure 4.

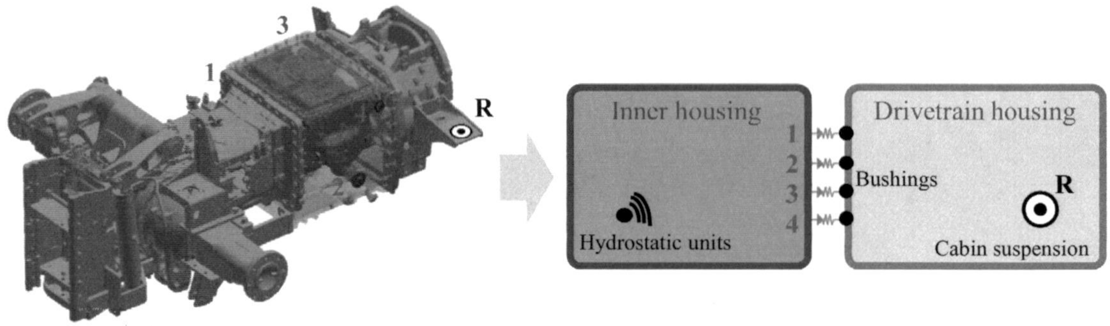

Figure 4. EMBS model of a tractor drivetrain as demonstrator for NTPA.

The main excitation source of the system are the hydrostatic units of the drivetrain. The hydrostatic units are mounted in a separate inner housing and connected to the drivetrain housing elastically via four elastomer bushings. In the NTPA, the paths from the bushing suspensions to the cabin suspension are considered. Therefore, the cutting plane is placed along the bushings. In the following study, the cabin suspension front right, marked with 'R' in Figure 4, is selected as the receiver point. The bushings transmit both translational and rotational forces along all DOFs. This results in six possible interface DOFs per bushing, in total 24 interface DOFs. In this example the total translational acceleration at the receiver is considered.

Using the methods shown in section 3, the sound contributions are calculated for the tractor drivetrain for a transient run-up of the vehicle speed. For verification of the calculation, the acceleration level at the receiver point measured in the reference model is compared with the synthesized total acceleration level generated by the NTPA using equation (1). The results are shown in Figure 5.

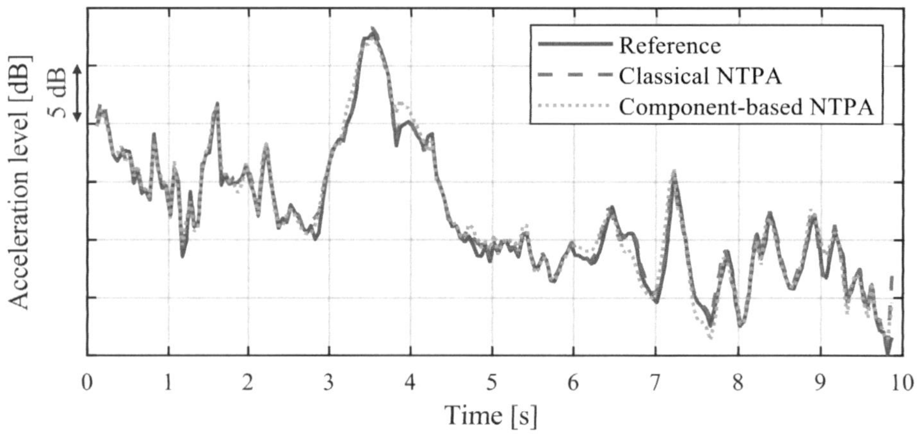

Figure 5. Comparison of the simulated and synthesized acceleration levels at the receiver.

The acceleration signals show a high degree of conformity. Small deviations of maximum 1-2 dB are visible. These deviations can be caused by effects of linearization or numerical inaccuracies. The deviations are relatively small and do not influence the qualitative results of the NTPA. Overall, the diagram shows a high accuracy of the calculation for both methods.

In the following the results of the NTPA are presented in the form of the sound contributions. As mentioned above, the sound contributions can be evaluated on different levels for visualization purposes. Figure 6 shows the sound contributions of the four bushings to the total sound for the classical NTPA. For comparison, the acceleration level of the reference model is presented in the first row.

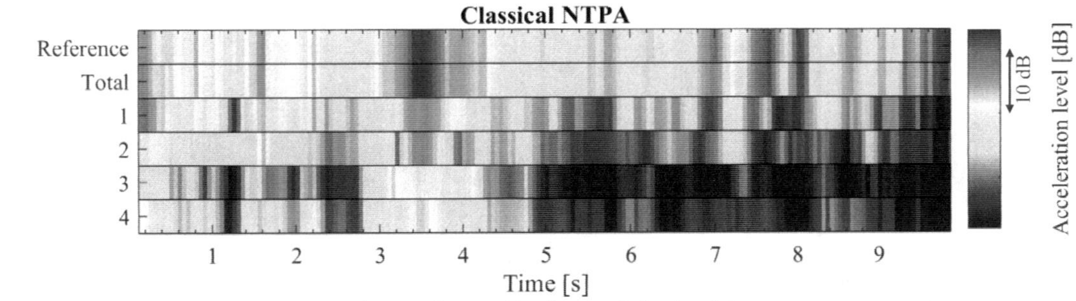

Figure 6. Classical NTPA: Total sound contributions of the bushings.

The color bars indicate the acceleration level of the individual paths. This visualization allows for a transparent identification and ranking of dominant paths. For example, it becomes clear that the peak at the time 3.5 s is mainly caused by bushings 1 and 2.

For a more in-depth investigation, the contributions of the bushings are divided into the six interface DOFs, Figure 7.

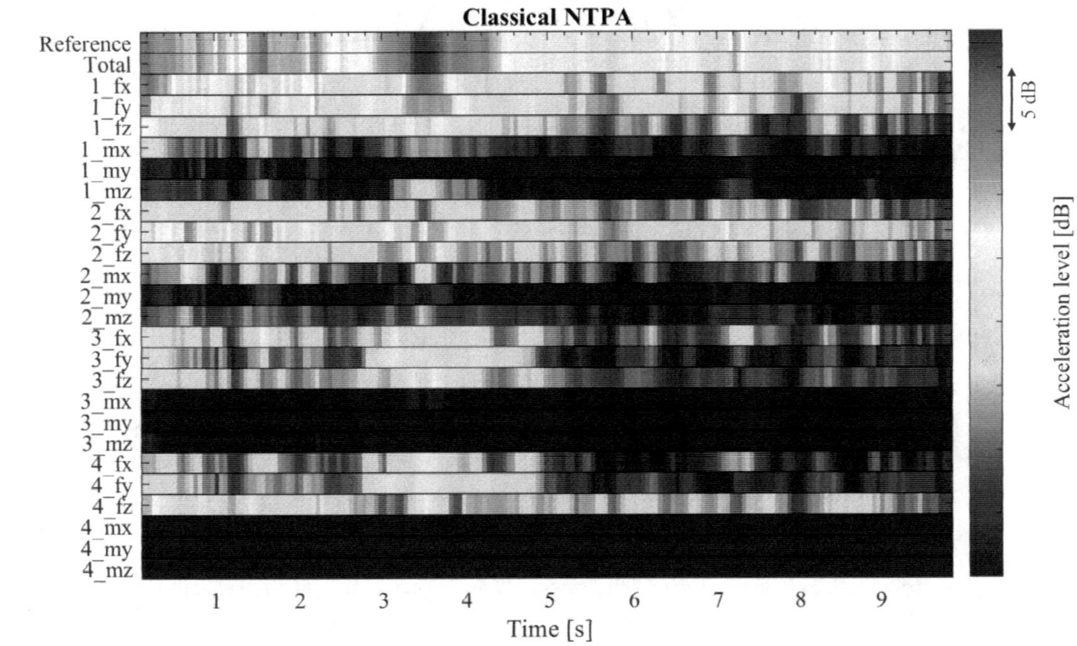

Figure 7. Classical NTPA: Sound contributions of the bushings divided into six interface DOFs.

It is visible that especially the paths in x- and y-direction of bushing 1 and y- and z-direction of bushing 2 have the highest contribution to the total sound.

The result of the component-based NTPA is shown in Figure 8.

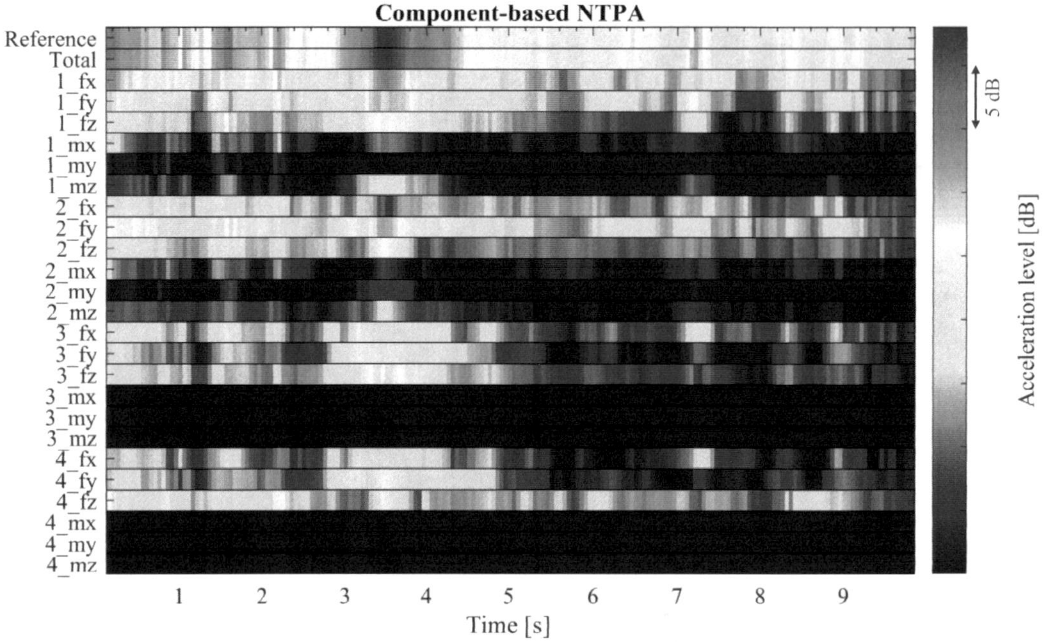

Figure 8. Component-based NTPA: Sound contributions of the bushings divided into six interface DOFs.

The results are similar to the results of classical NTPA, although not exactly the same. For example, the path of bushing 2 in z-direction is less dominant in component-based NTPA in comparison with classical NTPA with a difference of about 8 dB at the time 3.5 s. The reason for the differences in the sound contributions of the two methods will be discussed in the following section.

5. Discussion

The results presented in section 4 show that TPA can be performed on numerical models with the proposed calculation chains. The comparison between the simulated and synthesized acceleration levels at the receiver demonstrates the high accuracy of the calculation, as presented in Figure 5. Weak spots in the complex transfer behavior become clearly detectable by the visualization of the sound contributions in the color bar diagrams, Figure 6-8. These results can be used directly to derive optimization measures, such as a stiffening of the passive-side connection of the bushing.

The calculation time of the post-processing including the visualization takes only few minutes. Therefore, the routines developed in this study are well applicable in the PDP. Among them, the component-based NTPA with blocked-force determination offers another essential advantage: significant reductions of the calculation times of the time integration simulation. The time integration for this method was performed only on subsystem A and not on the overall system due to the substructuring strategy presented. Depending on the system, this can lead to significant reductions in calculation time. In the case of the tractor model, the calculation times for time integration were reduced by about 50%. Another advantage of the component-based method is that the blocked forces depend only on system A. In the case that changes or optimization measures are applied to system B, the blocked forces remain unchanged. For this reason, only an LSA calculation with its short calculation times (few seconds or minutes) must carried out to evaluate optimization measures. Thus, the sound contributions for a modified system can be generated within a few minutes.

As mentioned in the previous section, the sound contributions of the classical and component-based methods are similar but not equal for the system under study. While the total sound at the receiver u_3 is mathematically the same for both methods, the path contributions of the two methods are in general not equal. This is caused by the different way of substructuring and thus the different calculation order of equation (2). The difference in the mathematical calculation is derived in the following for one frequency grid point. For clarity, equation (2) is rewritten as

$$u_3 = A \cdot B \cdot C \tag{10}$$

with

$$A = Y_{32}^{B} \in \mathbb{C}^{1xn} \tag{11}$$

$$B = (Y_{22}^{A} + Y_{22}^{B})^{-1} \cdot Y_{22}^{A} \in \mathbb{C}^{nxn} \tag{12}$$

$$C = (Y_{22}^{A})^{-1} \cdot Y_{21}^{A} \cdot f_1 \in \mathbb{C}^{nx1} \tag{13}$$

The dimension n corresponds to the number of interface DOFs. In the case of classical TPA, entry i of A is multiplied by entry i of the matrix product of B and C for calculation of the sound contribution of path i:

$$u_{3_{\text{path i}}}^{\text{classic}} = A_i \cdot (B \cdot C)_i \tag{14}$$

For calculation of the sound contribution of path i in component-based TPA, entry i of the matrix product of A and B is multiplied by entry i of C:

$$u_{3_{\text{path i}}}^{\text{comp.}} = (A \cdot B)_i \cdot C_i \tag{15}$$

In general, equations (14) and (15) are not equal, and thus the path contributions of classical TPA and component TPA are not the same.

In the special case that B is diagonal, equations (14) and (15) and thus the path contributions of both methods are the same. B is diagonal when no crosstalk between the interface DOFs exists, because crosstalk is indicated by the side diagonal entries of B. Crosstalk means that a force at one interface DOF causes a response, e.g. movement or force, at another interface DOF. The level of crosstalk depends on the system and on the definition of the cutting plane. The special case of a system without crosstalk can also be illustrated as follows: an equivalent force acting at one interface DOF in the system AB causes only an interface force at the same interface DOF, whereas the interface forces at all other DOFs are zero.

However, real systems typically exhibit a certain amount of crosstalk. In the case of the tractor drivetrain presented in the previous section, this amount is relatively small, resulting in slightly different sound contributions for the two methods.

In systems with a high level of crosstalk, the results of the two methods can vary significantly, leading to different interpretations and optimization measures for the systems in question. Hence, the choice of method should be guided by the specific changes and optimization measures to be evaluated. For a fully filled matrix B, in classical TPA a sound contribution $u_{3_{\text{path i}}}^{\text{classic}}$ contains exactly one transfer path between an interface DOF and the receiver, and n forces acting within the interface DOFs. In contrast, for component-based TPA with a fully filled matrix B, a sound contribution $u_{3_{\text{path i}}}^{\text{comp.}}$ results from n transfer paths between the interface DOFs and the receiver, with exactly one equivalent force acting at one interface DOF. Thus, classical TPA can provide more transparent results for evaluating of single transfer path changes in the passive system. The results of the component-based TPA, on the other hand, can be more transparent for evaluating optimizations that aim to change a force dependent on the active system.

6. Summary and Outlook

This contribution shows the application of the classical and component-based TPA methods to numerical elastic multi-body system-simulation models and their comparison. The classical direct-force and component-based blocked-force TPA methods were chosen for achieving high accuracy with low computational effort even for the analysis of wide and transient operation ranges. Both methods are challenging for use in experimental TPA because of the direct way of force determination, but particularly suitable for the application to numerical models. The calculation chains developed for both methods were applied to a demonstrator model of a tractor drivetrain for investigating the structure-borne noise for a transient run-up of the vehicle speed. The NTPA allows for in-depth insights into the system behavior with low computational effort. The efficient and target-oriented analysis tool is suitable for being used in early phases of the product development process for identifying optimization potential in a transparent and structured way. It was shown that the component-based blocked-force method additionally offers a high potential for the reduction of simulation time. Since the classical and the component-based NTPA were carried out in parallel, a direct comparison of the sound contributions was possible, whereby slight differences were identified for the demonstrator model. Further investigations showed that these differences are caused by the different calculation order of the path contributions for the two methods. Both methods provide equal path contributions only for systems without crosstalk, which are barely found in reality. For systems with high crosstalk, on the other hand, a theoretical discussion is given about which method allows evaluating optimization changes with higher transparency.

In this contribution, the NTPA was presented on structure-borne sound level. In further works, the calculation chain of classical NTPA presented in this study has been extended in order to perform path evaluations on airborne sound level for numerical system-simulation models [10].

Furthermore, the NTPA methods are to be integrated into automated optimization processes. The component-based NTPA method is particularly suitable for this purpose due to its high potential for reducing simulation times. Especially for optimizations on the passive system, only FRF calculations are required per iteration, so that each iteration can be performed within few minutes.

References

[1] Drichel P, Jaeger M and Müller-Giebeler M 2019 *Mit elektrischem Antrieb und modellbasierter Systemanalyse nahezu lautlos in die Zukunft* ATZ Extra pp 52–57

[2] Jaeger M, Drichel P, Müller-Giebeler M, Schröder M, Berroth J, Behler G, Hameyer K, Jacobs G and Vorländer M 2019 *Erweiterung NVH Simulationsmodell: Erweiterung der Simulationsmöglichkeiten für maschinenakustische Untersuchungen an E-Motive-Antrieben im Kontext zur Fahrzeugstruktur* FVA Heft Nr. 1730 (Forschungsvereinigung Antriebstechnik e.V.)

[3] Pasch G, Jacobs G, and Berroth J K 2020 *NVH System Simulation of a Tractor with Hydrostatic-Mechanical Power Split Transmission* LANDTECHNIK 75(4) pp 301-15

[4] Rick S, Klein J, Wegerhoff M, Hameyer K, Vorländer M and Jacobs G 2015 *E-MOTIVE NVH-Simulationsmodell: Modellbildung zur NVH Simulation eines E-MOTIVE Antriebsstrangs* FVA Heft Nr. 1130 (Forschungsvereinigung Antriebstechnik e.V.)

[5] Klerk D de, Rixen D J and Voormeeren S N 2008 *General Framework for Dynamic Substructuring: History, Review and Classification of Techniques* AIAA Journal 46 pp 1169–81

[6] van der Seijs, Maarten V., Klerk D de and Rixen D J 2016 *General framework for transfer path analysis: History, theory and classification of techniques* Mechanical Systems and System Processing pp 217-44

[7] Vanhollebeke F, Peeters J, Vandepitte D and Desmet W 2015 *Using transfer path analysis to assess the influence of bearings on structural vibrations of a wind turbine gearbox* Wind Energy 18 pp 797–810
[8] Acri A, Offner G, Nijman E and Corradi R 2015 *NUMERICAL TRANSFER PATH ANALYSIS FOR THE ASSESSMENT OF AIR-BORNE NOISE IN INTERNAL COMBUSTION ENGINES* Proceedings of ICoEV 2015, International Conference on Engineering Vibration: Ljubljana, 7-10 September (Ljubljana: Faculty for Mechanical Engineering) pp 384–93
[9] Wegerhoff M, Sottek R and Amberg J 2018 *Numerical in-situ TPA of an elastic multibody dynamic model of a tractor gearbox* AAC 2018: Aachen Acoustics Colloquium/Aachener Akustik Kolloquium ed S Pischinger et al (Aachen, Germany: FEV Group GmbH) pp 177–88
[10] Wischmann S, Pasch G, Berroth J and Jacobs G 2021 *Acoustic Optimization of a Power Take-off Gear Box Using Numerical Transfer Path Analysis* IOP Conference Series, Material Science and Engineering

IOP Conf. Series: Materials Science and Engineering 1097 (2021) 012009 doi:10.1088/1757-899X/1097/1/012009

Investigations on the rolling bearing cage dynamics with regard to different operating conditions and cage properties

S. Schwarz[1], H. Grillenberger[2], S. Tremmel[3] and S. Wartzack[1]

[1] Friedrich-Alexander-Universität Erlangen-Nürnberg (FAU), Engineering Design (KTmfk), Martensstraße 9, 91058 Erlangen, Germany

[2] Schaeffler Technologies AG & Co. KG Industriestraße 1-3, 91074 Herzogenaurach, Germany

[3] Universität Bayreuth, Engineering Design and CAD, Universitätsstraße 30, 95447 Bayreuth, Germany

Abstract. The dynamic behavior of a cage can have a significant influence on the performance and the noise of a rolling bearing. In case of an unstable or high-frequency cage movement, vibrations are transmitted to the adjacent machine elements and the environment, which can influence the operating behavior of the whole system. In addition, a loud noise can often be perceived, which is referred to in the literature as cage rattling or squealing. In this paper, characteristics of different cage movements are investigated using multibody simulations and experimental investigations. For this purpose, essential properties of the fundamentally observable cage movement types "stable", "unstable" and "circling" are presented. The calculated cage dynamics and the type of cage motion are used to show dependencies between the operating conditions and the resulting cage movement such as inner ring rotational speed, bearing load or cage characteristics such as pocket clearance. Based on the simulations, interactions between the input parameters can also be determined. The results are used to identify operation-critical conditions and cage properties that lead to high cage dynamics. Finally, a comparison between the results of the multibody calculations and optical measurements is made. The optical measurements are performed using high-speed cameras. Reference markers fixed on the cage and digital image correlation allow the evaluation of the kinematics as well as the deformation of the cage. These results are compared with the simulation data to ensure a high quality of dynamics simulation.

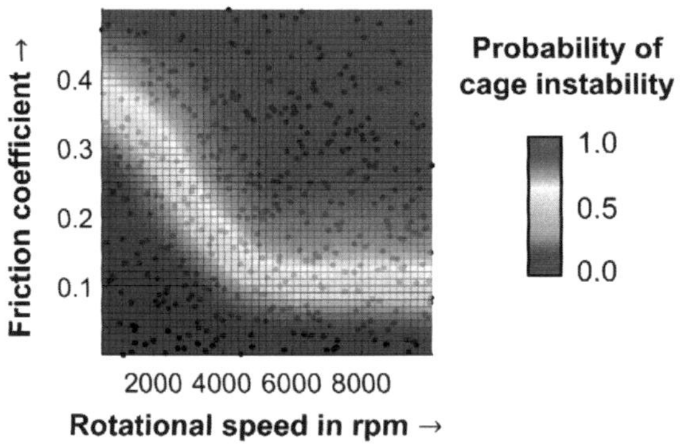

Dependency of unstable cage motion on the coefficient of friction in contact between cage/rolling element and speed of the inner ring. An interpolation between the simulation results shown as black dots is color highlighted.

IOP Conf. Series: Materials Science and Engineering 1097 (2021) 012009 doi:10.1088/1757-899X/1097/1/012009

1. Introduction

The demands on the performance of rolling bearings are continuously increasing. In addition to conventional functionalities such as shaft guidance or load rating, the focus is also on vibration and noise behavior. A possible cause of severe vibrations or noise can be the cage of the rolling bearing. Under certain conditions, the cage can perform very high-frequency movements, which can also be superimposed by severe deformations. In the literature this phenomenon is called cage rattling or cage instability [4; 9; 3]. In this case, the cage shows a high-dynamic movement which leads to an increase in contact forces, slip velocities and the frictional torque of the bearing. The severe deformations can also lead to high component stress on the cage and may significantly reduce cage lifetime. In addition high contact forces between the cage and rolling elements can affect the kinematics of the rolling elements and thus also lubricant film formation and bearing lifetime. In addition to this type of movement, which is classified as 'unstable', 'stable' and 'circling' movement patterns can also be observed [10]. During a circulating cage motion, the cage center of mass (cCoM) uses a major part given by the guidance clearance and the speed of the cCoM corresponds approximately to the rotational speed of the cage. A stable cage movement characterizes a very low dynamic. The cCoM of the cage moves comparatively slowly and the deformations of the cage are very low.
Research activities on cage dynamics are performed in the field of calculation as well as in real experiments. GUPTA used a dynamics simulation model to investigate the relationship between friction in the cage pocket and the resulting trajectory of the cage [5]. NOGI et al. determine a critical limit for the friction coefficient above which cage instability occurs. This limit was determined on the basis of dynamics simulations and depends on the cage mass, the friction in the contact between ball and raceway, cage-ball contact stiffness, speed of the cage and number of rolling elements [8]. There are also publications in the experimental area dealing with the dynamics of rolling bearing cages. The cage movement is often measured using optical methods (high-speed cameras). These are suitable for contactless determination of the cage movement even at high rotational speeds. ABELE and HOLLAND were able to observe a clear influence of accelerations of the rolling bearing and the resulting cage motion on the basis of measurement results [1]. ABELE, UNTERDERWEIDE and HOLLAND were also able to measure typical properties of unstable cage motion in experiments [2]. SCHWARZ, GRILLENBERGER and TREMMEL used high-speed cameras and digital image correlation to compare the deformations of unstable cage motions in calculation and experiment [9].
The previous investigations only selectively show the influence of the bearing load or cage geometry on the resulting dynamics. A comprehensive study is not available in this context. In the following, based on a comprehensive simulation study dependencies between parameters such as bearing load, rotational speed as well as cage pocket clearance and the resulting cage movement identified for a deep groove ball bearing. Thus, interactions of the investigated parameters with respect to the resulting cage dynamics can be determined, further improving the understanding of highly dynamic cage motions.

2. Main Section

2.1. Test and simulation conditions

A deep groove ball bearing with an inner diameter of 65 mm, an outer diameter of 140 mm and a glass fiber reinforced polyamide cage is used for these investigations.
The multi-body simulation software Caba3D (Computer Aided Bearing Analysis 3 Dimensional) is used to calculate the dynamics of the rolling bearing and in particular of the cage. Caba3D enables the determination of all kinematic and kinetic results of the rolling bearing components [7; 9]. In addition, a reduced finite element model is used to calculate the elastic deformation of the cage [7; 4]. The deformation is an elementary characteristic of the unstable cage motion and should therefore be taken into account in the simulation.
An important parameter regarding the cage dynamics is the friction force between cage and the rolling elements [5, 8]. This is considered in Caba3D according to COULOMB's friction model and consequently with a coefficient of friction μ. This method allows the simplified modelling of different lubrication conditions, e.g. dry or lubricated and is considered reasonably accurate, since it is not the friction force

itself but its effect on the cage dynamics that should be calculated within the scope of the simulation. In particular, for the unlubricated contact between the cage and the rolling element, COULOMB's friction model is considered adequate. Although the coefficient of friction shows tendencies for different lubrication conditions, there is no specific value for a given lubrication. Therefore, for the comparison between the simulation and the optical measurement, several simulations with different coefficients of friction are carried out.
A stereo camera system enables the measurement of the cage kinematics. The image section is captured at 8 000 fps. Reference markers are attached to the test rig, shaft and cage. By grouping these markers to form a component, all translation and rotation movements of these components can be determined using digital image correlation. In addition, the cage deformation is calculated by shifting the original pattern to the deformed pattern. Deformations can thus be determined on the basis of the deviations from the original pattern.
The Cage Dynamics Indicator (CDI) is calculated to assess the cage dynamics. This indicator forms the basis for quadratic discriminant analysis, which determines the class probability for the given cage movement [10].

2.2. *Influences on cage dynamics*

Latin Hypercube samplings were performed to determine the influence of friction coefficient (in the cage pocket), bearing load, inner ring rotational speed and pocket clearance on cage dynamics. The limits for the experimental design of the calculations are especially for the friction coefficient μ=0.5 high, in order to better analyze the influence of the parameters. In addition, increased coefficients of friction are to be expected especially in the contact of polyamide and steel without lubricant, which are in the range specified [6]. For each of the simulations, the motion type was determined using quadratic discriminant analysis and the CDI. Figure 1 illustrates the relationship between the probability of "unstable" class membership P_I as a function of friction coefficient, radial bearing load, load ratio (axial force divided by radial force) and pocket clearance based on an interpolated surface.

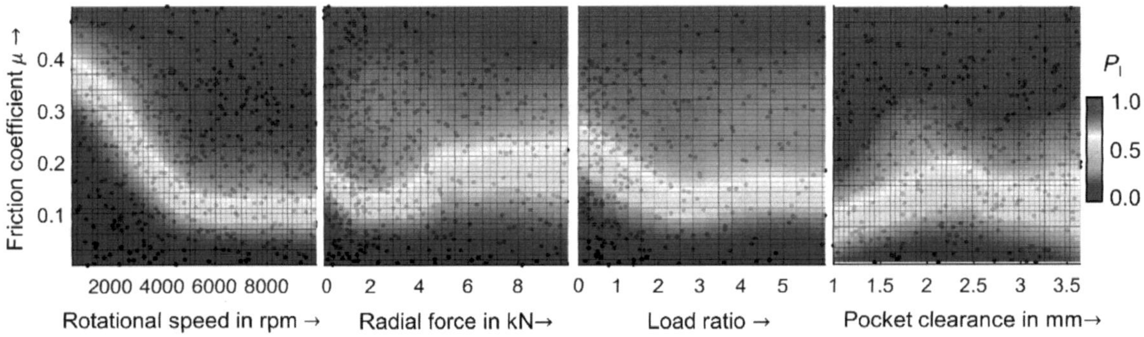

Figure 1. Class probability for the unstable cage movement as a function of the coefficient of friction in the cage pocket, bearing rotational speed, radial force, load ratio and cage pocket clearance. The black dots represent the specific conditions of the executed simulations.

The coefficient of friction in the cage pocket has the greatest influence on cage movement, which is why the other influencing factors are shown in combination. From a friction coefficient of 0.4, only unstable cage movements can be observed. Above a certain amount of frictional force, unstable cage movement is therefore to be expected independent of the operating condition. In contrast, no tendencies towards unstable cage movement can be observed at coefficients of friction of less than 0.1. This means that unstable cage movements can be avoided as good as possible by reliable lubrication of the contact between the cage and the rolling element, which ensures a coefficient of friction of less than 0.1. The rotational speed of the inner ring also influences the cage dynamics. Accordingly, figure 1 shows that an increase in speed can lead to unstable cage movements. From about 5 000 rpm, there is no change in the probability of unstable cage motion. With respect to the load, it should be noted that higher radial loads and load ratios of less than one tend to prevent unstable cage movements, since in these areas only higher friction coefficients lead to cage instability. From a load ratio of more than two, hardly any change in the influence on the cage dynamics can be detected. Finally, the cage pocket clearance also shows an influence on the cage dynamics. Small values of the pocket clearance increase the tendency to unstable cage movement. In addition, it is also possible to identify pocket clearances that are particularly suitable for reducing the probability of unstable cage movement.

2.3. *Comparison with optical measurements*

With the help of optical measurements, the quality of the simulations can be assessed and a range for the coefficient of friction can be identified which offers a good match with the simulation. Figure 2 shows the cage cCoM trajectories for the measurement and three simulations with different coefficients of friction. The simulation with the smallest coefficient of friction of 0.01 shows a circumferential cage movement, while the simulations with higher coefficients of friction of 0.3 and 0.5 show unstable cage dynamic. In addition, it is obvious that an increase in the coefficient of friction leads to unstable cage movement under otherwise constant operating conditions.

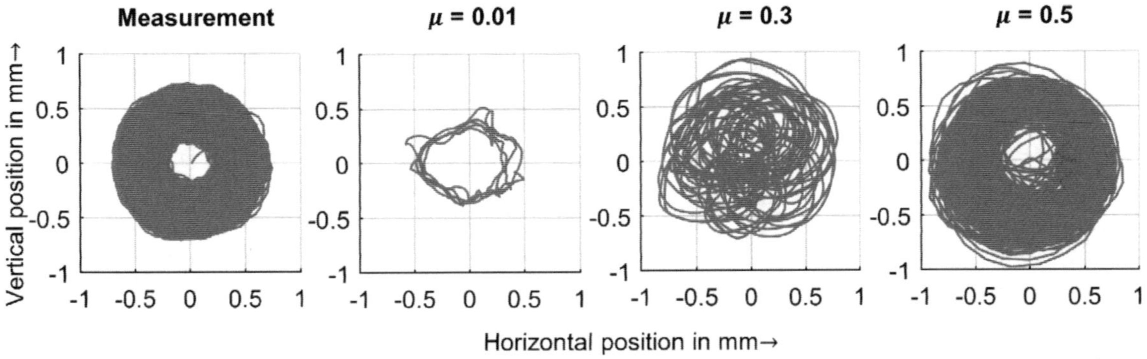

Figure 2. Trajectory of the cage cCoM for the optical measurement, and three simulations with different coefficients of friction. The speed of the inner ring is 1 500 rpm, the load on the bearing is 3 000 N in the axial and 1 500 N in the radial direction. No lubricant was used in the experiment.

In addition to the cage cCoM motion, the deformation of the cage can also be used for the comparison, see figure 3. As expected, comparatively high deformation values can be observed in the unstable cage movements, which lead besides loud noise to high and fast changing loads and a negative impact on the lifetime of the cage. In the optical measurement the highest amplitudes are present at a frequency of 477 Hz and 550 Hz. In comparison, the frequencies of the unstable cage movements in the simulation are somewhat higher. Thus, high amplitudes are present at a coefficient of friction both in the range of 453 Hz and 738 Hz. The differences are due to the glass fibre reinforced material of the cage, whose anisotropic material behavior is not considered in the simulation. Nevertheless, both the measured and the calculated cage dynamics at the coefficients of friction of 0.3 and 0.5 can be clearly classified as unstable. In principle, a high correlation between simulation and measurement can be determined. For a detailed evaluation of the agreement between calculation and measurement in the context of a validation, further measurements will be considered.

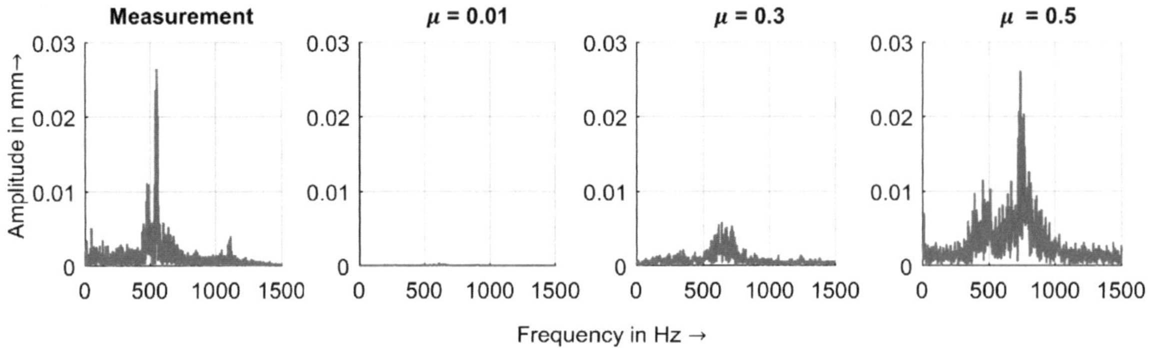

Figure 3 Deformation of the cage for optical measurement and three simulations with different coefficients of friction. The speed of the inner ring is 1 500 rpm, the load of the bearing is 3 000 N in axial direction and 1 500 N in radial direction. No lubricant was used in the experiment.

3. 4 Summary and conclusions

Based on a large number of simulations, dependencies between an unstable cage motion and the friction coefficient in the cage pocket, bearing load, bearing speed and cage pocket clearance are determined. Thus, it can be observed that for the investigated cage high coefficients of friction, high bearing speeds, load ratios greater than 2 and small pocket clearance support unstable cage motion. Finally, a comparison between optical measurements and dynamics simulations shows that the simulation results correlate very well with the measurement and thus with the actual cage movement. An unstable cage movement can be observed during the optical measurement, which can also be observed in the simulation. For the calculation, a coefficient of friction of 0.3 or higher must be defined, which can be considered plausible for the material pairing of polyamide and steel [6] in a contact without lubrication.

Based on these relationships, the operating conditions for the cage that lead to unstable cage motion can be identified for the analyzed rolling bearing. For example, by improving the lubrication condition in the contact between the cage and the rolling element, it is possible to reduce the friction and thus the dynamic behavior of the cage. Particularly at high speeds, sufficient lubrication should be ensured, as cage instability is stimulated under such conditions. Also, in the case of axial preload, this could be modified to vary the load ratio in order to avoid unstable cage movements. Adapting the cage geometry to influence the guidance and pocket clearance would be a suitable approach for affecting the resulting cage dynamics. In this way, the dynamic properties of the rolling bearing cages can already be taken into account by the simulation during the design process.

References

[1] Abele, E.; Holland, L.: Image-Based Movement Analysis of Bearing Cages of Cylindrical Hybrid Roller Bearings. In: Journal of Tribology 139(6), pp. 061101

[2] Abele, E.; Unterderweide, F.; Holland, L.: Experimentelle Untersuchungen zur stabilen und instabilen Käfigbewegung eines Schrägkugellagers. In: VDI Wissensforum (ed.): VDI-Fachtagung. Gleit- und Wälzlagerungen. Gestaltung - Berechnung - Einsatz. Düsseldorf, Germany: VDI Verlag, 2017.

[3] Ghaisas, N.; Wassgren, C. R.; Sadeghi, F.: Cage Instabilities in Cylindrical Roller Bearings. In: Journal of Tribology 126(4), pp 681–689, 2004.

[4] Grillenberger, H.; Hahn, B.; Koch, O.: Elastic Cage Instability in Rolling Element Bearings. Simulation and Test. In: Society of Tribologists and Lubrication Engineers. 70th STLE Annual Meeting and Exhibition, 2015.

[5] Gupta, P. K.: Frictional Instabilities in Ball Bearings. Tribology Transactions (1987) nr. 31(2), pp 258-268.

[6] Habig, K.-H.; Fischer, A.: Tribomaterialien. Tribotechnische Werkstoffe. In: Czichos, H.; Habig, K.-H. (ed.): Tribologie-Handbuch. Wiesbaden: Springer Vieweg, 2015, pp 283-416.

[7] Hahn, B.; Vlasenko, D.; GAILE, S.: Detailed Cage Analysis Using Caba3D. In: Society of Tribologists and Lubrication Engineers: 73rd STLE Annual Meeting and Exhibition. Minneapolis, 2018.

[8] Nogi, T.; Maniwa, K.; Matsuoka, N.: A Dynamic Analysis of Cage Instability in Ball Bearings. In: Journal of Tribology 140(1), pp. 011101, 2018.

[9] Schwarz, S.; Grillenberger, H.; Tremmel, S.: Investigations on cage dynamics in rolling bearings by test and simulation. In: Society of Tribologists and Lubrication Engineers: 74th STLE Annual Meeting and Exhibition. Nashville, USA, 2019.

[10] Schwarz, S.; Grillenberger, H.; Tremmel, S.: Kennzahl zur Identifikation der Wälzlagerkäfigdynamik. In: VDI Wissensforum (ed.): VDI-Fachtagung. Gleit- und Wälzlagerungen. Gestaltung – Berechnung – Einsatz. Düsseldorf, Germany: VDI Verlag, 2019.

Acknowledgments

The results and insights published in this paper were obtained within the projects "KILL VIB – Reduction of vibration and noise emissions due to cage instabilities in rolling bearings" (AZ-1233-16) and “MeLD – Machine Learned Dynamics” (AZ-1398-19), funded by the Bavarian Research Foundation. The authors thank the Bavarian Research Foundation for financing and supporting their work.

IOP Conf. Series: Materials Science and Engineering 1097 (2021) 012010 doi:10.1088/1757-899X/1097/1/012010

Influence of Model Detail Level on the Prognosis Quality of the Perceived Sound of an Electrified Passenger Car Powertrain

P Drichel[1,4], M Jaeger[2,5], M Müller-Giebeler[3,6], J Berroth[1], M Schröder[2], G Behler[3], G Jacobs[1], K Hameyer[2], M Vorländer[3]

[1] Institute for Machine Elements and Systems Engineering (iMSE), RWTH Aachen University, Schinkelstraße 10, 52062 Aachen, Germany
www.imse.rwth-aachen.de
[2] Institute of Electrical Machines (IEM), RWTH Aachen University, Schinkelstraße 4, 52062 Aachen, Germany
www.iem.rwth-aachen.de
[3] Institute of Technical Acoustics (ITA), RWTH Aachen University, Neustraße 50, 52066 Aachen, Germany
www.akustik.rwth-aachen.de
[4] Email: pascal.drichel@imse.rwth-aachen.de
[5] Email: markus.jaeger@iem.rwth-aachen.de
[6] Email: mark.mueller-giebeler@akustik.rwth-aachen.de

Abstract. Complex and resource intensive multi-physical models which integrate the domains of electrotechnical, structural dynamics and acoustics are needed to predict the noise, vibration and harshness (NVH) behavior of an electric passenger car drive train. Because of the highly limited resources in product development process of cars, engineers request efficient models with a good balance between accuracy and model creation as well as calculation effort. Knowing about which components and effects influence the prognosis is essential to adjust this balance between accuracy and effort. Therefore, in this paper a method is presented to determine the influence of the model fidelity level on the prognosis quality of the perceived drive train-related airborne sound under the use of hearing tests to consider the human noise perception. The work is based on a high fidelity multi-physical NVH of a small, series produced electric vehicle drivetrain which is validated in structure- and air-borne-noise against test bench measurements. The influence of model detail level on the prognosis quality is analyzed using the aforementioned method taking into account four exemplary variant calculations of the proposed model: (1) Power electronics harmonics vs. ideal sine electric machine current, (2) electric machine with vs. without production deviations, (3) modal damping parameters derived from component measurements vs. modal damping taken from literature, and (4) numeric vs. analytical sound calculation method.

1. Introduction

Distinctive design and noise characteristics of vehicles are important, brand-specific differentiators in terms of aesthetic and emotional criteria. Hence, designing a sound that matches the vehicles' character and corresponds to customer expectations has always been a central task. In this process, simulation is an important tool. [1]

For the prognosis of an electric car drivetrain induced sound, typically multi-physical modelling approaches with a variety of model fidelity levels are used. This includes different methods for coupling the domains of electronics, structural dynamics and acoustics involved in the generation and transmission of noise, as well as different domain-specific models. The challenge is the efficient interdisciplinary combination of different, highly specialized models and computation methods that work in the time or frequency domain, as well as the definition of their interfaces [2].

The focus of past work regarding electrical machines, performed by electrical engineers, has often been the aspect of exciting forces [3], which includes field and force calculation and for more sophisticated approaches also a circuit simulation. However, state-of-the-art NVH simulation chains [4, 5] combine this aspect with the structural dynamic behavior. This can e.g. be achieved by methods of transfer path analysis as in [5], or by calculating excitation forces at the stator teeth and at the bearing points for a subsequent forced-response analysis as in [6]. All referenced approaches share the neglection of feedback effects. To include these, the electrotechnical and structural dynamics domain both have to be calculated in the time domain.

In structural dynamics domain in practice methods like torsion vibration [14], (elastic) multibody simulation (EMBS) [7, 8] or finite element method (FEM) models [9] are used. Mechanical engineers tend to focus on structural dynamics using precalculated electromagnetic force excitation spectra and neglect electric current feedback effects. [7, 8, 9]

The coupling between the domains of structural dynamics and acoustics is typically classified as weak [10]. This is due to the low reaction of the air on the drive train and, in principle, enables the structural dynamic and acoustic calculation to be done consecutively [11]. The surface velocities required for the acoustic calculation at each mesh node can be reconstructed by using the time-dependent modal velocity in combination with the recovery matrix, which establishes the connection between modal and nodal velocity [12]. However, the amount of data to be processed requires skillful handling, which is why occasionally either analytical radiation models are used [7, 8] or the system is often cut at the bearings. In this case, the bearing forces are calculated in time domain transformed into the frequency domain and the airborne noise is calculated using a finite element model of the drive train housing and boundary element model of the air envelope [14]. With this approach the interaction between the shaft system and housing [15] is neglected.

The frontloading effort for creating, parameterizing and calculating such Multiphysics models is high. Thus, models with an appropriate level of accuracy are required. However, today it is hard to judge the expediency of models, because the effect of different model fidelity levels on the predicted noise behaviour of a complete electric drivetrain has been investigated rarely in literature.

In [16] the influence of electrical machine stator modelling on the spatially averaged square of housing surface velocity of an electric drivetrain for the response Order 1 of the electric machine (EM) is analyzed. They found out that the transversely isotropic material model does not affect the general trend in the response but does show an increase in magnitude.

To what extent detailed modelling improves the prognosis quality of the perceived sound compared to simple modelling approaches is not investigated in literature. Hence, in this paper a method is presented to determine the influence of the model fidelity level on the prognosis quality of the perceived drive train-related noise with hearing tests considering the human noise perception. Therefore, a high fidelity multi-physical NVH model of a small, series-produced electric vehicle drivetrain is presented which is validated in structure- and air-borne-noise. The influence of the model detail level on the prognosis quality is analyzed using the method of four exemplary variant calculations of the proposed model.

IOP Conf. Series: Materials Science and Engineering 1097 (2021) 012010 doi:10.1088/1757-899X/1097/1/012010

2. Approach

For setting up the approach to determine the influence of the model fidelity level on the prognosis quality of the perceived drive train-related noise, a central question is, what base model should be chosen? Should it be a more realistic, but complex model which will be simplified as part of the investigation? Or should a more artificial, but simpler one be chosen, which is increased further in complexity? To answer this question, one has to be aware of aspects of the interactions of component models and human perception of sound. In context of human perception of sound, one important aspect is masking. In simple terms, masking occurs when a tone can no longer be heard due to the level of other noise sources. Furthermore, the interaction of model components describes the dependence of the behavior of one component model on the other. For example, in a gearbox model the complexity of a bearing model (linear vs. nonlinear) leads to a different meshing situation and thus, to a different noise spectrum. Being aware of these phenomena, a complex model with a high degree of realism and a noise spectrum as complete as possible is chosen as the base model.

Therefore, in the first step of the overall approach (Figure 1), a transient high-fidelity multi-physics model is built (section 3) and validated (section 4). It's including most of the often-discussed components and effects which are relevant for the sound of electric drivetrains. For example, this includes power electronics and control, a fully coupled structural dynamics model between shafts and housing, as well as an accurate acoustic radiation model. With this, feedback in the current and the loads between the electrical and structural dynamics model and production deviations like geometric and magnetic eccentricity are considered. This model is the starting point for the analysis of several model simplifications (section 5.1). Aware that noise quality is fundamentally a complex multidimensional perceptual attribute to whose assessment various acoustic characteristics contribute. The MUSHRA hearing test is used to obtain an integral one-dimensional comparison quantity (section 5.2.). For the detailed objective characterization of individual dimensions of the auditory perception additional common psychoacoustic parameters are determined.

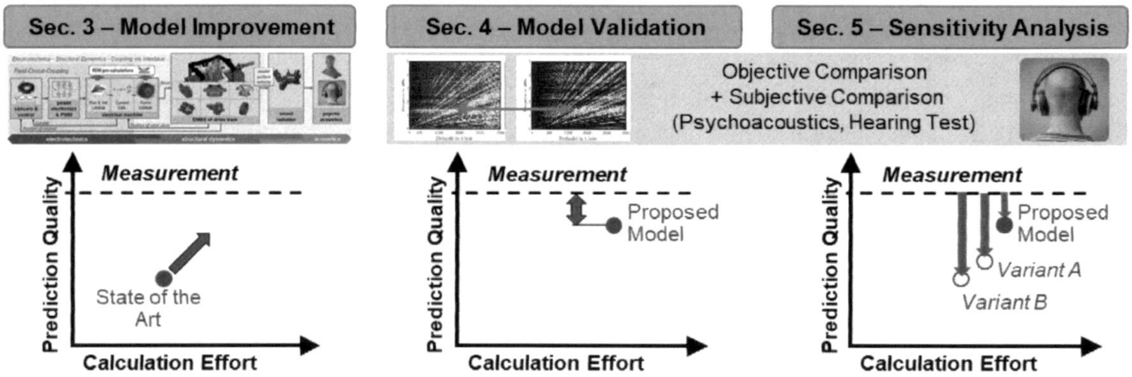

Figure 1. Approach to determine the influence of the model fidelity level on the prognosis quality of the perceived drive train-related noise with hearing tests considering the human noise perception.

3. Multi-Physical modelling approach

On the top level, the simulation chain consists of three parts according to the three domains: The electrotechnical domain, the structure dynamics domain, and the acoustics domain (Figure 2).

The electrotechnical model is a circuit simulation including an advanced representation of a permanent magnet synchronous machine (PMSM). Via the interface described in section 3.2 it is physically, not numerically, strongly coupled to the structure dynamics domain, which is an EMBS. Output of the EMBS is the surface velocity, from which the radiated sound is calculated in the acoustics domain. This interface (section 3.4) is unidirectional, since the effect of the airborne sound on the structural dynamics is small and thus negligible.

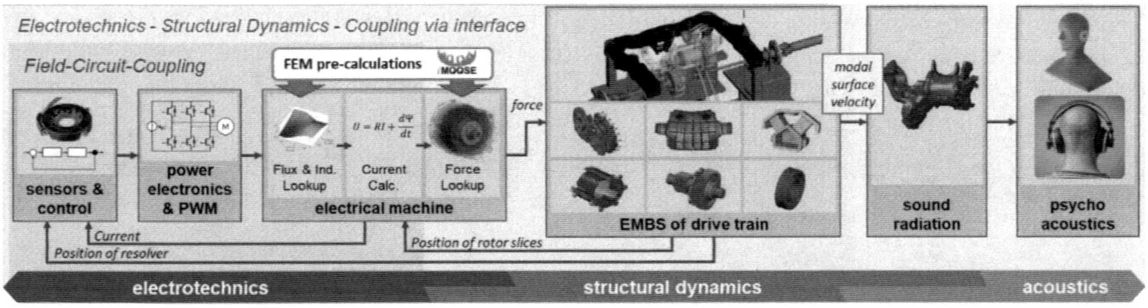

Figure 2. Overview of the simulation chain.

3.1. *Electrotechnical Domain*

The electrotechnical approach is extensively described in [17], here a brief summary is given. According to Maxwell's stress tensor, magnetic forces are proportional to the square of the magnetic flux density. Its determining parameters, also considering the dominant non-idealities, are

1. The excitation from the current, whose calculation is described in section 3.1.1.
2. The excitation from the magnets. The scattering of the permanent magnet remanence is in the order of ±5 % according to the manufacturer's data sheet, which is supported by our own measurements. It is considered in the field calculation. Geometric deviations like magnet position are not addressed in this paper but can also easily be considered.
3. The magnetic conductance of the airgap as primary magnetic resistance. The dominant influence on the airgap is the eccentricity due to displacement or bending of the shaft. In various analytical [18, 19] and numerical [20] studies, dynamic eccentricity has been identified as the cause of sidebands and is therefore also modeled here.

3.1.1. *Field- and Current Calculation*

The field calculation and the circuit simulation including the power electronics are connected according to the approach from [5] and [21], hence decoupled due to the difference of the time constants. It is extended to the multi-slice method as shown in [22]. The field of the PMSM with buried magnets is calculated stationary, subsequently it is only dependent on the state variables rotor angle γ and current id and iq. The eccentricity extends the state space of the field calculation in 2D by the eccentricity radius r_{ecc} and eccentricity angle φ_{ecc}. The calculation is performed in advance in parallel using the FEM software *iMOOSE* from the Institute of Electrical Machines. The flux linkages Ψ_d, Ψ_d, Inductances L_d, L_q and tooth forces (see section 3.1.2) are stored in characteristic maps.

3D effects are approximated by means of the multi-slice method (see [22]). To reduce calculation effort the same Look-Up tables (LUTs) can be used for all slices, in this case differences in production deviations between slices can be approximated by rotation angle offsets many times a pole pitch.

The symmetry of the PMSM reduces the number of field calculations required. To get forces at all teeth, which differ because of the eccentricity, a full 2D model is used. The stator geometry shows symmetry by the slot pitch, which has to be covered by the eccentricity angle dimension. If magnetization deviations are considered, the rotor has no symmetry making a full rotation necessary for the rotor angle dimension, otherwise a pole pitch is sufficient.

3.1.2. *Tooth Forces and Eccentricity*

The nodal forces from the FEM calculation must be prepared for the modal representation of the EMBS with its limited number of force application points, and aliasing must be avoided. The order reduction from [23] summarizes the nodal forces on each tooth tip and transforms them into a tooth-specific coordinate system (tangential / radial / axial). The axial force is estimated for the continuously skewed machine from the skew angle and the assumption that the tan. force acts orthogonal on the tooth flank.

To remove aliasing problems while containing a good execution speed. the LUT of tooth forces has to be processed as illustrated in [17]. In short, the rotation dimension is Fourier-decomposed and high orders are neglected. Then it gets resampled, and finally some data is discarded to prevent redundant but mismatching forces which disturb the EMBS. The resulting tooth forces are illustrated in Figure 3.

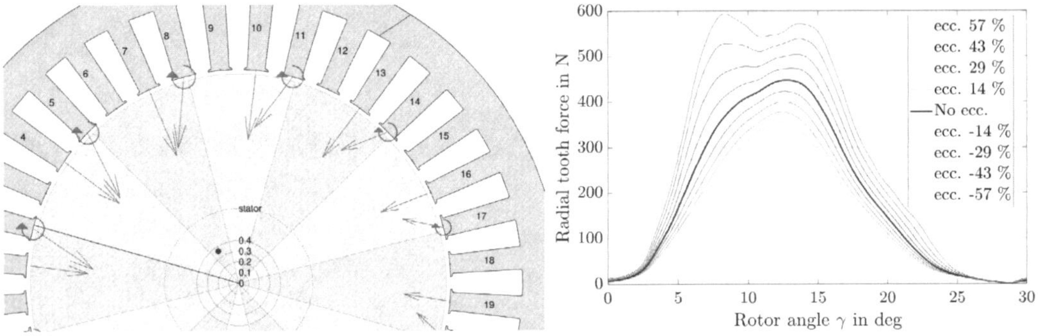

Figure 3. Tooth forces at 80 Nm. *Left*: Tooth forces with radial, tangential and torque component at one exemplary rotational step, which has a rotor displacement of 0.37 mm (53 % of the 0.7 mm airgap) to the top left. *Right*: Radial force over all rotational steps at one tooth, shown for different airgap widths due to rotor displacement. Positive eccentricity represents smaller airgap at this tooth (and larger on the opposite one), negative eccentricity represents a larger airgap at the regarded tooth.

3.2. *Interface between Electrotechnical and Structural Dynamics Domain*

The electrotechnical domain and the structural dynamics domain both run simulations in the time domain with variable step-size solvers to account for nonlinearities. The circuit simulation is implemented in MATLAB Simulink (MathWorks). The EMBS is implemented in Multi-Body Simulation framework called SIMPACK, which features an interface to Simulink. The interface works by a fixed rate data transmission, which, as compromise between execution speed and frequency resolution, is chosen to 60 kHz here. To avoid aliasing in the interface, it is necessary to lowpass the high frequency current, but with little phase lag. A second-order Butterworth-filter (corner 20 kHz) is chosen as a compromise between steepness and phase lag.

In each time step the circuit simulation provides the machine current i_d and i_q to the EMBS, which in return feeds back the rotor angle γ of each rotor slice. The force lookup is performed in the EMBS, which has the advantage of the EMBS being able to run with an ideal current and therefore without the circuit simulation, effectively modeling a perfect controller. For the force lookup, via symmetries the machine state is reduced to a state which is mapped in the LUT [17].

3.3. *Structural Dynamics Domain*

The structural dynamics NVH model includes the drivetrain components electric machine with coolant, gearbox with differential, side shafts, support arms, subframe and coolant supply (Figure 2).

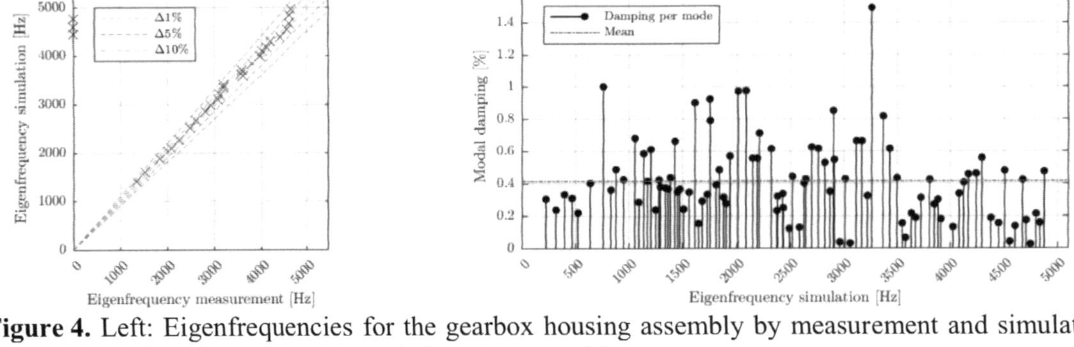

Figure 4. Left: Eigenfrequencies for the gearbox housing assembly by measurement and simulation; Right: Modal damping of the drivetrain housing assembly.

The components are idealized with elastic bodies and coupling elements (force elements and joints). Component modes up to a frequency of 8 kHz are considered within the model. The structural dynamic properties of rotor and stator, which are characterized by a layered structure, are modeled with the help of material models with direction-dependent properties and parameterized using methods of multi-scale modeling (representative volume element - RVE) and measurements. In addition, the load-dependent, non-linear stiffnesses of the deep groove and tapered roller bearings as well as the excitation behavior of the gears are integrated in this environment. Joints are used to model the kinematic behavior of tripod and constant velocity joints of the side shafts.

Elastic component models have been extensively validated with modal analysis and impedance measurements. The results for housings (Figure 4) and rotor (Table 1) are presented here as an example.

Table 1. Mode shapes and eigenfrequencies of the rotor by measurement and simulation.

Mode		Measurement		Simulation		Deviation
#	**Type**	**Shape**	**Frequency**	**Shape**	**Frequency**	
1./2.	Bending		3741 Hz		3753 Hz	0,32 %
3.	Torsion		4022 Hz	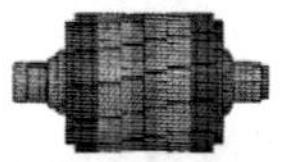	3919 Hz	-2,56 %
4.	Axial		5500 Hz		5601 Hz	1,84 %

Furthermore, the imbalance condition of the rotor is measured in the bearing planes and considered in the modally reduced EMBS model of the rotor. The investigated rotor fulfils the requirements of balance quality 2.5 as the usual balance quality for high-speed electrical machines.

3.4. Interface between Structural Dynamics and Acoustics Domain

The calculation of the sound radiation by means of a state-of-the-art Boundary-Element-Method-simulation (BEM) requires (spatially and temporally) high-resolution export of the surface velocities on the drive train which leads to a practically unmanageable amount of data in this case. Therefore, a minimum data representation under the use of the ATV (Acoustic Transfer Vector) Method [12] is applied. The ATV between the housing velocities and the resulting sound pressures at selected receiver points represent the transfer properties of the air volume taking into account the radiating housing geometry [24]. Once the ATV matrices have been determined for all frequencies, the calculation of the resulting sound pressures for any load case is performed as a simple vector product of the velocity distribution on the surface mesh and the calculated transfer vectors. The transient surface velocities required for each node of the acoustic model are calculated per time window, transformed into the frequency domain, and then multiplied with the ATV-Matrices. After corresponding calculation for all time windows of a simulated run-up, inverse transformation and superposition in the time domain are performed. Detailed description can be found in [25].

3.5. Acoustics Domain

A BEM model considering all partial elements of the entire drive train geometry as radiating enveloping surface is used to calculate the ATV-Matrices (Figure 5). The mesh with nearly 100000 nodes is suitable for the calculation of air borne sound up to a frequency of 16 kHz.

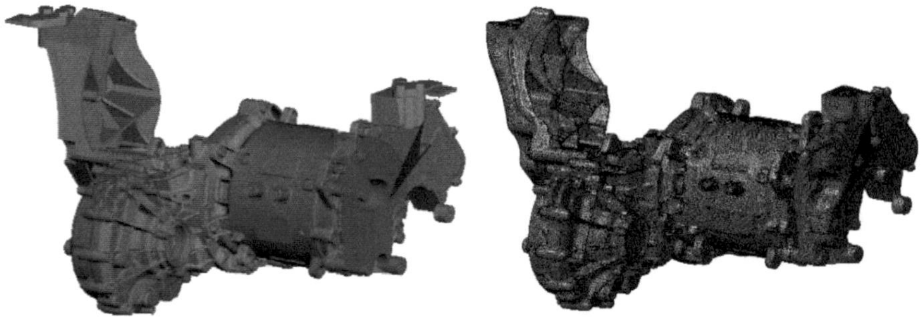

Figure 5. Left: Approximation of gearbox and stator housing by means of replacement cylinders; Right: Boundary element housing model (mesh) in LMS Virtual.Lab.

4. Validation

Simulation results for the electric machine current show good accordance to the measurement and represent the machine harmonics (-5, +7, -11, +13, etc.) good as well as the harmonics from the pulse width modulation (PWM) of the inverter (Figure 6). However, the PWM harmonics amplitude is too low due to the neglected nonlinear behavior of the machine inductance which decreases at higher frequencies. This effect requires some kind of hysteresis in the time domain and therefore is difficult to implement. Therefore some of the PWM harmonics are not visible in the simulation.

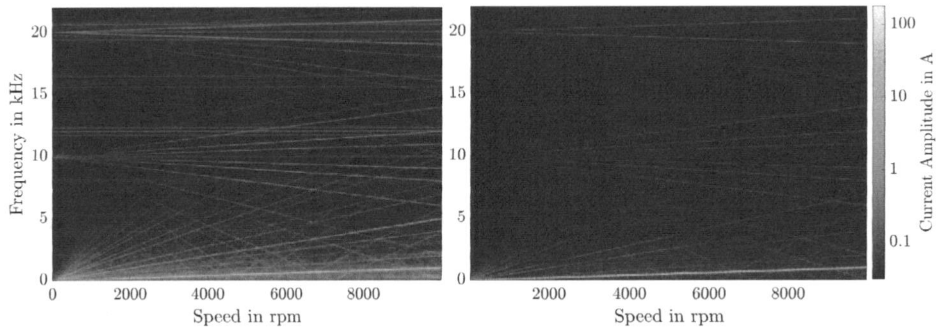

Figure 6. Comparison between the measured current (left) and the simulated current (right).

In the following sound pressure level (SPL) (Figure 8) and acceleration at an exemplary point orthogonal on the electrical machine surface (Figure 7) are investigated.

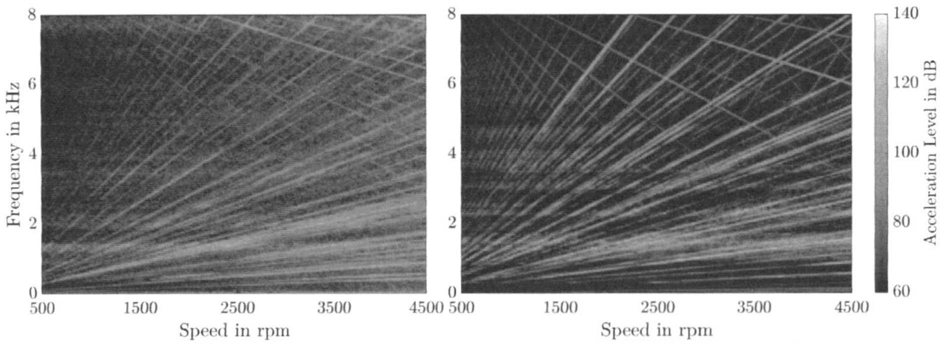

Figure 7. Surface acceleration radial on the housing of the electrical machine; Left: Measurement. Right: Simulation. Runup at 80 Nm.

A good agreement is achieved, especially regarding the main harmonics. Peaks in acceleration level resulting from natural frequencies, especially in the lower frequency range up to 5 kHz, are well represented in the simulation. The allocation of the harmonics to their origin reveals that the gearbox of the examined drive train dominates the NVH behavior compared to the acoustically unobtrusive PMSM. Furthermore, in the simulation the rotor of the electric machine has a very low dynamic eccentricity. Therefore, sidebands do arise solely from the 'magnetic' eccentricity.

Deviations between measurement and simulation can be observed: E.g., PWM harmonics are missing, probably due to the three parallel paths of the winding on 120° of the motor circumference each, which allow for minor equalizing currents. In addition to that, underrepresented grinding noises in simulation could be caused by the missing excitation of the side shaft models. Furthermore, the gearbox oil filling and therefore the mass and damping influence is ignored. And in the context of acoustic radiation calculation surrounding passive components are neglected. Nevertheless, the reference model far exceeds the state of the art and is rated suitable for valuable variant simulations.

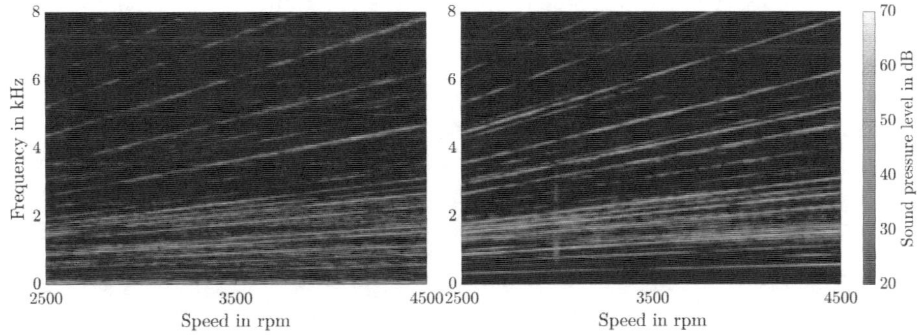

Figure 8. Sound pressure level 1 m above the center of the drive train; Left: Measurement; Right: Simulation.

5. Sensitivity analysis

5.1. Variant models and objective results

Within this section the variant models of validated simulation chain are presented and compared against the reference simulation in terms of SPL (Figure 9).

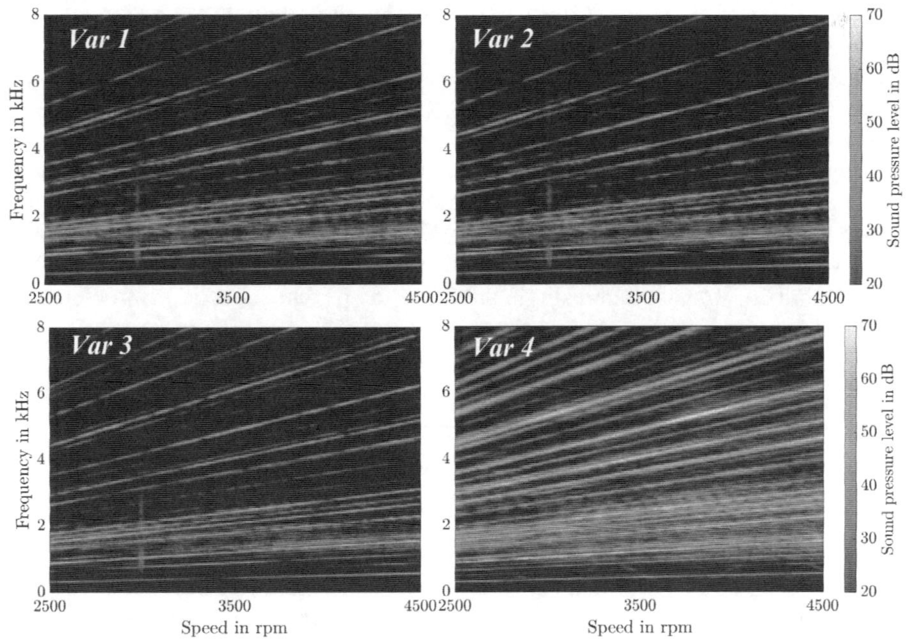

Figure 9. Sound pressure level 1 m above the drive train center, runup at 80 Nm; Top left: Variant 1, ideal current; Top right: Variant 2, ideal electric motor; Bottom left: Variant 3, 2 % modal damping; Bottom right: Variant 4, analytical sound radiation model.

5.1.1. Variant 1: Ideal Current

The first variant assumes an ideal controller which keeps the current perfectly constant at any time. This is done by feeding the force lookup with a fixed current combination. As expected, the PWM harmonics are missing in this variant. The motor current harmonics are also not present, but even in the reference simulation they are so small that they cannot be separated from other excitation orders.

5.1.2. Variant 2: No Production Deviations in Electric Machine

Variant two assumes an ideal electrical machine to determine the influence of its production deviations. Within this model, the rotor eccentricity has no effect on the electromagnetic force calculation and all magnets properties are set equal. As expected, a small difference to the reference simulation can be seen (Figure 12). As stated above, the drive trains NVH behavior is dominated by the two transmission stages, the PMSM is fairly quiet even with its production deviations. To some extend the PWM harmonics are reduced in this variant.

5.1.3. Variant 3: Damping from Literature

The third variant tunes all modal damping ratios of the housing group in the EMBS to 2 %, a value from a standard range given in literature [26]. The result shows that the sensitivity of the surface acceleration for the modal damping is rather low. The measured values of the housing group scatter between 0.02…1.5 % with a mean of 0.43 %, thus with a 2 % value the mean surface acceleration level is slightly reduced.

5.1.4. Variant 4: Analytical Sound Radiation Model

The existing velocity data is sampled on the housing parts (stator and gear housing), which are approximated as cylinder bodies, both in axial and in circumferential direction (Figure 6 left). According to the underlying theory, only those modes on cylindrical bodies contribute to the radiated airborne sound whose wavelength is greater than or equal to the considered airborne sound wavelength [27]. Since at least two interpolation points per wavelength are necessary for the correct description of these modes, the maximum interpolation point distance required for an upper cut-off frequency of 16 kHz is about 1.1 cm. On the ends of the cylinder bodies, which are regarded as ideal piston radiators, the velocity is scanned at the same spatial distance. The detailed mathematical derivation of the formulas used in the analytical approach can be found in [7].

5.1.5. Psychoacoustic Evaluation

The following psychoacoustic parameters where computed based on the sound signal as prescribed in the given source, and averaged over the duration of the runup. They are listed in Table 2: Loudness (DIN 45631/A1), Sharpness (DIN 45631/A1; DIN 45692), Roughness (Hearing Model Sottek), Tonality (Hearing Model Sottek) and Psychoacoustic Annoyance (after Fastl & Zwicker).

Table 2. Time-averaged single values of the psychoacoustic parameters

	Loudness [sone]	**Sharpness [acum]**	**Roughness [asper]**	**Tonality [tuHMS]**	**Annoyance []**
Measurement	28,1	2,1	0,045	1,67	32,1
Reference model	29,3	1,85	0,029	2,48	30,6
Ideal current	29,3	1,85	0,029	2,48	30,6
Ideal E-Machine	28,7	1,87	0,032	2,42	30,2
Literature damping	26,0	1,94	0,027	2,29	28,0
Analytical radiation	66,2	2,35	0,035	4,63	84,9

Overall, it can be stated that the loudness level is basically well met by the reference simulation. The deviations between measurement and reference model which are discussed at the end of section 4. , among other things friction related excitations and additional sidebands, make the measurement sound less tonal and rougher. This is due to more broad-band noise, while the amount of tonal noise is comparable to the simulation. The higher sharpness results probably from noise components in the higher frequency range where the simulation underestimates the SPL due to the different reasons stated in the previous sections (nonlinear inductivities, eigenfrequencies in the EMBS up to 8 kHz, etc.). Of the variant simulations, the analytical radiation stands out by noticeable deviations in all psychoacoustic parameters, rendering this approach unsuitable. The other variant simulations are just about in the order of magnitude of the just noticeable differences (JNDs) for steady state signals reported in literature [28].

5.2. Hearing test

In order to investigate the dependence between the prediction quality of the noise behavior and the model detailing as well as the general prediction quality of the system model considering the human auditory perception, a "Multi-Stimulus Test with Hidden Reference and Anchor" (MUSHRA) test with 30 participants was performed. In this test, different noise stimuli are evaluated by the test participants in relation to a reference on a continuous scale from 0 (completely dissimilar) to 100 (absolutely identical). Also presented are the so-called hidden reference (an unlabeled duplicate of the reference signal) and an anchor signal in order to promote the best possible utilization of the scale range and/or disqualify unreliable or unqualified listeners. In the present case, the reference stimulus, a noise measurement on the powertrain test bench, was compared with the reference simulation and the four abstracted model variants. As can be seen from the averaged similarity value of 98.5, all test persons assigned by far the highest similarity value to the hidden copy, as expected. While 4 of the 5 tested model detail levels with scores between 50.1 and 56.0 are perceived as very similar, only the variant with the analytical radiation approach with an average value of 26.9 deviates significantly from this, which according to the participant survey was due in particular to the substantially different loudness of the radiation model variant implemented here. On the one hand, a comparatively large variance in the evaluations can be seen, which means that the absolute similarity of individual simulations to the test bench measurement is assessed differently by the test persons individually. The slightly different mean value and the somewhat larger variance indicate that the model variant literature attenuation of the 4 very similar noises could have been perceived differently at least by some subjects. However, a post-hoc Tukey multiple comparison test could show that there are no significant differences between the groups 'reference model, 'ideal current', 'ideal electric machine' and 'literature damping', whereas the group averages of 'measurement', 'radiation model' and 'anchor signal' differ significantly from all other six groups. (Figure 10)

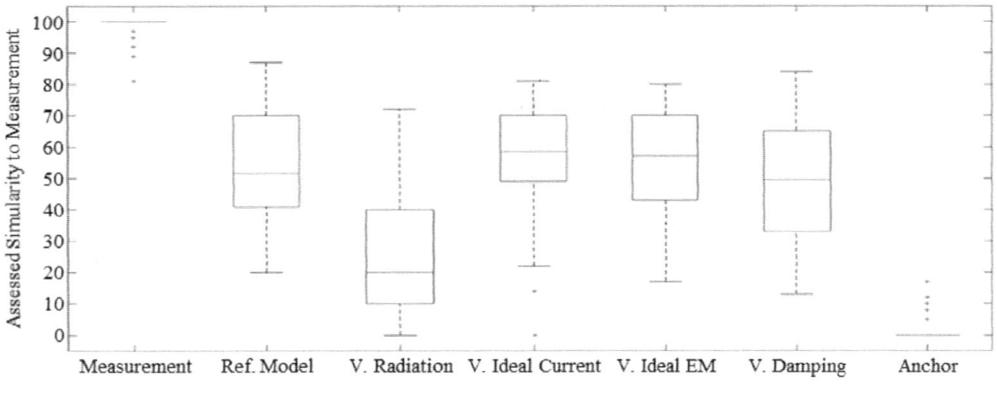

Figure 10. Boxplot of the similarity evaluations of all test persons in the listening test.

6. Conclusion and Outlook

In this paper a method to analyze the sensitivity of the NVH behavior of an electrical drive train is proposed, implemented and exemplarily applied. It was developed in a joint 2.5-year project of the three contributing institutes.

Central part of this method is a combined implementation of state-of-the-art simulation techniques and modelling approaches from other researchers to an extensive, multi-domain simulation chain with the goal to model as many possibly relevant effects as feasible in the project. To achieve this, efficient interfaces between the electrotechnical and structural dynamics domain, as well as between the structural dynamics and the acoustics domain were developed. The simulation chain includes active and retroactive effects between the electrotechnical and structural dynamics domain, various production deviations and a sophisticated sound radiation calculation. The validation of this simulation chain versus the measurement shows a good agreement, but also some remaining discrepancies which leave scope for improvement.

This NVH simulation chain is used to exemplarily evaluate four variants and their influence: The current harmonics from the controlled electrical machine, its production deviations, the modal damping of the housing group, and the sound calculation method. The results are compared objectively via Campbell diagrams as well as via the MUSHRA hearing test to evaluate the perception of the NVH behavior by the subjects. For variant one, two and three a low sensitivity is determined which partly can be attributed to the NVH behavior being dominated by the transmission, while the analytical sound radiation calculation of variant 4 shows a high sensitivity and is deemed unsuitable.

The method can be used to further analyze the sensitivity of the NVH behavior on this drive train and can be applied to other electrical drive trains. Improvements are possible regarding the frequency range, friction excitations, PWM harmonics and by further reducing the calculation effort, especially by a more efficient sound radiation calculation.

7. Acknowledgement

This research was supported by the program for Cooperative Industrial Research (IGF funding no. 18764 N) with funding from the German Federal Ministry for Economic Affairs and Energy (BMWi) on the basis of a decision by the German Parliament. The authors thank the Forschungsvereinigung Antriebstechnik e.V. (FVA) for the realization and support from experts of the FVA network. The authors further thank the German Research Association (DFG) for the funding and support of the interdisciplinary research in the post graduate program mobilEM (GRK 1856). Some simulations were performed with computing resources granted by the RWTH Aachen University under project rwth0263.

Supported by: Federal Ministry for Economic Affairs and Energy on the basis of a decision by the German Bundestag

8. References

[1] Genuit, K.: Sound-Engineering im Automobilbereich. Methoden zur Messung und Auswertung von Geräuschen und Schwingungen.Springer-Verlag Berlin Heidelberg, 2010.

[2] Beitelschmidt, Michael: Durchgängige Integration standardisierter NVH-Berechnungen in den Entwicklungsprozess von Schienenfahrzeugantriebssträngen, 2014.

[3] Gieras, J. Wang, C. Lai, J.: Noise of polyphase electric motors, CRC Press Group, 2006.

[4] Valavi, M., Nysveen, A., Nilsen, R., Le Besnerais, J., Devillers, E. (2017): Analysis of magnetic forces and vibration in a converter-fed synchronous hydrogenerator. In IEEE energy conversion congress and exposition (ECCE).

[5] Franck, D., Herold, T., Hameyer, K. (2013): Transient acoustic simulations of electrical drive-trains. In Conference on Acoustics, AIA-DAGA 2013.

[6] Knaus, O., Schneider, J., Klarin, B. (2018): Simulation in der Entwicklung automotiver E-Maschinen. ATZ Extra (S. 32–37). Berlin: Springer. 05/2018.

[7] Rick, S.; Wegerhoff, Matthias; Klein, J.; Hameyer, K.; Jacobs, Georg; Vorländer, Michael: E-MOTIVE NVH-Simulationsmodell. Modellbildung zur NVH Simulation eines E-MOTIVE Antriebsstrangs. Forschungsvereinigung Antriebstechnik (FVA), 2015 (1130).

[8] Wegerhoff, Matthias: Methodik zur numerischen NVH Analyse eines elektrifizierten PKW Antriebsstrangs. Doktorarbeit. RWTH Aachen, 2017.

[9] Schwarzer, M. (2017): Structural Dynamic Modeling and Simulation of Acoustic Sound Emissions of Electric Traction Motors. Ph.D. Dissertation, TU Darmstadt.

[10] Norton, Michael P.; Karczub, Dennis G.: Fundamentals of noise and vibration analysis for engineers. 2. ed., transferred to digital print. Cambridge: Cambridge Univ. Press, 2007.

[11] Dassault Systemes: Abaqus 2019 Documentation, 2019.

[12] Gerard, F.; Tournour, M.; El Masri, N.; Cremers, L.; Felice, M.; Selmane, A.: Acoustic transfer vectors for num. modeling of engine noise. In: Sound and Vibration 36 (7), 2002.

[13] Falkenberger, S.; Neher, J.; Graf, B.; Wender, B.: Experimental and computational studies of the sound radiation of gearboxes. In: ISMA 2014, International Conference on Noise and Vibration Engineering. Proceedings. CD-ROM. Leuven: KU Leuven, 2014.

[14] Bosse, Dennis: Motor-Getriebe-Akustik bei hohen Drehzahlen. Diss. RWTH Aachen, 2018.

[15] Klein, Andreas: Interaktion der Antriebsstrang- und Gehäusedynamik bei Industriegetrieben. RWTH Aachen, 2007.

[16] Michon, M., Holehouse, R., Shahaj, A., Jafarali, H., Janakiraman, V.; System Interactions Affecting NVH Performance of an Electric Vehicle Drivetrain, SAE Technical Paper Series 2019, DOI: 10.4271/2019-01-1545, 2019.

[17] Jaeger, M.; Drichel, P.; Schröder, M., Berroth, J., Jacobs, G.; Hameyer, K.: Die Kopplung elektrotechnischer und strukturdynamischer Domänen zu einem NVH-Systemmodell eines elektrischen Antriebsstrangs, Elektrotechnik & Informationstechnik https://doi.org/10.1007/s00502-020-00802-z, 2020.

[18] Jordan, H. Schroeder, R.-D. Seinsch, H. O.: Zur Berechnung einseitig magnetischer Zugkräfte in Drehfeldmaschinen. AfE. 63 (1981) 117-124.

[19] Zhang, M., Macdonald, A., Tseng, K.-J., Burt, G. M.: Magnetic Equivalent Circuit Modeling for Interior Permanent Magnet Synchronous Machine under Eccentricity Fault, Power Engineering Conference (UPEC), 2013.

[20] Lundin, U., Wolfbrandt, A. (2009): Method for Modeling Time-Dependent Nonuniform Rotor/Stator Conf. in Electrical Machines. IEEE Trans. on magnetics, vol. 45, no. 7, 2009.

[21] Herold, T.; Lange, E.; Hameyer, K.; *System Simulation of a PMSM Servo Drive Using Field-Circuit Coupling,* IEEE Trans. on Magnetics, vol. 47, issue 5, pp. 938-941, May 2011.

[22] Jaeger, M.; Rick, S.; Hameyer, K.; Current Simulation of a Controlled PMSM Incl. Skew and Torsional Rotor Vibrations, 2018 XIII Int. Conf. on Elect. Machines (ICEM), 2018.

[23] Herold, T., Franck, D., Schröder, M. *et al.* Transientes Simulationsmodell für die akustische Bewertung elektrischer Antriebe. *Elektrotech. Inftech.* **133,** 55–64 (2016).

[24] Drichel, P.; Jaeger, M.; Müller-Giebeler, M; Jacobs, G.; Hameyer, K.; Vorländer, M.; Mit elektrischem Antrieb und modellbasierter Systemanalyse nahezu lautlos in die Zukunft, ATZ Extra 24, 52–57 (2019). https://doi.org/10.1007/s35778-019-0050-2, 2019.

[25] Jaeger, M.; Drichel, P.; Müller-Giebeler, M.: FVA 682 II Erweiterung NVH Simulationsmodell Sachstandsbericht 05/2018, 2018.

[26] Williamson, S.; Flack, T. J.; Volschenk, A. F.; *Representation of skew in time-stepped two-dimensional finite-element models of electrical machines*, IEEE Industry Applications. Society Annual Meeting, Oct. 1994.

[27] Williams, E.G.; Fourier Acoustics. Academic Press, 1999.

[28] Zwicker, E.; Fastl, H.: *Psychoacoustics. Facts and models.* 3. ed., [repr.]. Berlin, Heidelberg [u.a.]: Springer, 2010.

Active vibration control of slewing drives with gear backlash

M Spiegelhauer and B Schlecht

Technische Universität Dresden, Institute of Machine Elements and Machine Design (IMM), Chair of Machine Elements, Münchner Platz 1-3, 01187 Dresden, Germany

Abstract Geared transmissions with high reduction ratios are widely utilized in drive technology. Common to these drives is the occurrence of considerable gear backlash. If operating scenarios with repeated reversals of the rotational direction or changing load directions occur, the backlash will be traversed. To overcome the gear play, rotation angles of several hundred degrees can be required at the motor. As a result, unwanted vibrations can occur and high dynamic loads may arise. It is shown that an active vibration control strategy based on full state feedback (LQG) allows to reduce both the dynamic overload and the time for reversing the drive. The model-based design approach relies on two angular measurements only. Multibody system simulations of an exemplary drivetrain are used to verify the possible improvements.

1. Introduction

A multitude of processes relies on the precise rotational motion of a machine around its vertical axis. Typical examples are wind turbines, radio telescopes as well as a variety of machines in material handling technology, e.g. tower cranes. Their slewing motion or yaw motion is usually carried out by means of electrical drives and multi-stage gearboxes with high transmission ratios.

Due to the inevitable elasticity of shafts, couplings and gears, these drive systems are prone to low-frequency torsional vibrations.

One source of excitation is associated with the special operating conditions of slewing drives: As a result of the operating scenario with repeated reversals of the rotational direction and changing load directions at the gearbox output, the backlash of all gear stages is compulsorily traversed. After each backlash traverse, the tooth flanks collide with a certain speed difference. This excites the system's natural frequencies and causes high dynamic loads in the gearbox.

Large backlash and high acceleration capabilities of electric machines promote the occurrence of substantial speed differences during collision of the tooth flanks.

The angular backlash of a single gear stage rarely exceeds a few degrees. However, in typical drive configurations with multi-stage gearboxes, cumulated backlash angles of significantly larger dimensions frequently occur. In particular, the angular backlash of the low speed gear stages – in conjunction with the high gear ratio in between – can lead to substantial rotation angles of several hundred degrees at the motor shaft, that are required to overcome the gear play.

Backlash itself degrades positioning accuracy and the associated vibration-induced overload contributes significantly to fatigue and wear of drivetrain components, which should be minimized. [1,2]

For positioning drives with explicit requirements for highest precision – used e.g. in machine tools, industrial robots, radar equipment – zero or low backlash can be guaranteed by taking constructive measures. Possible approaches are the use of special gearbox concepts (strain wave gearing, cycloidal drives), or the optimization of conventional gear pairs. Among these are the use of spring-loaded gears, the precise manufacturing of the centre distance combined with tight gearing tolerances (manufacturing effort) and the use of two braced drives. [2,3]

For the majority of slewing drives in heavy industry and materials handling technology, the focus is on providing very high torques with the most robust and simple drive technology possible, so that the occurrence of greater gear backlash is accepted.

In order to achieve a precise tracking of rotational speed or angular position, feedback control systems are utilized.
Linear proportional-integral (PI) control based on motor speed feedback forms the de-facto industry standard for slewing drives [1]. The simple (on-site) parameterization and the acceptable robustness justify its widespread use. However, the capability to actively dampen torsional vibrations is limited, even if no backlash is present [4]. Different authors propose the additional feedback of the shaft torque, which is referred to as "disturbance observer" [5], "Gear Torque Observer" [6] or "Brandenburg Observer" [1]. This measure should prevent limit cycles, but since it actually decouples motor and load, improvements are limited to the motor side. The control no longer affects the load side, where only friction remains to counteract vibrations [7].
Since backlash introduces two distinct modes in a drive (contact mode and backlash mode), control by means of two different controllers, each active in either contact or backlash mode would suggest itself. The switching is realized by measuring the motor and load positions and comparing them to the backlash size [8]. In [7] a switching controller is presented, where strong control action enforces a fast reestablishment of flank contact. Another concept [9,10] relies on the computation of an optimal trajectory, allowing to smoothly reach the former non-operating flank.
In [11] the switching between two different control concepts is proposed: A linear-quadratic regulator (LQR) is enabled in contact mode to achieve an active vibration control. A model predictive control (MPC) algorithm is used in backlash mode for an optimal traverse. However, all states are assumed to be measured and no disturbances are taken into account. The concept of MPC is further refined in [12], where a nonlinear system model valid in both contact and backlash mode is presented. The experimental validation reveals the sensitivity to parametric uncertainty.
Common to these approaches is the necessity to precisely know the backlash size and especially avoid underestimation. The analyses known to the authors are concerned with backlash sizes smaller than 20°.

In this article, a modular control strategy based on a model-based design approach is presented, which effectively limits the torsional load of elastic drivetrains subjected to large backlash. The reduced dynamical overloads prevent mechanical failure of load-bearing machine elements and thus increase the drive's reliability. This can be achieved by active damping of torsional vibrations and a novel backlash traverse strategy. The proposed control strategy facilitates practical implementation due to the necessity to provide only two incremental encoders.
First, the system dynamics of an exemplary slewing drive are analysed using measured data and a fully flexible multibody system simulation model. The latter allows to derive a simple reduced order model, suitable for designing a full-state feedback controller and a state estimator for the nonlinear system. Robustness and achievable load reduction are investigated in subsequent steps. The mathematical formulation is based upon a drive with a single motor for the sake of clarity. An extension to multi-motor drives is straightforward.

2. Measurements and multibody system simulation of an exemplary slewing drive

In the field of bulk material handling, bucket wheel excavators are utilized if large amounts of goods have to be conveyed continuously. The slewing drive of such a large-scale equipment is shown in figure 1. The drive is prone to low-frequency torsional vibration. The angular backlash under nominal conditions (i.e. no excessive wear) corresponds to a rotation angle of $\alpha = 560°$ at the motor shaft.
The input shaft speed and the shaft torque at the gearbox input were recorded under representative operating conditions during a measurement campaign (see figure 2). As the measured data in figure 3 illustrate, similar speed and torque signals occur in all reversal cycles. A proportional-integral (PI) speed controller is currently implemented to regulate the slewing motion. The direction of rotation is reversed

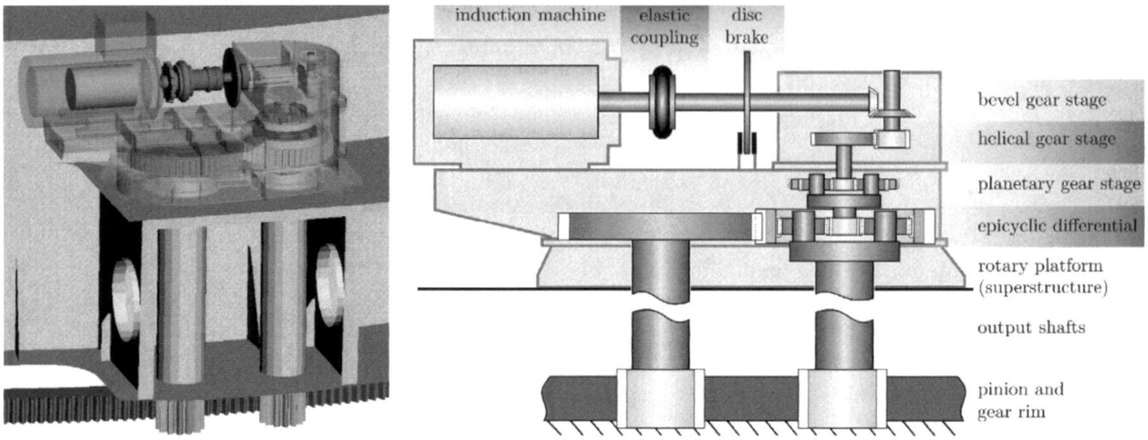

Figure 1. Schematic view of the slewing drive's gearbox (right) and multibody system simulation model (left)

by following a jerk-limited speed profile. As a consequence of a defensive setpoint setting, the reversing time is currently approx. 16 s. Nevertheless, the measured data reveal a severely increased load, which can be attributed to the gear backlash:
As soon as the load direction or the rotational direction is reversed, the tooth flanks previously transmitting the torque disengage and the backlash of all gear stages is run through until the former non-operating flanks touch. The increase in load is thereby caused by two independent mechanisms:

1. The collision during the initial contact of the former non-operating flanks. The magnitude of the mean torque throughout the impact approximately follows from the change in momentum $M = \Delta L/\Delta t$ with the angular momentum $= J\omega$.
2. The excitation of the lowest torsional vibration mode of frequency f_0, leading to higher loads subsequent to the reestablishment of the flank contact.

The current PI speed control structure is neither capable to effectively limit the speed difference during impact nor the subsequent torsional overload.

Based on the system's technical drawings, a fully-flexible multi body system (MBS) simulation model has been set up using the program environment SIMPACK. The MBS model, including its validation process, is described in [13] as well as an in depth-analysis of the effects associated with gear backlash.

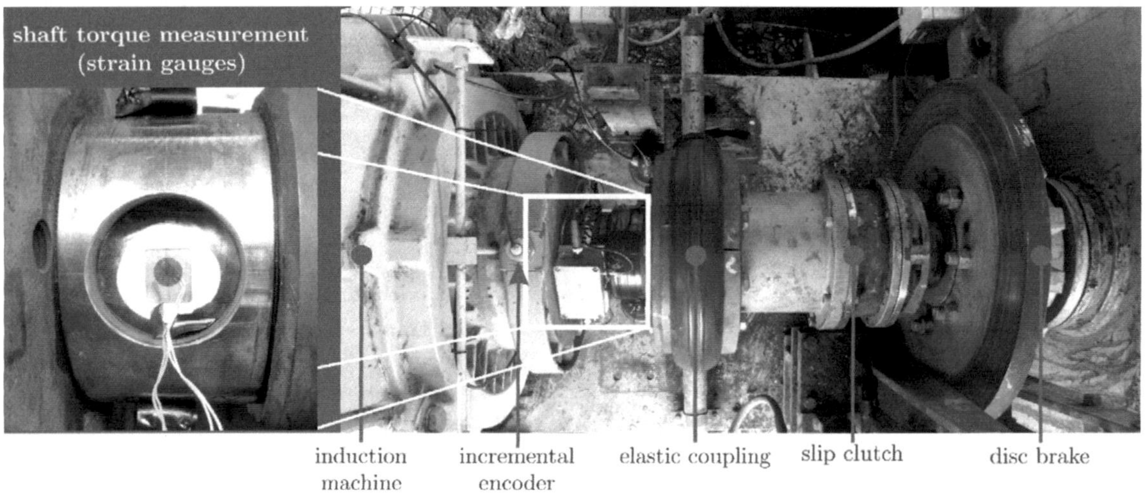

Figure 2. Input shaft assembly of the gearbox with measurement setup for shaft torque and motor speed

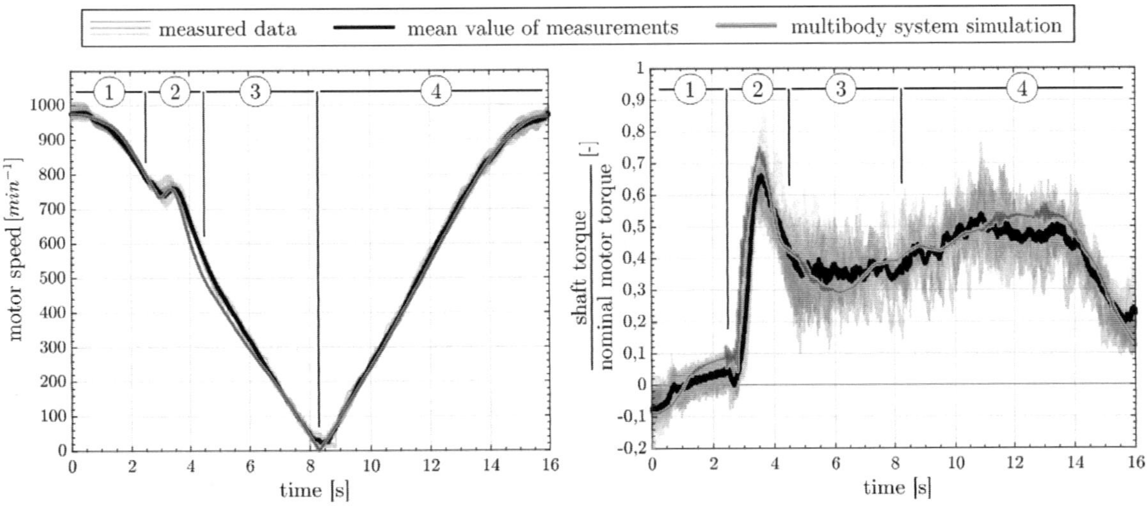

Figure 3. Measured data from multiple reversals of the rotational direction compared to the multibody system simulation results – the absolute value of the motor speed is shown to facilitate the visual comparison

As the simulation results of one reversal of the rotational direction in figure 3 indicate, there is a good agreement between measurement and simulation. The detailed MBS model will serve to evaluate the possible improvements of a new control strategy being presented in the subsequent sections.
But first, the model allows to derive a simple low-order system representation suitable for the intended model-based controller design. Analyzing the distribution of the system's rotational inertia reveals that the input shaft assembly (consisting of induction machine, coupling and brake disc) and the excavator superstructure (load side) represent approximately 99 % of the drivetrain's total rotational inertia. Consequently, the gear stages and shafts may be modelled as an idealized massless torsion spring. This leads to the two-mass system depicted in figure 4, with a torsional shaft stiffness $k = 585$ Nm/rad and the rotational inertias $J_1 = 12$ kgm² and $J_2 = 26$ kgm² of motor and load, respectively. In figure 5 the transfer functions of the two-mass system are compared to the detailed MBS model. A sufficient agreement can be seen in the low-frequency range up to 10 Hz. The torsional natural frequency $f_0 = 1{,}34$ Hz of the two-mass system corresponds to that of the MBS model.
As the torsional damping of most drives is small and hardly known exactly, it is neglected in the design phase but included in the final MBS simulations (section 4) and the robustness assessment (section 3.3) as uncertain parameter. This allows to extend the two-mass system by the known dead-zone backlash model (see figure 4), which is valid for shafts without damping only [14]. Consequently the angular difference between motor and load

$$\varphi_\mathrm{d} = \varphi_1 - \varphi_2 = \varphi_\mathrm{s} + \varphi_\mathrm{b} \tag{1}$$

consists of the elastic twist of the drivetrain φ_s (proportional to the torque transmitted) and the position in the backlash φ_b. The latter is bounded by the backlash size α and remains constant at either of both limits in case of flank contact.

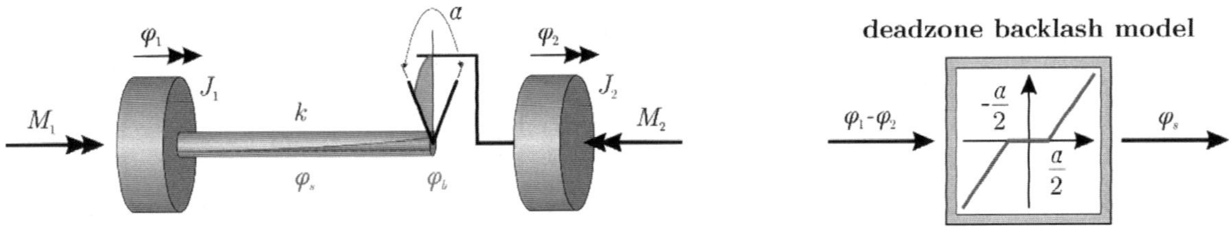

Figure 4. Dynamic model used for the control design (left) and nonlinear characteristic representing the backlash (right)

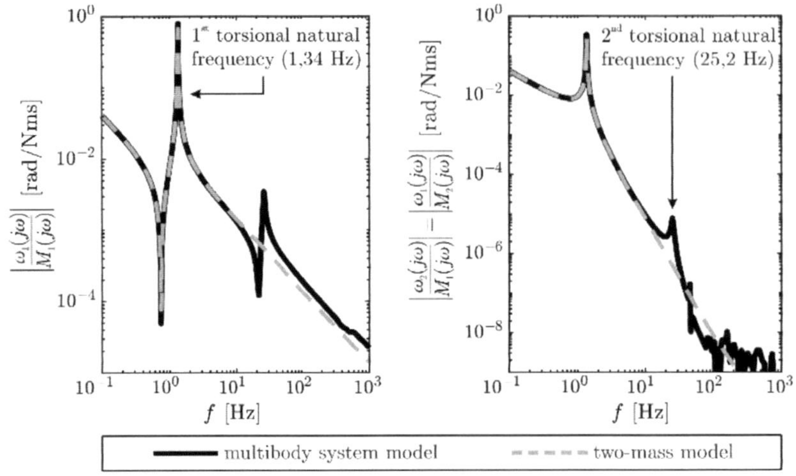

Figure 5. Frequency response functions (point-to-point and transfer mobility) of the detailed MBS model and the derived two-mass system

3. Active vibration control strategy

Due to the described shortcomings of the widespread PI control, a different control structure is pursued. Figure 6 summarizes the modules of the proposed control system and the interfaces to the actual drive system.

The main objective is to actuate the drive by means of a motor torque $M_1(t)$ that ensures tracking of the slewing speed reference $w(t)$ without excessive overload due to backlash-induced vibrations. The torque setpoint value is slightly delayed by the induction machine's underlying vector control. The latter can be modelled as first-order delay with a time constant T_{VC} of several milliseconds.

The feedback control relies on measurements at the drive. Usually the rotation angle φ_1 of the motor is measured, additionally the slewing angle φ_2 of the load side is assumed to be available. The limited resolution of incremental encoders causes a quantization error of the measurements.

The utilized linear quadratic regulator (LQR) is described in section 3.1. Subsequently, section 3.2 is dedicated to the state estimation process using a Kalman filter. A robust and practical approach for a fast and yet smooth traversal of the backlash gap is presented in section 4.

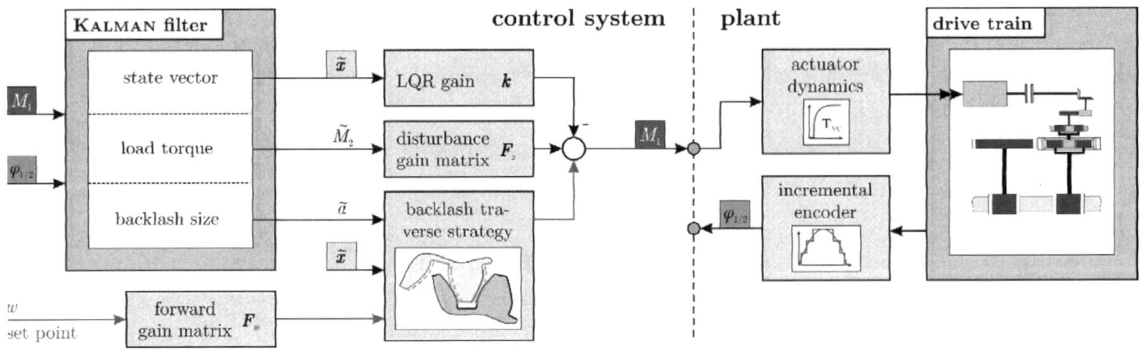

Figure 6. Proposed control structure for slewing drives subject to large backlash

3.1. LQ-optimal state feedback

In contrast to the PI control, which is based solely on the feedback of motor speed, additional kinematic quantities are fed into the control algorithm here. This full state feedback will enable an active damping of torsional drivetrain oscillations. [4]

The starting point is the state space representation of the system model shown in figure 4

$$\dot{\boldsymbol{x}}(t) = \boldsymbol{A}_{\mathrm{LQ}}\boldsymbol{x}(t) + \boldsymbol{b}_{\mathrm{LQ}}u(t) \quad \text{with} \quad \boldsymbol{A}_{\mathrm{LQ}} = \begin{bmatrix} 0 & 0 & -k/J_1 \\ 0 & 0 & k/J_2 \\ 1 & -1 & 0 \end{bmatrix} \quad \text{and} \quad \boldsymbol{b}_{\mathrm{LQ}} = \begin{bmatrix} 1/J_1 \\ 0 \\ 0 \end{bmatrix}. \tag{2}$$

Herein the state vector $\boldsymbol{x} = \left[\omega_1\ \omega_2\ \varphi_{\mathrm{s}}\right]^T$ includes the angular velocities of motor and load, as well as the shaft twist. The system matrix $\boldsymbol{A}_{\mathrm{LQ}}$ represents the dynamics in terms of the rigid body motion and the torsional mode of vibration, when the tooth flanks engage, i.e. in contact mode. The input vector $\boldsymbol{b}_{\mathrm{LQ}}$ describes, how the motor torque $u(t) = M_1(t)$ is acting on the drive.
In order to arbitrarily manipulate the system's poles – in particular the damping of the torsional mode – full state feedback is considered. This leads to the linear control law $u(t) = -\boldsymbol{k}\boldsymbol{x}(t)$. The gain vector $\boldsymbol{k}$ is chosen to minimize the quadratic cost functional

$$J(t) = \int_0^t \boldsymbol{x}(\tau)^T \boldsymbol{Q}\boldsymbol{x}(\tau) + u(\tau)^T r\, u(\tau)\, d\tau \tag{3}$$

and thus defines an LQ-optimal controller [15]. The weighting matrix $\boldsymbol{Q}$ and the scalar value r serve as tuning parameters to achieve favourable control characteristic, as they balance the relative importance of states deviating from zero and the resulting control action.
As the rotational speeds of motor and load are required to track some reference slewing speed value und reach the desired steady state value, a forward gain matrix

$$F_w = \left(\boldsymbol{c}_{\mathrm{LQ}}\left(\boldsymbol{b}_{\mathrm{LQ}}\boldsymbol{k} - \boldsymbol{A}_{\mathrm{LQ}}\right)^{-1}\boldsymbol{b}_{\mathrm{LQ}}\right)^{-1} \tag{4}$$

is introduced [16]. The output vector $\boldsymbol{c}_{\mathrm{LQ}} = [1\ 0\ 0]^T$ defines the state to be tracked. To maintain the requested rotational velocity, even if a non-zero load torque is acting, the disturbance gain matrix

$$F_z = F_w \boldsymbol{c}_{\mathrm{LQ}}\left(\boldsymbol{b}_{\mathrm{LQ}}\boldsymbol{k} - \boldsymbol{A}_{\mathrm{LQ}}\right)^{-1}\boldsymbol{b}_z \tag{5}$$

is included [16]. The input vector $\boldsymbol{b}_z = \left[0 - 1/J_2\ 0\right]^T$ describes how the load torque acts on the system. The resulting overall control law including full state feedback, setpoint tracking and disturbance rejection becomes

$$u(t) = -\boldsymbol{k}\boldsymbol{x}(t) + F_w w(t) + F_z z(t)\,. \tag{6}$$

The realization of the control structure relies on the availability of the state vector $\boldsymbol{x}(t)$ and the load torque $z(t) = M_2(t)$. While both rotational velocities may either be measured directly or processed from angle measurements, the determination of the elastic shaft twist φ_{s} is more involved. As stated in equation 1, the backlash nonlinearity imposes an uncertainty (φ_{b}) to the shaft twist. The magnitude of the backlash can substantially exceed the shaft twist angle, thus requiring special attention during the state estimation process described below.

3.2. State estimation

The estimation of the state vector can be accomplished by means of a soft-switching continuous-time Kalman filter. A system model in state space formulation

$$\dot{\tilde{\boldsymbol{x}}}(t) = \boldsymbol{A}_{\mathrm{KF}}\tilde{\boldsymbol{x}}(t) + \boldsymbol{b}_{\mathrm{KF}}u(t) + \boldsymbol{v}_d \tag{7}$$

forms the basis. The state transition is described in terms of a deterministic portion, capturing the system's internal dynamics and the effects of external inputs, e.g. the actuation due to control. Effects that are not captured by the estimation model, but are present in the actual physical system, are modelled as stochastic state disturbances $\boldsymbol{v}_d$. This includes unmodelled high-frequency dynamics or forced vibration due to some excitation.
Additionally, measurements taken from the actual system

$$\boldsymbol{y}(t) = \boldsymbol{C}_{\mathrm{KF}}\boldsymbol{x}(t) + \boldsymbol{v}_n \tag{8}$$

are also assumed to be corrupted by measurement noise $\boldsymbol{v}_n$. All stochastic processes are supposed to have white noise characteristics with covariance matrices $\boldsymbol{V}_d$ and $\boldsymbol{V}_n$ for simplification [17]. In the described application, measurements $\boldsymbol{y} = \left[\varphi_1\ \varphi_2\right]^T$ are taken at the motor and at the load of the drive.
The specific state vector of the estimation model used in equation 7 is based upon the two mass system shown in figure 4.

$$\tilde{\boldsymbol{x}} = \left[\widetilde{\omega}_1\ \widetilde{\omega}_2\ \widetilde{\varphi}_1\ \widetilde{\varphi}_2\ \widetilde{\varphi}_b\ \widetilde{M}_2\right]^T \tag{9}$$

All estimates are denote by the tilde. In contrast to the state-space model utilized in section 3.1 for controlling the system, the filter model involves both measured states. Due to the special requirements of a system containing backlash, also the angular position in the backlash gap φ_b is included. Finally, the Kalman filter allows to estimate the load torque acting on the drive. [8]
The state vector used by the controller is obtained from the estimated states using a Boolean matrix

$$\boldsymbol{x}_\mathrm{LQ} = \begin{bmatrix} 1 & 0 & 0 & 0 & 0 & 0 \\ 0 & 1 & 0 & 0 & 0 & 0 \\ 0 & 0 & 1 & -1 & -1 & 0 \end{bmatrix} \tilde{\boldsymbol{x}}_\mathrm{KF}\,. \tag{10}$$

The dynamical properties of the drivetrain are expressed in terms of the system matrix

$$\boldsymbol{A}_\mathrm{KF,co} = \begin{bmatrix} 0 & 0 & -k/J_1 & k/J_1 & k/J_1 & 0 \\ 0 & 0 & k/J_2 & k/J_2 & -k/J_2 & -1/J_2 \\ 1 & 0 & 0 & 0 & 0 & 0 \\ 0 & 1 & 0 & 0 & 0 & 0 \\ 0 & 0 & 0 & 0 & 0 & 0 \\ 0 & 0 & 0 & 0 & 0 & 0 \end{bmatrix} \tag{11}$$

which is valid if torque is transmitted, i.e. contact of tooth flanks is established. If motor and load are decoupled during backlash, the system dynamics change, leading to a different system matrix

$$\boldsymbol{A}_\mathrm{KF,bl} = \begin{bmatrix} 0 & 0 & 0 & 0 & 0 & 0 \\ 0 & 0 & 0 & 0 & 0 & -1/J_2 \\ 1 & 0 & 0 & 0 & 0 & 0 \\ 0 & 1 & 0 & 0 & 0 & 0 \\ 1 & -1 & 1/dt & -1/dt & -1/dt & 0 \\ 0 & 0 & 0 & 0 & 0 & 0 \end{bmatrix}. \tag{12}$$

No torque will be transmitted and thus the shaft twist becomes zero. This enables to augment the calculation of the backlash position, as $\widetilde{\varphi}_1 - \widetilde{\varphi}_2 = \widetilde{\varphi}_\mathrm{b}$ holds true. The additional parameter dt governs the dynamics of the correction of $\widetilde{\varphi}_\mathrm{b}$. The actuation $\boldsymbol{b}_\mathrm{KF} = [1/J_1 \quad [\boldsymbol{0}]_{1x5}]^T$ due to the input torque is unaffected by the backlash nonlinearity.

The hybrid characteristics of drives subject to backlash with two distinct modes (contact and backlash) leads to a switching observer, as proposed in [8]. The switching between the two models (equation 11 and equation 12) may be realized by comparing the angle difference $\varphi_1 - \varphi_2$ to the backlash size.
In order to avoid estimation errors associated to harsh switching between different dynamic models as in [8], a different approach is presented. A smooth transition of state estimates can be accomplished by means of a Kalman filter

$$\dot{\tilde{\boldsymbol{x}}} = \boldsymbol{A}_\mathrm{KF}\tilde{\boldsymbol{x}} + \boldsymbol{B}_\mathrm{KF}u + \boldsymbol{K}_\mathrm{KF}(t)(\boldsymbol{y} - \boldsymbol{C}_\mathrm{KF}\,\tilde{\boldsymbol{x}}) \tag{13}$$

with a time-varying gain matrix [17]

$$\boldsymbol{K}_\mathrm{KF}(t) = \boldsymbol{P}(t)\boldsymbol{C}_\mathrm{KF}^T\boldsymbol{V}_n^{-1}(t)\,. \tag{14}$$

In addition to the usual time-dependency of the estimation error covariance $\boldsymbol{P}$, expressed as [17]

$$\dot{\boldsymbol{P}}(t) = -\boldsymbol{P}(t)\boldsymbol{C}_\mathrm{KF}^T\boldsymbol{V}_n^{-1}(t)\boldsymbol{C}_\mathrm{KF}\boldsymbol{P}(t) + \boldsymbol{A}_\mathrm{KF}\boldsymbol{P}(t) + \boldsymbol{P}(t)\boldsymbol{A}_\mathrm{KF}^T + \boldsymbol{V}_d(t) \tag{15}$$

IOP Conf. Series: Materials Science and Engineering 1097 (2021) 012011 doi:10.1088/1757-899X/1097/1/012011

here also the noise covariances $\boldsymbol{V}_n(t)$ and $\boldsymbol{V}_d(t)$ are explicitly chosen to vary. After every switching event between the two models $\boldsymbol{A}_{\mathrm{KF,co}}$ and $\boldsymbol{A}_{\mathrm{KF,bl}}$ a transition phase is defined, during which the noise covariances gradually approach its final values for contact mode or backlash mode. This effectively reduces the impact of the filter's fast correction term $\boldsymbol{K}_{\mathrm{KF}}(\boldsymbol{y} - \boldsymbol{C}_{\mathrm{KF}}\,\tilde{\boldsymbol{x}})$ and puts emphasis on the slower system model.

The estimation performance of the proposed soft-switching Kalman-Filter can be analysed be means of an MBS simulation of the drive. Due to the separation principle, the simulation of the state estimator does not rely on the use of the state-feedback controller [15]. Thus, the simulation data of the PI-controlled load case described in section 2 is used. Only the motor and load angle, as well as the control input are provided to the Kalman filter. The estimation results and the actual signals taken from the MBS simulation are compared in figure 7.
It is important to note, that the simple internal model of the Kalman filter (two-mass system) is distinct from the detailed MBS model. The latter one includes the realistic distribution of inertia across the drive components, as well as the individual backlash of each gear mesh. Moreover, the MBS-model introduces higher frequency dynamics and disturbances from gear meshing frequencies. Finally, the angle "measurements" are artificially quantized to emulate the effect of a limited angular resolution of incremental encoders with 5000 pulses per revolution (ppr) at the motor and 50 ppr at the load side.
As the results in figure 7 indicate, both rotational velocities are estimated correctly. The reduced resolution of the load angle limits the feasible dynamics of the load speed estimation.
The backlash traverse during the simulation is apparent in the lower right diagram. Even during this phase the filter provides a reliable estimate of the elastic shaft twist φ_s (top right diagram). Herein the small step at $t = 2\ s$ is caused by the switching from contact mode to backlash mode. The higher frequency signal components present in the MBS simulation (e.g. due to resonance with the second torsional natural frequency at $t = 12\ s$) are effectively suppressed.

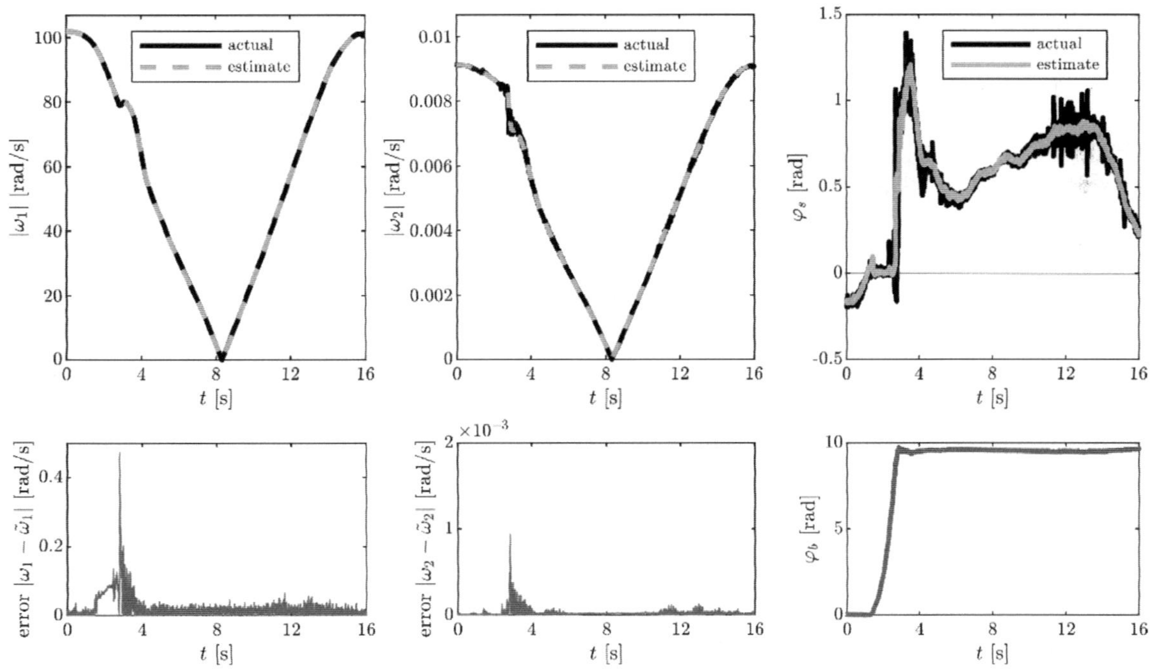

Figure 7. Estimated states and actual values during a reversal of the rotational direction, simulated by means of the MBS model under PI control

IOP Conf. Series: Materials Science and Engineering 1097 (2021) 012011 doi:10.1088/1757-899X/1097/1/012011

3.3. Active damping capability and robustness

One goal of the presented control concept is to increase the damping of the first torsional mode, which can be excited by load fluctuations or the gear meshing frequency of the low speed gear stages. Figure 8 shows the amplitude of the gearbox' shaft torque, when torque fluctuations of defined frequency act at the load side. The standard PI controller is compared to the full state feedback control based on estimated states (LQG control). Apparently, the shaft torque magnitude can be decreased by up to 50 %. The maximum improvement is achieved if the load torque fluctuations match the system's first torsional natural frequency.

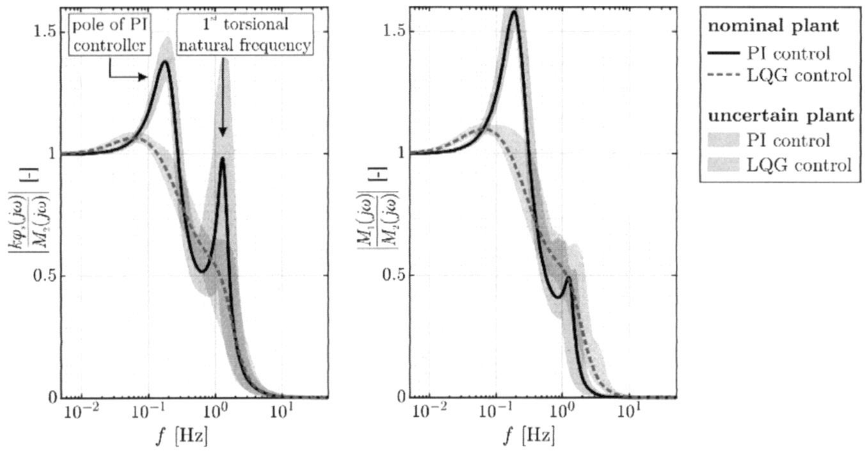

Figure 8. Frequency response functions indicating the active vibration damping capabilities (left) and the associated control effort (right) for PI control and LQG control considering parametric plant uncertainty

Special attention has to be drawn to the effects of parametric uncertainty. It is well known, that LQR and Kalman filter have guaranteed stability margins, while their combination (LQG) does not [18]. Any model imperfection and especially time delays can severely degrade control performance or even cause instability. Therefor the parameters of the controlled system are complemented by margins of uncertainty (table 1). The multitude of frequency response functions resulting from the uncertain parameters are represented by shaded areas in figure 8. The load sensitivity functions in the left diagram suggest, that considerable damping can be accomplished even if the control model significantly deviates from the actual system. The transfer functions from load disturbance to control effort in the right diagram reveal the higher bandwidth of the LQG controller, but indicate no excess control actions due to the load variations.

Table 1. Parametric uncertainty of the drivetrain

parameter	nominal value	relative uncertainty
rotational inertia J_1	12 kg m^2	±20 %
rotational inertia J_2	26 kg m^2	±20 %
torsional stiffness k	585 Nm rad^{-1}	±50 %
damping constant d	0.5 Nm s rad^{-1}	−100 % + 1000 %
time constant T_{VC}	10 ms	±100 %

4. Backlash traverse strategy

The second objective of the slewing speed control is to reduce the overload and the excitation that originate from the backlash traverse. As stated in section 2, the speed difference between load and motor during flank contact is decisive for these effects and thus should be limited.

A prerequisite for the proposed approach is the coarse knowledge of the backlash size. It may either be gained from measurements, estimated using a separate Kalman filter [8] or processed from the signal $\widetilde{\varphi}_b$ introduced in section 3.2. Furthermore, the angular velocities and positions of motor and load are required.

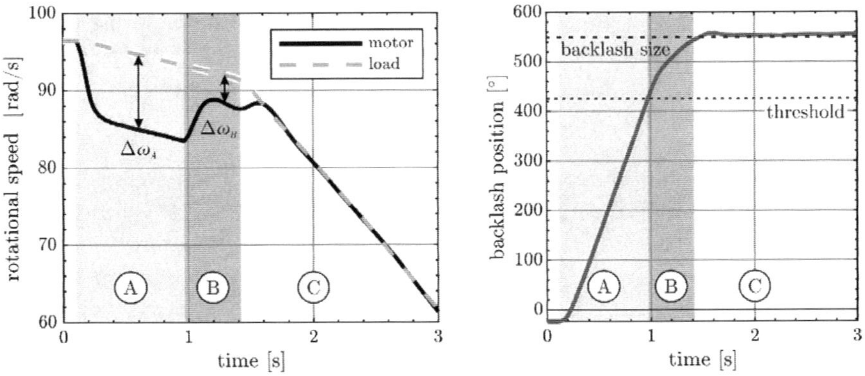

Figure 9. Reference speed profile for backlash traverse (left) and corresponding backlash position (right)

The backlash traverse strategy shown in figure 9 consists of three phases:

Ⓐ Fast backlash traverse

First, the motor speed reference value is reduced by $\Delta\omega_A$ relative to the load speed ω_2. The choice of the speed difference depends on the admissible control effort and the desired duration of the backlash traverse.

Ⓑ Preparation of flank contact

If the backlash position φ_b exceeds a threshold value, the speed difference between motor and load is reduced to $\Delta\omega_B$ in preparation of the expected flank contact. The speed difference is chosen to traverse the remaining backlash gap sufficiently fast and yet limit the ensuing overload.

Ⓒ Contact phase

After re-establishing the flank contact in all gear meshes, the targeted final slewing speed can be reached.

Finally, the overall control structure consisting of the LQG control and the backlash traverse strategy is used in an MBS simulation for reversing the drive's rotational direction to demonstrate its functionality. In figure 10 the motor speed and shaft torque are shown and compared to the present PI controller. Currently, the process takes 16 s due to a defensive jerk-limited reference speed profile. A faster reversal is preferable. However, the simple adjustment of the PI controller's reference speed profile to reduce the duration for a motor speed reversal to 12 s would result in an unacceptable strain in the drivetrain (see figure 10). Moreover, the PI control is unable to reach the target speed without overshooting, thus entering backlash again.

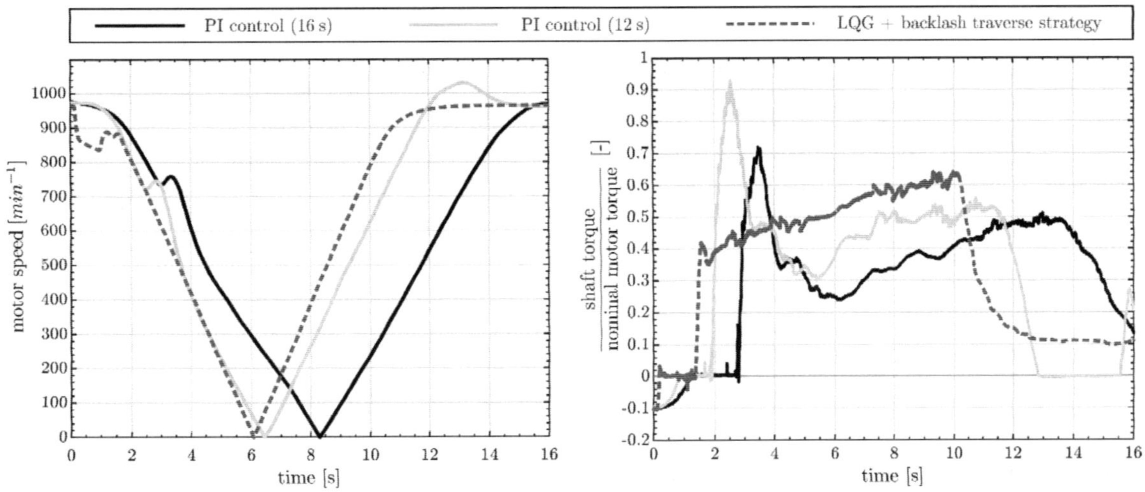

Figure 10. MBS simulation of three different reversals of the rotational direction – the absolute value of the motor speed is shown to facilitate the visual comparison

In contrast, the proposed control is capable of reducing the time for reversing the rotational direction by 25 %, while almost completely avoiding torque overshoot. The increased shaft torque in the second half of the simulation results from the higher acceleration torque necessary for the faster reversal.

5. Conclusion

The present article provides a practical solution to two common problems in drive technology. Geared transmissions with high reduction ratios cause both large backlash and elasticity in drivetrains. As a result, unwanted vibrations can occur and high dynamic loads may arise. A modular control strategy is proposed, which is based on full-speed feedback in order to actively dampen torsional vibrations. A soft-switching state estimator is presented, allowing to reconstruct the necessary states for feedback based on two measured kinematic quantities. The overload resulting from the abrupt contact of tooth flanks after a backlash traverse is reduced by means of a simple and robust strategy. MBS simulations and a robustness assessment in the frequency domain proof the concept's effectiveness and reliability. The practical implementation is facilitated due to the limited sensor demand and the moderate computational cost. Future work will cover the realization on an actual drive system.

6. References

[1] Nordin M and Gutman P-O 2002 Controlling mechanical systems with backlash - A survey *Automatica* **38** 1633–1649

[2] Linke H, Börner J and Hess R 2016 *Cylindrical gears* (München: Hanser)

[3] Schlecht B 2009 *Maschinenelemente 2* (München: Pearson)

[4] Thomsen S 2013 *Drehzahlregelungsverfahren für Antriebssysteme mit schwingungsfähiger Last unter Berücksichtigung verschiedener mechanischer Antriebsparameter* phdthesis (CAU Kiel, Technische Fakultät)

[5] Brandenburg G 1989 Einfluss und Kompensation von Lose und Coulombscher Reibung bei einem drehzahl- und lagegeregelten, elastischen Zweimassensystem *at* **37** 23–31

[6] Dhaouadi R, Kubo K and Tobise M 1994 Analysis and compensation of speed drive systems with torsional loads *IEEE Trans. Ind. Applic.* **30** 760–766

[7] Boneh R and Yaniv O 1999 Reduction of limit cycle amplitude in the presence of backlash *J. Dyn. Sys., Meas., Control* **121** 278–284

[8] Lagerberg A and Egardt B 2007 Backlash estimation with application to automotive powertrains *IEEE Trans. Control Sys. Tech.* **15** 483–493

[9] Schäfer U 1993 *Entwicklung von nichtlinearen Drehzahl- und Lageregelungen zur Kompensation von Coulomb-Reibung und Lose bei einem elektrisch angetriebenen, elastischen Zweimassensystem* phdthesis (TU München, Lehrstuhl für Elektrische Antriebstechnik)

[10] Ezal K, Kokotovic P V and Tao G 1997 Optimal control of tracking systems with backlash and flexibility *Proc. of the 36th IEEE Conf. on Decision and Control* (San Diego) pp 1749–1754

[11] Lagerberg A and Egardt B 2005 Model predictive control of automotive powertrains with backlash *IFAC Proc. Vol.* **38** 1–6

[12] Rostalski P, Besselmann T, Barić M, Belzen F V and Morari M 2007 A hybrid approach to modelling, control and state estimation of mechanical systems with backlash *Int. J. Control* **80** 1729–1740

[13] Spiegelhauer M, Schlecht B 2020 On the influence of gear backlash in speed controlled slewing drives *World of Mining* **72** 200–204

[14] Nordin M, Galic J and Gutman P-O 1997 New models for backlash and gear play *Int. J. Adaptive Control and Signal Processing* **11** 49–63

[15] Stengel R 1994 *Optimal Control and Estimation* (New York: Dover)

[16] Unbehauen H 2000 *Regelungstechnik III – Identifikation, Adaption, Optimierung* (Braunschweig: Vieweg)

[17] Simon D 2006 *Optimal State Estimation - Kalman, H_∞ and Nonlinear Approaches* (Hoboken: Wiley-Interscience)

[18] Skogestad S and Postlethwaite I 2005 *Multivariable Feedback Control: Analysis and Design* (Wiley)

IOP Conf. Series: Materials Science and Engineering 1097 (2021) 012012 doi:10.1088/1757-899X/1097/1/012012

Acoustic Optimization of a Power Take-off Gear Box Using Numerical Transfer Path Analysis

Stefan Wischmann[1,2], Gerwin Pasch[1], Joerg Berroth[1] and Georg Jacobs[1]

[1]Institute for Machine Elements and Systems Engineering, RWTH Aachen University, Schinkelstraße 10, 52062 Aachen, Germany

[2]stefan.wischmann@imse.rwth-aachen.de

Abstract. In this contribution, a method is proposed which allows for a target-oriented structural optimization of drive train systems using transfer path analysis on a multi body system simulation with respect to the airborne sound. It combines the existing approaches of multi body simulation (MBS) and experimental transfer path analysis (TPA) with sound radiation calculations thereby creating a new tool chain to not only identify critical operation points but also critical transfer paths of an MBS model. This approach combines the advantages of elastic MBS (i.e. information about modes and forces acting between different components in critical operating points) with the advantages of TPA (i.e. identifying critical paths from the excitation to the receiver) thereby enabling the designer to numerically deduct and evaluate target-oriented structural optimizations already in early stages of the development process. It therefore reduces development times and costs by reducing iteration loops and the manufacturing of physical prototypes. The method is demonstrated on a power take-off (PTO) gear box, of which the sound pressure level was reduced by 6 dB within three virtual optimization loops that each only take a few hours by applying the proposed method.

1. Introduction

The continuous shortening of product lifecycles has created a demand for faster time to market and thus for faster and more efficient development processes. Therefore, methods of virtual product development (VPD) have become increasingly established, as they reduce development time and cost by incorporating numerical methods into the development process, thereby reducing the number of iteration loops with expensive, physical prototypes. Current trends in VPD include the integration of simulation models of different domains and detail levels into seamless multi-domain system models which is referred to as model-based systems engineering (MBSE) and allows for the evaluation and optimization of features of complex systems. Within this framework, elastic multi body simulation (eMBS) is used to analyze the structural dynamic properties of drive train systems. eMBS models have enabled the design engineer to efficiently evaluate and quantify the effect of design changes on the dynamic behavior at system level, which can be quantified in terms of structure borne and airborne sound that both play an important role for the customer acceptance of a product. However, the eMBS method alone does not allow for the target-oriented deducting of optimization measures to the structure as this requires the identification of critical transfer paths of numerical models which still remains a challenge. TPA is a well-known method for experimentally identifying these critical paths on physical prototypes. As this method however is a purely experimental approach, it leaves the advantages of the simulation unused. The aim of this work is therefore to propose a method that enables the design engineer to identify optimizations to an existing virtual prototype with regard to the airborne sound by combining eMBS and TPA and extending it by incorporating the sound radiation calculation into the TPA-formulation.

Based on the state of the art described in chapter 2 an approach to incorporate the experimental approach of transfer path analysis (TPA) into the simulation domain is presented in chapter 3. Special focus will be directed towards the analysis of airborne sound as the most important NVH quantity. By doing so, critical transfer paths can be identified and optimization measures deduced. The method is demonstrated on a power take-off (PTO) gear box in chapter 4. It yields a 6 dB optimization of the total sound pressure level within three virtual optimization loops. Conclusions and outlook will be presented in chapter 5.

2. State of the art

Two fundamentals are essential to understanding the presented work: The widely used experimental TPA-Method as well as the calculation of airborne sound radiation as part of system level simulation models based on the eMBS method. Both are presented in greater detail in what follows:

2.1. *Transfer path analysis*

In the experimental world, TPA is a widely established method for identifying critical transfer paths for different operation points [1–4]. Its basic concept is shown in figure 1. A structural dynamic system is divided into an active system which contains the excitations $\boldsymbol{f_1}$ (e.g. from gear boxes, combustion or electrical engines or road-tire noise), denoted “A” in the following, and a passive system, denoted “B”, that contains the structure linking the active system to a receiver point (e.g. the driver’s ear). These two systems are connected at discrete interface points, where forces are transferred from the active to the passive subsystem.

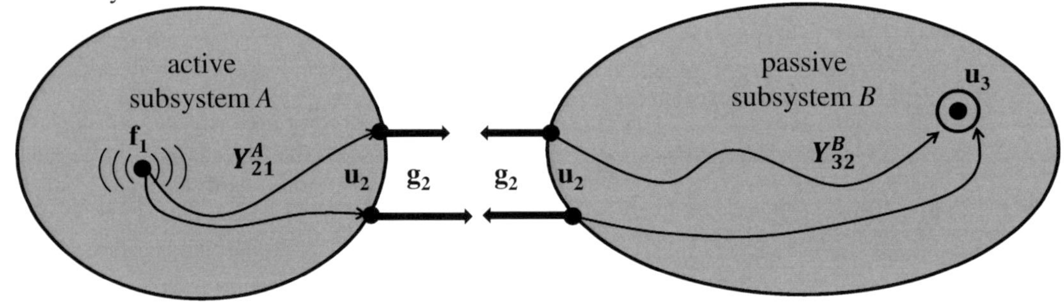

Figure 1. General concept of TPA

The receiver response, which can be a displacement, velocity, acceleration or sound pressure level, denoted $\boldsymbol{u_3}$, can be calculated depending on the interface forces $\boldsymbol{g_2}$ acting between the two subsystems when assembled and the frequency response functions (FRF) of the passive System $\boldsymbol{Y^B_{32}}$ as shown in equation (1).

$$\boldsymbol{u_3} = \boldsymbol{Y^B_{32}} \cdot \boldsymbol{g_2} \tag{1}$$

The multiplication of an interface force in one degree of freedom (DOF) with the corresponding FRF to the receiver, results in the transfer path contribution. As these interface forces are typically difficult to measure in an experimental setup, different methods have been developed for reconstructing the interface forces [5, 6]. The advantage of the simulation however is, that the interface forces can directly be obtained from time-domain analysis (e.g. a rpm runup) so that the experimental methods (e.g. matrix-inverse methods) don’t have to be applied on a simulation model.

While some efforts have been made to extend TPA-methodology into the simulation domain, their application remains limited to structure borne sound [7, 8] or requires the computational intensive export of all surface node’s velocities in the time-domain for every interface force and every time step [9]. A computationally efficient extension to airborne sound in the frequency domain therefore is necessary to deduce optimization measures for a system's NVH behavior.

2.2. *Calculating airborne sound pressure levels*

The toolchain for calculating airborne borne sound levels is depicted in figure 2. [10]

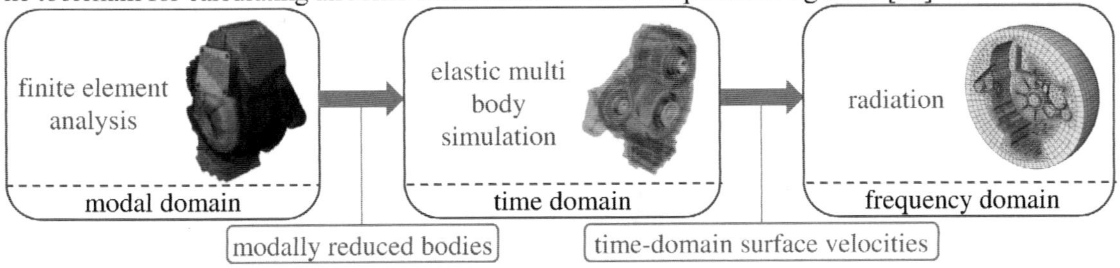

Figure 2. Current tool chain for calculating airborne sound radiation.

Based on modeling the components in a finite element (FE) environment, a modal reduction [11] is carried out which allows for the flexible modeling of these components in an eMBS environment. After defining the interaction between these components (in form of force elements like bearings and gear mesh behavior as well as boundary conditions like external torque and rotational speed) a time-domain analysis of the entire system is carried out. The system model has to be solved in the time domain and cannot be represented in the frequency domain as nonlinearities (e.g. nonlinear bearing stiffnesses and clearances, gear mesh excitations and load depending gear mesh stiffnesses) have to considered. The time-dependent deformation of the bodies (in form of elongation of modal states) is subsequently exported. Together with the recovery matrix of the flexible body, which links the modal states to nodal DOFs for a selected node set on the radiating surface, the time-dependent surface velocities can be calculated [10]. To calculate the sound pressure level at a receiver point, an acoustic transfer vector (ATV) matrix needs to be generated, which links the surface node velocities to the sound pressure levels at the receiver point [12]. This is done in an FE-environment in the frequency domain by modeling the surrounding air as a fluid and exciting this fluid individually at every surface node in all translational directions. By using a fast Fourier transform (FFT) to transform the surface into the frequency domain and multiplying these with the ATV matrix, the sound pressure level at the receiver point is obtained.

3. Problem formulation and solution approach

The existing approach does not allow for the tracking of transfer paths from the excitation at the interface points to the response at the receiver point, as the nodal surface velocities are exported as time-domain values and only represent the response of the entire system, see figure 2. Therefore, information about the transfer paths is not included in the radiation calculation. For a TPA to be carried out, both the time-domain values of the interface forces and the transfer paths from the interfaces to the receiver points (formulated in a frequency domain) are necessary. As the interface points are within the eMBS model and as the transfer paths from the interface points to the surface velocities are not available in the frequency domain in current radiation calculations, a TPA cannot be executed based on the toolchain of figure 2. Therefore, the design engineer can extract no knowledge about transfer paths and optimization methods out of the airborne sound signal. Hence, a new modeling approach is suggested, see figure 3.

IOP Conf. Series: Materials Science and Engineering 1097 (2021) 012012 doi:10.1088/1757-899X/1097/1/012012

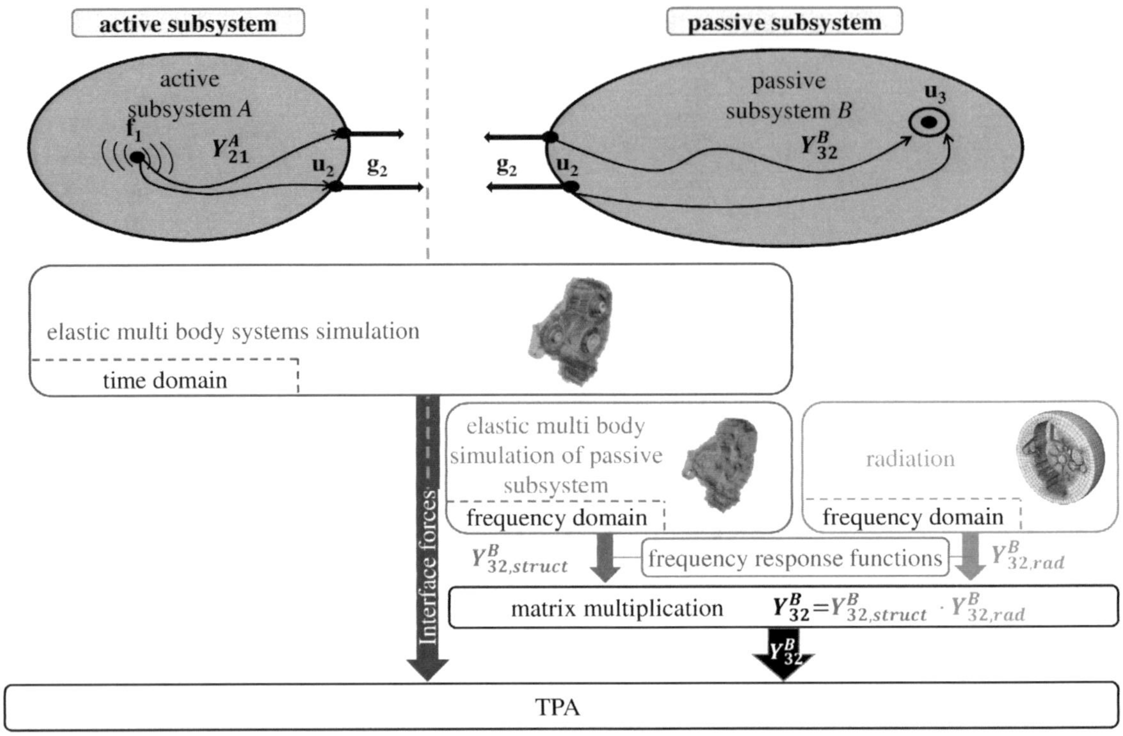

Figure 3. Proposed method and toolchain for performing TPA on airborne sound

The general concept of the proposed method is to set up a model structure compliant to the TPA formulation. This requires the interface forces of the assembled model to be exported as time-domain values and the passive system (containing both the structural dynamic model and the radiation calculation) to be modeled in the frequency domain. By combining the passive system models' frequency responses (structural dynamic and radiation), a TPA can be subsequently carried out. In detail, this leads to the following adaptions to the original toolchain for airborne sound (see figure 2):

First of all, instead of using the surface velocities as a result of the system model (marked in blue) in time-domain, the time-domain values of the interface forces are used as this is the necessary input for the calculation of the TPA. By doing so, the discontinuity between the different simulation models is shifted to the interface points to match the model structure required for a TPA.

In order to model the transfer from the interface points to the receiver point (i.e. for the modeling of the passive system), two separate models are necessary, both of which are formulated in the frequency domain. The radiation is still calculated using the ATV-matrices (marked in green). However, to link the interface forces to the surface velocities, an additional model is required (marked in magenta). This model is the structural dynamics model of the passive system. It is excited at the interface points and yields the frequency response functions from the interface forces to the surface velocities. To minimize the data output, the modal states are exported and using the recovery matrix of the passive subsystem the full transfer functions are calculated. The required frequency response matrix of the entire passive system (consisting of the structural passive system and the radiation calculation) is eventually calculated by multiplying the FRF from the interface forces to the surface nodes' velocities $\boldsymbol{Y}^{\boldsymbol{B}}_{32,struct}$ with the ATV matrix $\boldsymbol{Y}^{\boldsymbol{B}}_{\boldsymbol{32,rad}}$ according to equation (2).

$$\boldsymbol{Y}^{\boldsymbol{B}}_{32} = \boldsymbol{Y}^{\boldsymbol{B}}_{32,struct} \cdot \boldsymbol{Y}^{\boldsymbol{B}}_{\boldsymbol{32,rad}} \tag{2}$$

4. Application to a power take-off gearbox

The proposed method is applied to a PTO gearbox, depicted in figure 4. It consists of an input shaft, an auxiliary PTO shaft, an intermediate shaft (also allowing a PTO), an output shaft and the surrounding housing. Between these shafts, gear stages transmit the power. A receiver point is defined at the rear side of the housing, at a distance of 0,5 m from the surface.

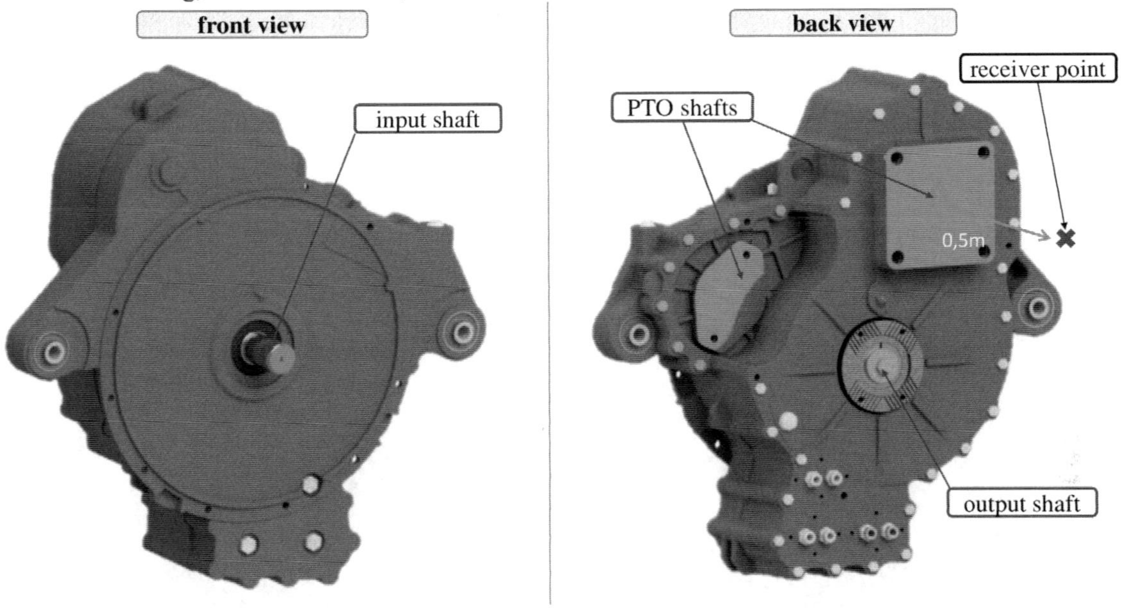

Figure 4. Demonstrator for airborne sound TPA calculation: power take-off gearbox

In order to perform a TPA on this system, the boundaries between active and passive subsystem have to be defined (see also figure 1). The active subsystem contains the shafts and the gear stages, see figure 5. The passive system contains the housing and the radiation model. The 8 bearings (two per shaft) form the interface points.

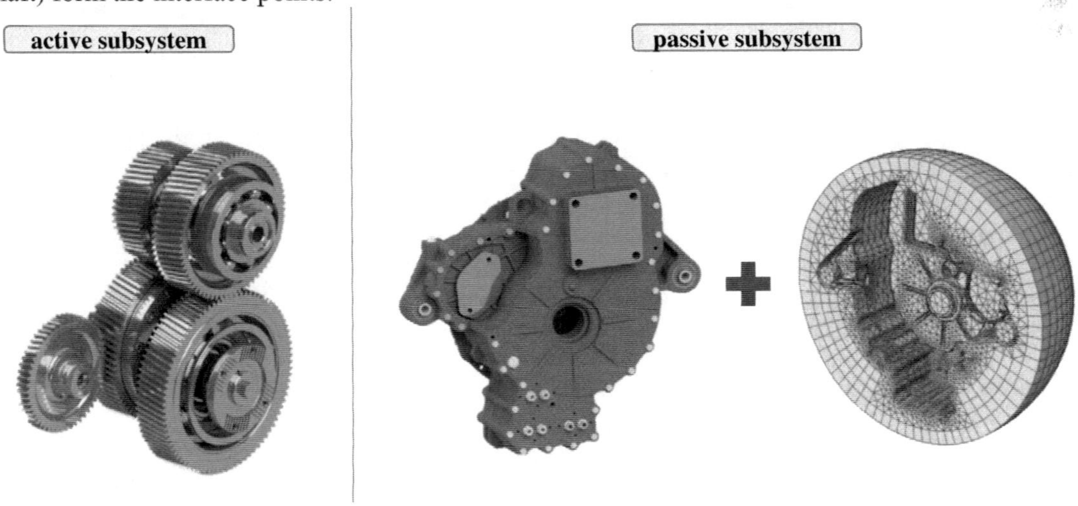

Figure 5. Subsystem definition of the power take-off gearbox

According to the method described in chapter 3, three different models are generated: First, an eMBS model of the entire mechanical system is created (highlighted in figure 3 in blue), which yields the time-domain values of the interface forces. The model is loaded with a constant torque and an rpm runup is conducted in the time domain. Bearing stiffnesses are taken into account by the Hertz'ian contact theory, bearing clearances are also considered. The gear mesh excitations are modeled based on the load-dependent Weber/Banaschek model [13]. Secondly, a model of the housing is created in an eMBS environment (highlighted in figure 3 in magenta). The housing is modeled linearly. The material properties of the cast iron are identified by fitting the model to a conducted experimental modal analysis. This model's purpose is to yield the frequency domain transfer paths from the bearings to the radiating surface. Lastly, an FE model to calculate ATV matrices is created. The air is modeled as a linear fluid and the boundary nodes are copied to the housing model, where they are tied to the radiating surface. For reference, the modeling method according to figure 2 is also applied – the surface velocities are exported from the systems model and used to calculate a reference sound pressure level at the receiver point. Measurement data were available for the validation of the model.

The measured sound pressure level is compared to the calculated one (according to the modeling tool chain described in figure 2), see figure 6.

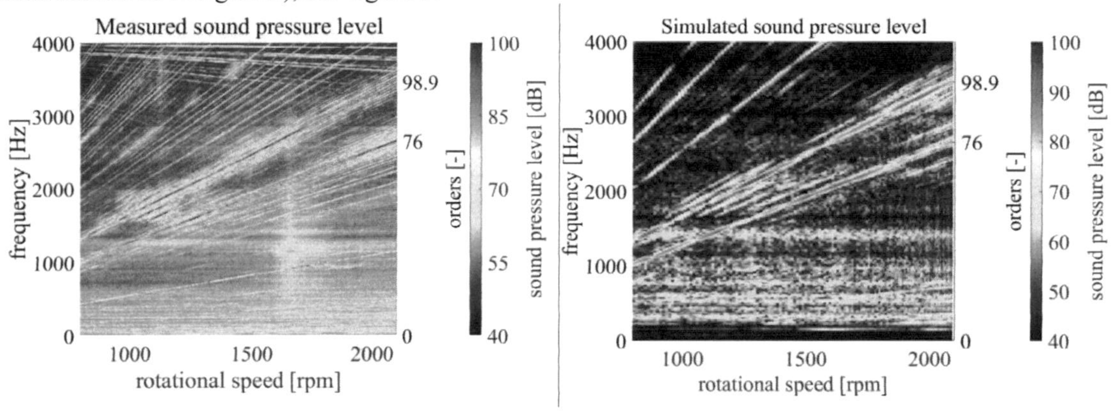

Figure 6. Comparison between measured (left) and simulated (right) sound pressure levels

First of all, the comparison in the waterfall diagram shows a good match in the frequency content. Both first gear stage orders (76th and 98.9th) can clearly be identified in both the simulation results and the measurement results. The measurement results additionally show pulse-width-modulation-orders, induced by the testbench, which are not considered in the simulation. Looking at the amplitudes of the waterfall diagrams it can be deduced that several resonance peaks are clearly identified. However, the simulation proved to be about 10% too stiff. The first resonance occurs at 1180 rpm in the measurement and at 1300 rpm in the simulation and is accompanied by sidebands modulating the exciting order of the second gear stage. Both peaks show similar amplitude values of around 100 dB. Also, the second dominant resonance of the second gear stage at 1600 rpm can be identified in the simulation, again at slightly higher rotational speeds of about 1650 rpm. The resonance peaks of the first gear stage order (76th) at 1300 rpm are also represented in the simulation. However, several additional peaks in the simulation can be found (e.g. at 2000 rpm in the second gear stage or 1800 rpm in the first gear stage) which seem to be significantly more damped in the real system. The well matching rpm band between 1000 rpm and 1500 rpm is used for further validation of the proposed method. The sum levels between simulation and measurement are compared in figure 7. It can clearly be seen that while the absolute values are differencing slightly, the three peaks at ca. 1100 rpm, 1300 rpm and 1500 rpm are all found in the simulation as well. The model therefore can be considered as qualitatively valid and will be used for the further analysis.

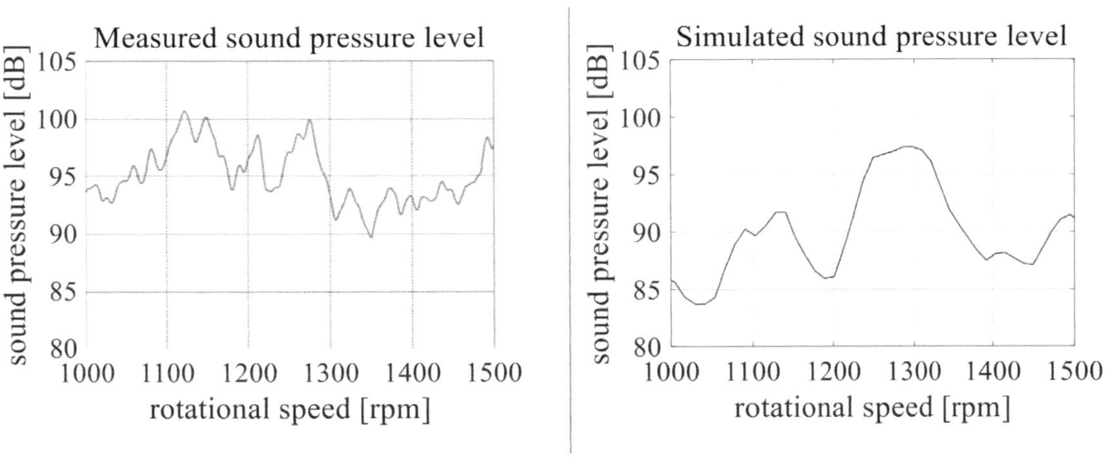

Figure 7. Comparison between Sound pressure level of measurement and simulation

To validate the proposed method, a comparison between the sound pressure levels calculated according to figure 2 and the sound pressure levels created from superimposing all transfer paths is shown in figure 8. The reconstructed signal matches the original signal both in frequency content and in absolute amplitudes very well, so that it can be deduced, that no transfer paths have been missed or miscalculated.

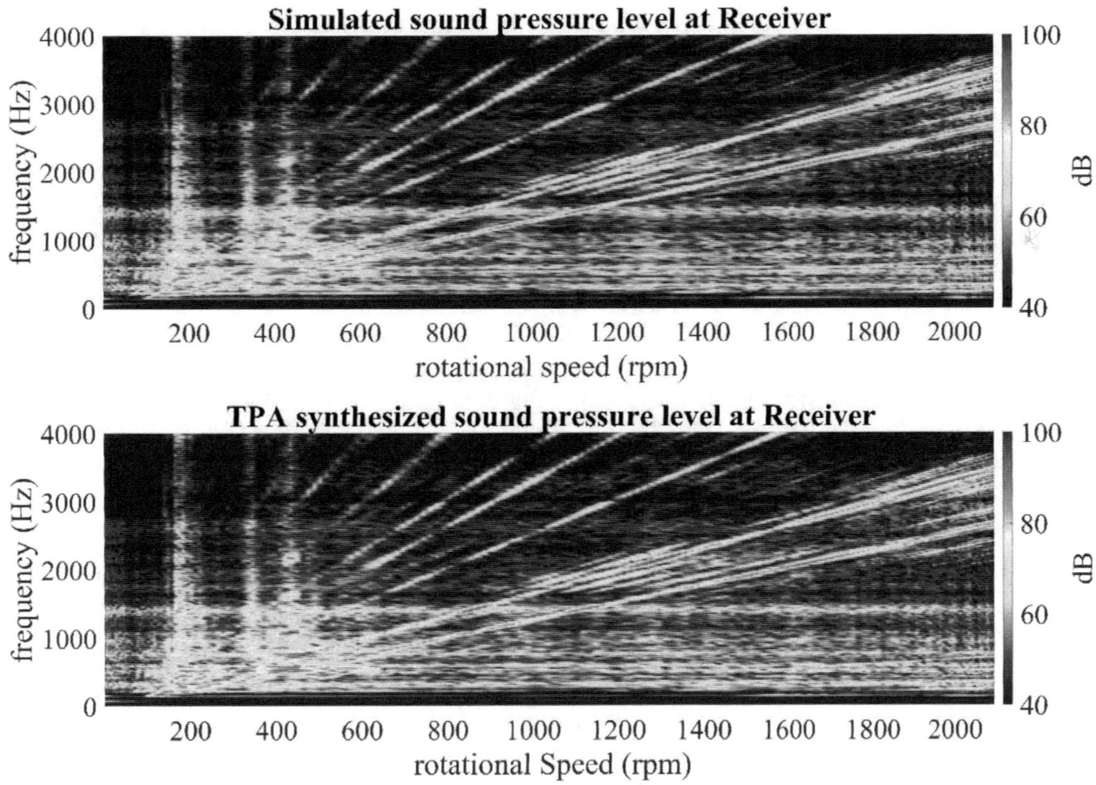

Figure 8. TPA-Method Verification: Comparison of simulated sound pressure levels at the receiver point (top) with TPA synthesized levels (bottom)

Now, the transfer path contributions can be calculated by multiplying the ATV matrix with the FRF matrix of the structural dynamic passive system and with the force of each interface DOF. The path contributions are depicted in figure 9. Each row shows the aggregated path contributions of one bearing to the sound pressure level. It can clearly be seen that the two tapered bearings supporting the output shaft are the dominant factors for the sound pressure level in the rpm range of up to 1200 rpm, while at higher rpm values the intermediate shaft cylinder bearing becomes dominant. It is also evident, that the two bearings supporting the auxiliary PTO shaft are not contributing to the sound pressure level. This is to be expected, as no power is transmitted through this shaft in the chosen verification setup.

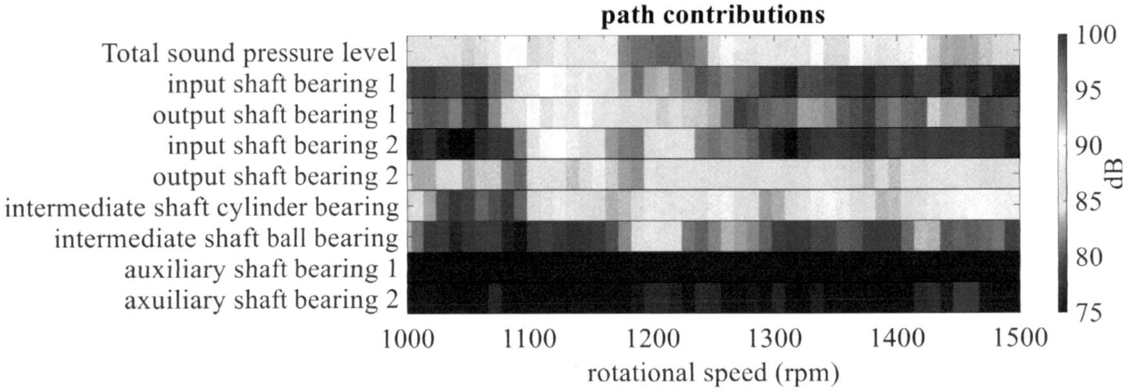

Figure 9. Path contributions of the eight bearings to the sound pressure level of the power take-off gearbox

4.1. Target-oriented optimization – Strength of the proposed method

The proposed method allows for the identification of critical transfer paths and therefore enables the design engineer to deduct optimization measures. If necessary, the identification of optimization measures can be assisted by the deformation shapes of the radiating surface in the critical operation points, which can directly be obtained from the eMBS model. In what follows, optimization measures will be derived for the power take-off gearbox and will be evaluated based on the effect on the airborne sound pressure level.

Based on the critical transfer paths identified in figure 9, the optimization measure has been derived to increase the mass of the housing at the bearing support of the output shaft by increasing the thickness of the support, see figure 10, left. This adds a mass of 2,5% to the housing's original weight. The sum levels of the sound pressure level at the receiver point are shown in figure 10 on the right. While a significant reduction of 5 dB lower peak values can be achieved in the rpm range of up to 1200 rpm, only little optimization is to be seen at higher rpm values, as the output shaft bearings are not the critical paths at this point. Therefore, in a second optimization, additional mass was added to the cylinder bearing of the intermediate shaft as well. It can clearly be seen how especially at rotational speeds higher than 1200 rpm a significant decrease in peak values of 6 dB can be achieved.

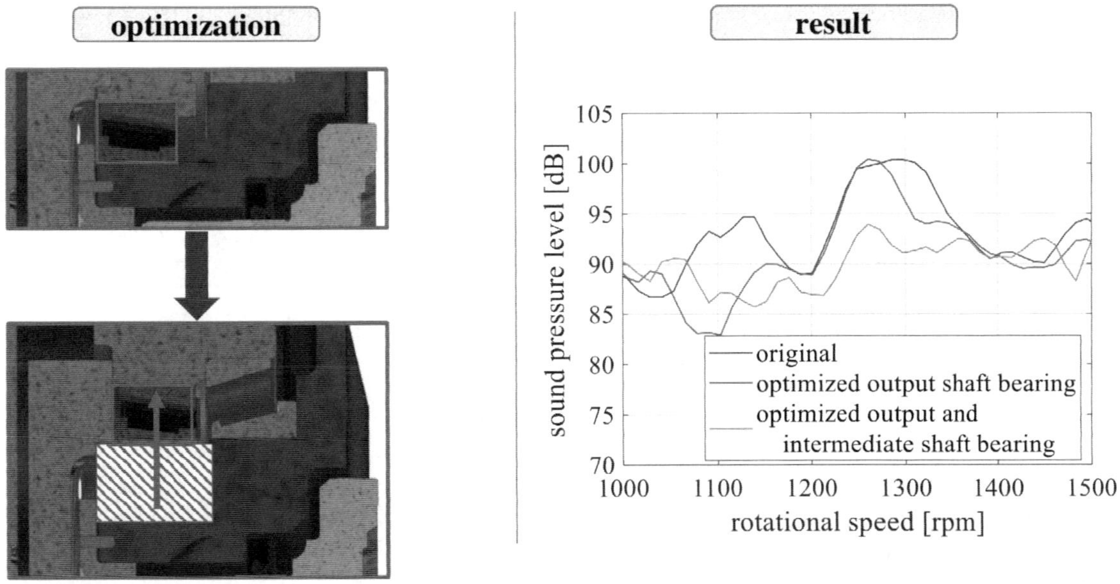

Figure 10. Optimization measure (left) and result (right) based on the transfer path analysis

This shows how with using TPA on numerical models a target-oriented optimization of individual peaks can be carried out, as transfer paths for every operational point can be identified and compared. By closing the gap of TPA towards airborne sound in the numerical domain, a direct evaluation with the most dominant NVH quantity at relevant receiver points as the driver's ear can be carried out for the first time.

5. Summary and Outlook

In this work, a method has been proposed to use transfer path analysis (TPA) to identify critical transfer paths to airborne sound in an elastic multi body system simulation (eMBS). It incorporates a newly developed toolchain which overcomes the constraint of exporting time-domain velocities of the eMBS model, which doesn't allow conclusion to the critical transfer path. Instead, an additional model of the passive structure in the frequency domain allows for the interface forces to be exported from the system model and the frequency response function of the entire passive system (consisting of the passive structure and the radiation properties) to be calculated. Thus, the transfer path contributions to the airborne sound can be obtained. This allows for the target-oriented optimization of the NVH behavior on a virtual prototype, thereby reducing development time and cost. The method is demonstrated on a power take-off (PTO) gear box and yields a 6 dB decrease in the sound pressure level.

Further work an experimental TPA needs to be done with regard to rotational degrees of freedom. Existing methods for deducing interface forces need to be extended to include the information about torques and moments acting on the interface.

References

[1] van der Seijs M V, Klerk D de and Rixen D J 2015 General framework for transfer path analysis *Mech. Syst. Signal Process.* **68-69** 217–44

[2] van der Seijs M V, Pasma E, de Klerk D and Rixen D 2014 A robust Transfer Path Analysis method for steering gear vibrations on a test bench *Proc. of ISMA 2014 International Conference on Noise and Vibration Engineering (Leuven)* (Heverlee: Sas P) pp 4027-40

[3] Plunt J 2005 Finding and fixing vehicle NVH problems with transfer path analysis *Sound and Vibration* **39** 12–6
[4] Diez-Ibarbia A, Battarra M, Palenzuela J, Cervantes G, Walsh S, De-la-Cruz M, Theodossiades S and Gagliardini L 2017 Comparison between transfer path analysis methods on an electric vehicle *Applied Acoustic* **118** 83–101
[5] Dom S, Geluk T, Janssens K and van der Auweraer H 2014 Transfer Path Analysis: Accurate Load Prediction beyond the Traditional Mount Stiffness and Matrix Inversion Methods *SAE Technical Papers* (Florianópolis, November 4th – 5th, 2014) (São Paulo: SAE Brasil)
[6] Oktav A, Yılmaz Ç and Anlaş G 2017 Transfer path analysis: Current practice, trade-offs and consideration of damping *Mech. Syst. Signal Process.* **85** 760–72
[7] Wegerhoff M, Sottek R and Amberg J 2018 Numerical in-situ TPA of an elastic multibody dynamic model of a tractor gearbox *Aachen Acoustics Colloquium* (Aachen, November 26th – 28th, 2018) pp 177–87
[8] Pasch G, Jacobs G and Berroth J 2021 Numerical Transfer Path Analysis for NVH System-Simulation Models *IOP Conference Series, Material Science and Engineering*
[9] Acri A, Offner G, Nijman E and Corradi R 2015 Numerical transfer path analysis for the assessment of air-borne nois in internal combustion engines *Proc. Of ICoEV 2015 International Conference on Engineering Vibrations* (Lubljiana, September 7th – 10th, 2015) pp 384–93
[10] Drichel P, Jaeger M, Müller-Giebeler M, Hameyer K, Jacobs G and Vorländer M 2019 Erweiterung NVH Simulationsmodell. Erweiterung der Simulationsmöglichkeiten für maschinenakustische Untersuchungen an E-Motive-Antrieben im Kontext zur Fahrzeugstruktur (Heft Nr. 1730) (Forschungsvereinigung Antriebstechnik)
[11] Bampton M C, Craig JR and Roy R 1968 Coupling of substructures for dynamic analyses *AIAA Journal* **6** 1313–9
[12] Gerard F, Tournour M, El Masri N, Cremers L, Felice M and Selmane A 2002 Acoustic transfer vectors for numerical modeling of engine noise *Sound and Vibration* **36** 20–25
[13] Weber C and Banaschek K 1953 Formänderung und Profilrücknahme bei gerad- und schrägverzahnten Rädern *Schriftenreihe Antriebstechnik* **11** 1–17

Fault detection for sliding bearings using acoustic emission signals and machine learning methods

F. König[1], G. Jacobs[1], A. Stratmann[1], D. Cornel[1]

[1] RWTH Aachen University, Institute for Machine Elements and Systems Engineering, Schinkelstrasse 10, 52062 Aachen, Germany

Florian.koenig@imse.rwth-aachen.de

Abstract. Driven by the potential application of sliding bearings under wear- and fatigue-critical operating conditions, i.e. in planetary gearboxes for wind turbines or automotive engines with start-stop systems, the reliability and lifetime prognosis of heavy loaded sliding bearings under low rotational speeds is an emerging field of research. The application of machine learning (ML) offers a great potential for all kinds of engineering applications when physical models are not feasible due to their complexity. This study showcases the application of ML to wear and fatigue fault detection and lifetime prognosis for sliding bearings using acoustic emission signals.

1. Introduction

Changing operating strategies such as start-stop systems for internal combustion engines and new sliding bearing applications such as gearboxes in wind turbines lead to an increased proportion of sliding bearing operation under mixed friction conditions, which limits the service life of a sliding bearing. Furthermore, the continuous trend of downsizing leads to higher loads, which potentially increases the risk of fatigue in the sliding layer of a bearing [1–5]. In many sliding bearing applications, the wear and fatigue reliability of sliding bearings is a potential bottleneck for the design [6,7]. As shown in Figure 1, the study showcased that abrasive and adhesive wear as well as fatigue are the main causes (~ 80 %) of bearing failure [8].

Wear occurs for instance during starting and stopping procedures [9] or in scenarios with low shaft speeds in combination with high loads [10]. Depending on the system's operating conditions, abrasive wear (two-body abrasion), adhesive wear or fatigue may occur that limit the remaining useful life (RUL) [8,11].

Additionally, the RUL is influenced by abrasive wear caused by solid particle contamination in the oil (three-body abrasion) [12–15]. The RUL prediction of sliding bearings subjected to wear critical operating conditions is possible with physical models and empirical wear laws if the underlying wear mechanism is known. However, a combination of the aforementioned wear mechanisms is often observed which hinders the direct use of such sophisticated approaches outside the laboratory. The important task is to detect the actual fault that determines the RUL [16,17].

For an estimation of the fatigue lifetime, failure hypotheses can be elaborated. In the case of a fatigue failure, a broad distribution of the expected lifetime can be expected, which makes the fatigue lifetime prediction inaccurate.

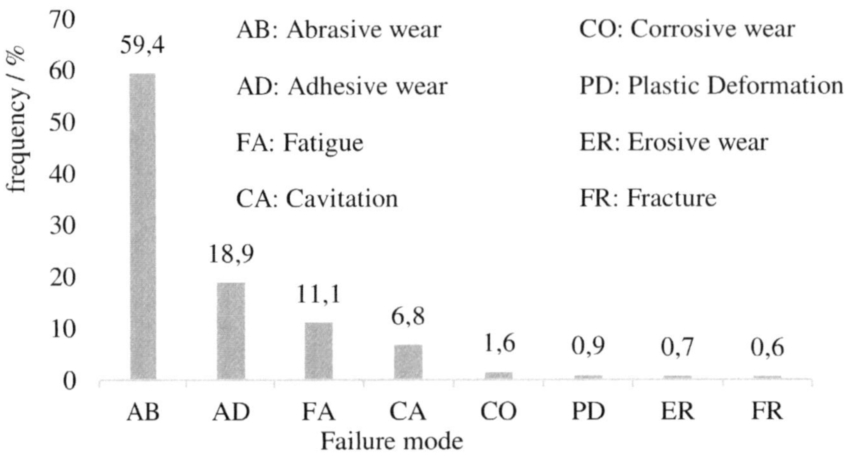

Figure 1. Frequency of failure causes of combustion engine sliding bearings, case study on 616 sliding bearings [8]

Due to the multi-variate, unpredictable failure of sliding bearings, the condition monitoring for sliding bearings is an important aspect. For this purpose, multiple sensor signals can be used, which are juxtaposed in Figure 2 with respect to early detection capabilities.

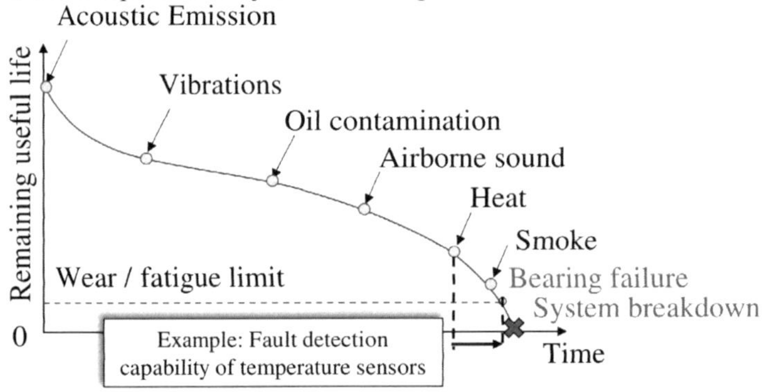

Figure 2. Ranking of sensor signals regarding their early detection of damage [18]

In recent years, Acoustic Emission signals (AE) have been identified to allow an early detection of faults that would cause a bearing failure. AE is the phenomenon of radiation of acoustic (elastic) waves in solids that occurs when a material undergoes irreversible changes in its internal structure, for example as a result of crack initiation/propagation or plastic deformation due to temperature gradients or external mechanical forces. AE signals can be measured with a high-frequency transducer as shown in Figure 3.

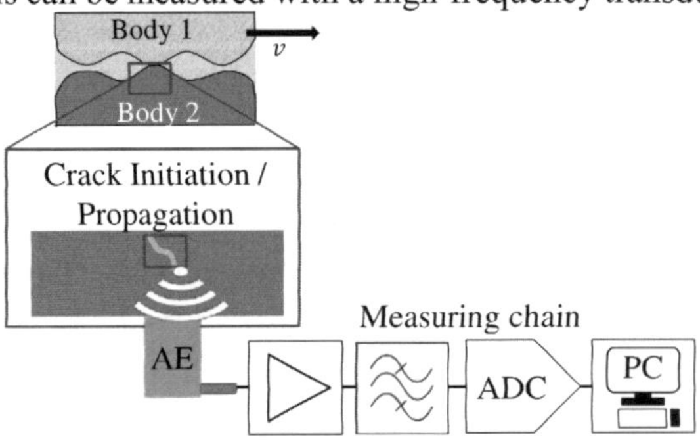

Figure 3. Description of AE measuring system. The AE signal is amplified, bandpass filtered and then undergoes analog-digital conversion. Then the digitized AE signal can be analysed.

Previous studies showed, the correlation between AE signals and different wear and fatigue mechanisms as depicted in Figure 4.

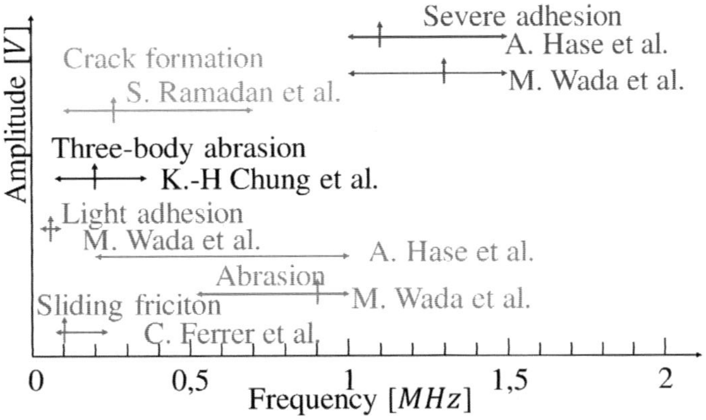

Figure 4. Relation between AE frequency and wear, redrawn from [19,20]

However, the AE data post-processing and analytical approach of extracting useful performance indicators from AE signals is not feasible for human kind. Characteristic frequencies overlap for multiple failure modes and the AE amplitude, viz. signal intensity, strongly depends on the transfer path between the fault source and the sensor position.

The recent advancements in the domain of machine learning (ML) make the development of advanced monitoring systems with ML methods for online wear and fatigue prediction more prominent. With ML, the potential AE signal patterns can be detected without human bias.

The ML problem for online wear and fatigue prediction can be defined in a similar way as the standard procedure for condition monitoring and diagnostics of machines according to ISO 13379 [21]. The consecutive tasks of the machine diagnosis can be translated into ML tasks, which is shown in Figure 5.

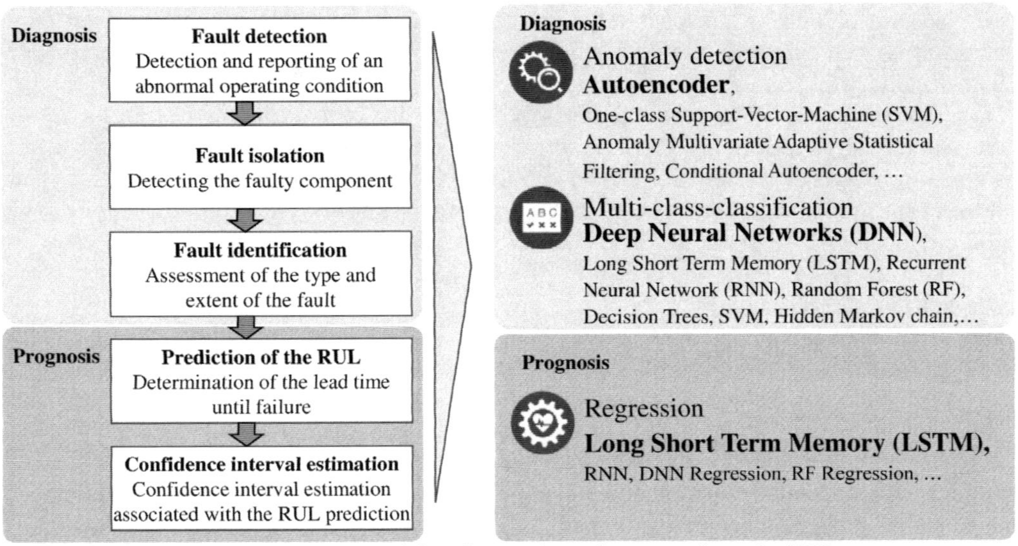

Figure 5. Typical ML methods used for diagnosis and prognosis tasks [22]

From this overview, it can be concluded that supervised learning ML methods on artificial neural networks (ANN) are commonly used for diagnosis and prognosis, due to their accuracy, robustness and speed [23]. In this study, an approach using ML algorithms for wear- and fatigue related failure diagnosis is presented. Here, abnormal conditions in sliding bearings were detected utilizing an autoencoder, before multi-class classification with a deep neural networks (DNN) was employed identify the present fault. Two different setups were used to provoke multi-variate wear failures as well as sliding layer fatigue failure. In the case of wear, a gradual degradation can be expected. This gradual degradation shall be modelled and thereby monitored with regression modelling utilizing long short-term memory ANNs (LSTM).

2. Materials and methods

In this study, we recapitulate a number of research papers that deal with the wear monitoring and diagnostics and prognostics with ML methods [15,24,25]. Furthermore, we apply a ML-based condition monitoring to the fatigue monitoring of dynamically loaded sliding bearings.

2.1. Experiments

Multi-variate wear failures were provoked by variation of the operational conditions of the test rig. An operation of the bearing under high loads and low speeds was used to generate two-body abrasive wear. For the simulation of adhesive wear events, the oil pump was shut off. In order to achieve three-body-abrasive wear, the oil was contaminated with solid particles (norm dust). A detailed description of the operational conditions can be found in [24].

For the fatigue tests, the bearings were subjected to multiple million oscillating-load cycles at a frequency of f_{Load} = 49 Hz and a stationary speed. The dynamic load was applied in multiple steps until a final load level was reached.

2.2. Specimens

The wear experiments were performed with a turned 30 mm diameter, 15 mm width bronze sliding bearing CuSn12NiC-GCB with a roughness of Ra 1 µm (Rq 1.5 µm, Rz 4 µm). The material of the shaft sleeve was hardened steel AISI 52100 (100Cr6) with 62 HRC. In accordance to automotive crankshaft journals, the roughness was about Ra 0.25 µm (Rq 0.45 µm, Rz 1 µm) [ISO/CD 27507]. The relative clearance was 1.67 ‰. the bearings were lubricated with the additive-free mineral oil "FVA 2" with a viscosity grade of ISO VG 32 and a density of 850 $kg \cdot m^{-3}$. The kinematic viscosities at 40 °C and 100 °C were 32 $mm^2 \cdot s^{-1}$ and 5.35 $mm^2 \cdot s^{-1}$, respectively. For the purpose of comparability, the material-lubricant combination was chosen similar to previous studies on sliding bearing behaviour. All specimens were cleaned with acetone prior testing.

The fatigue experiments were performed with centrifugally casted sliding bearings (Gleitlager-technik Weißbacher, Alpen, Germany) with 2 mm thick babbitt sliding layer made of SnSb8Cu4. The shaft was made from temper-hardened 42CrMo4 with and hardness of 56 HRC. The bearing diameter and width were 70.01 mm and 28 mm. The shaft diameter was manufactured to 69.96 mm, which results in a relative clearance of 2 ‰. The shaft and bearing were cleaned with Acetone prior to each test. In this study, a commercial turbine oil (Teresstic™ T 100, Exxon Mobil) was used. The specimens are exemplarily shown in Figure 6.

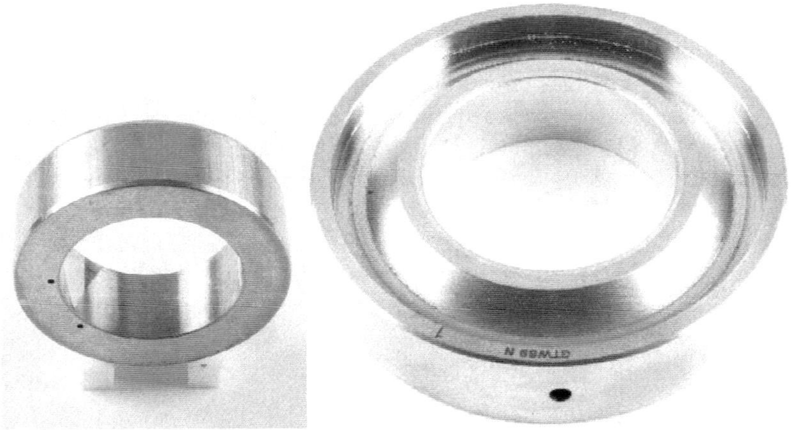

Figure 6. Bearing specimens used for wear (left) and fatigue tests (right)

2.3. Acoustic Emission (AE) technique

During the tests, AE measurements were conducted with a NANO30-sensor (MISTRAS, Princeton Jct, NJ, USA). The sensor has a resonance frequency of 300 kHz and a bandwidth of 125 kHz – 750 kHz. To ensure a coupling between the sensor head and the bearing housing, thread seal paste (Neo-Fermit Universal 2000/1, Fermit, Vettelschoß, Germany) was used. Before each experiment, the coupling between sensor head and housing was verified with the procedure according to DIN EN 13477-2:2010 or so-called Hsu-Nielson approach, where the AE response of pencil-lead breaks on the bearing housing was observed.

2.4. Machine learning algorithms

2.4.1. Anomaly detection

For the purpose of anomaly detection, AE features (AE RMS, Standard deviation, AE energy, number of counts) were processed from the original AE data (transient volt-age signal). The AE features of a sliding bearing operating under 'normal' hydrodynamic conditions are used as a reference state for the anomaly detection. The anomaly detection is performed with an autoencoder, which is a special type of ANN. In this ANN, the input AE features represent the input as well as the output. In a training phase with the hydrodynamic conditions, the required node weights of this ANN were acquired. If the trained ANN is supplied with other data from 'normal' hydrodynamic operation, the input and output values should still be comparable. In contrast, an anomaly is present when a large difference between input and output values is observed.

2.4.2. Multi-class failure classification

Following the fault detection on the basis of an autoencoder, the underlying fault has to be identified Therefore, these anomalies were classified into different classes of lubrication regimes and wear mechanisms / fatigue stages utilizing a DNN. GoogLeNet, the DNN in this study, is a pretrained convolutional neural network that is 22 layers deep commonly used for classification and detection of images. Continuous wavelet transformations (CWT) of the positive AE signals were used as input. The use of CWT can be seen in contrast to existing solutions, which focus on the extraction of statistical features from the time domain signal as well as time-frequency domain signal. The dependency on the feature extraction step limits the potential of traditional machine learning methods and imposes a strong inductive bias in the learning process The CWT reduced the feature extraction and analysis efforts to detect and classify AE events.

2.4.3. *Wear regression*

For the purpose of wear regression prognosis, the ML approach has to predict a sequence. This can be seen in contrast to the previous problem, where scalar values are feed into an ANN to obtain one or more scalar value(s). The data-driven wear prognosis of a component is commonly performed with a long short-term memory (LSTM), which is a special time of artificial recurrent neural network (RNN). In an LSTM, the output of previous time steps can be used as an input for the next time step. In contrast to conventional RNN, a LSTM can learn long-time dependencies in an efficient way. In this study, the wear volume was predicted with long short-term memory (LSTM) than can be used for transient sensor data. The LSTM was trained and validated with experimental and simulation results.

3. Results and discussion

3.1. *Anomaly detection*

The AE events from the wear and fatigue experiments were subjected to anomaly detection. Figure 7 shows an example of these anomalies and the hydrodynamic state or in other words, the abnormal and normal AE hits.

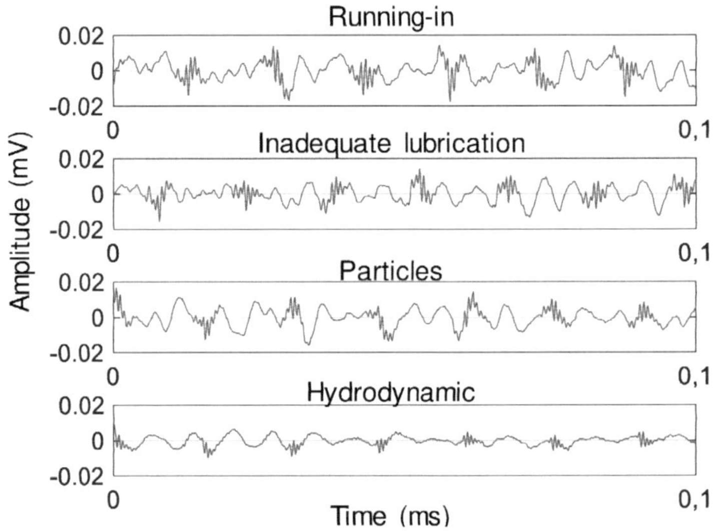

Figure 7. Anomaly detection in wear experiments [24]

In the transient AE data of the anomalies, no obvious difference can be overserved that could be used for classification. Furthermore, the AE signal shows a locality in time and frequency domain, which means that the commonly employed statistical feature extraction in time or frequency domain would lead to a loss of transient information. In contrast, this information is still present in the CWT of the raw AE hits, which elucidates the potential difference in the AE signals for different wear mechanisms. The CWT derived from an AE hit associated with hydrodynamic operation in shown in Figure 8a. In that case, recurrent peaks occur at a frequency of 130 kHz, which can be associated with sliding friction . In the case of running-in (Figure 8b), the frequency of the AE signal is spread between frequencies of 40 kHz and 600 kHz, which agrees with previous studies.

A similar observation can be made in the cases of oil starvation (Figure 8c) and particle contamination (Figure 8d). However, the signals show a rather similar pattern with a frequency range of 40 kHz up to 700 kHz, consequently an attribution of an unknown AE hit to one of the three mechanisms is not possible for a human being.

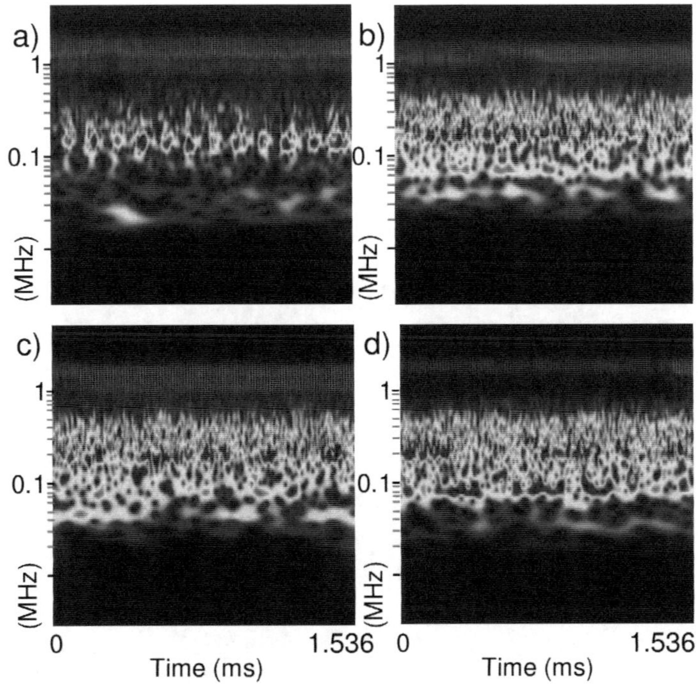

Figure 8. CWT of the AE hit signals from different tests a) Hydrodynamic operation b) Running-in, c) Oil starvation, d) Particle contamination. [24]

In a similar fashion, the fatigue monitoring was performed. Examples of the CWT of different stages of the fatigue experiment are shown in Figure 9. Here, it can be seen that the overall intensity of the AE signal is increasing up to the point of imminent fatigue failure. Also, it can be observed that the frequencies of a fatigue failure are assumed in a range of 50 kHz up to 500 kHz, which agrees quite well with the state-of-the-art knowledge as summarized in Figure 4.

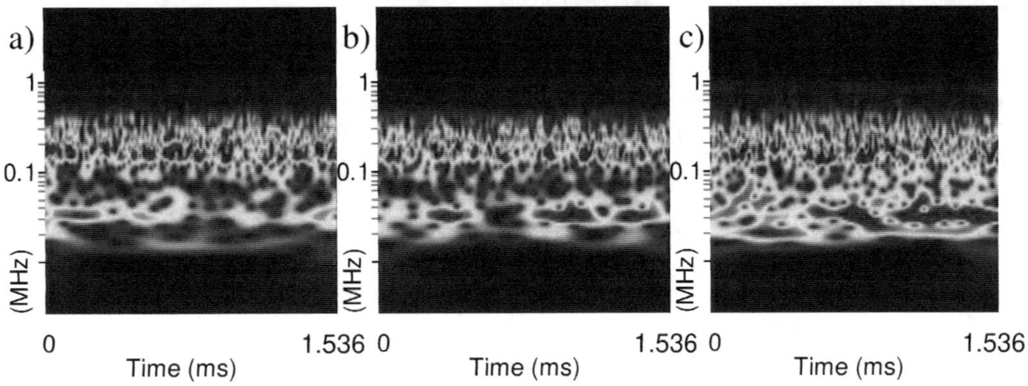

Figure 9. CWT of the AE hit signals from different stages of the fatigue experiment (a) Hydrodynamic operation (b) Incubation period, (c) Imminent fatigue failure.

3.2. Multi-class classification

3.2.1. Wear modes

The DNN approach was employed for classification of wear critical AE events. The results are summarized in a confusion matrix in Figure 10.

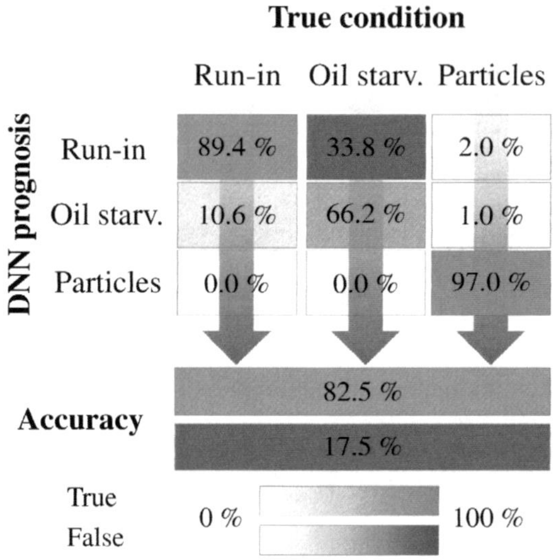

Figure 10. Confusion matrix of the wear mode classification problem [24]

As can be seen, the trained DNN correctly predicted running-in with a high accuracy of 89.4 %. However, only 66.2 % of the AE hits that originated from the experiments on oil starvation were correctly classified by DNN. The remaining 33.8 % were incorrectly classified as running-in, which can be explained by the gradual drop of the oil pressure and the oil-inlet flow. During this stage of operation, the contact conditions are closely related to steady-state mixed-friction conditions that are observed during running-in. Consequently, the DNN yet has a limited accuracy for the classification of oil starvation events. In the case of particle contamination, a high accuracy above 97 % was demonstrated. The overall accuracy was therefore at 82.5 %.

3.2.2. Fatigue failure modes

A similar approach was adopted to the fatigue testing, which is summarized in Figure 11. In this case, however, the normal state was defined of one additional anomaly to ensure that the possibility of a false classification is reduced.

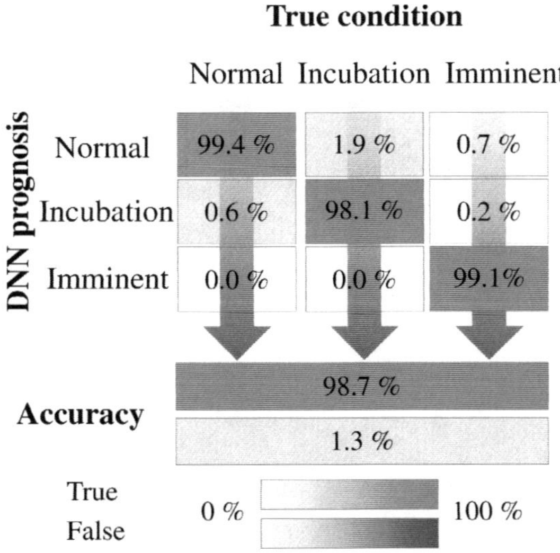

Figure 11. Confusion matrix of the fatigue failure classification problem

As can be seen, the trained DNN correctly predicted the normal state with a high accuracy of 99.4 %. 98.1 % of the incubation periods were correctly classified by the DNN. The wrong classifications were exclusively part of the normal state, which shows that the transition between incubation period and imminent failure was successful. In the case of imminent failure, the accuracy was 99.1 % with a short percentage incorrectly classified as incubation as well as normal state. Thus, the overall accuracy was 98.7 %, which clearly demonstrates the combination of AE and DNNs for fatigue monitoring. It should be noted that the imminent failure was predicted roughly 22 minutes before the actual thermally induced failure due to the breakdown of the hydrodynamic film. Consequently, a fatigue induced system failure can be prohibited.

3.2.3. Wear regression

In the case of wear monitoring, the gradual degradation of a bearing has to be considered. Wear has been predicted with a long short-term memory (LSTM) than can be used for transient sensor data. The method and the results were previously published here [25]. The LSTM was trained and validated with experimental and simulation results from [10]. A good agreement between experimental results and ML based wear prediction was demonstrated for Running-in Type A ($v= 0{,}41$ m/s, $\bar{p} = 5$ MPa, $\vartheta = 40$ °C), which corresponds to mild wearing-in, see. Figure 12. For Running-in Type B ($v= 0{,}41$ m/s, $\bar{p} = 8$ MPa, $\vartheta = 64$ °C), the overall scatter of the wear volume after an experiment was higher. However, as further experiments under similar operational conditions show, the general trend is clearly met for both operational conditions [26].

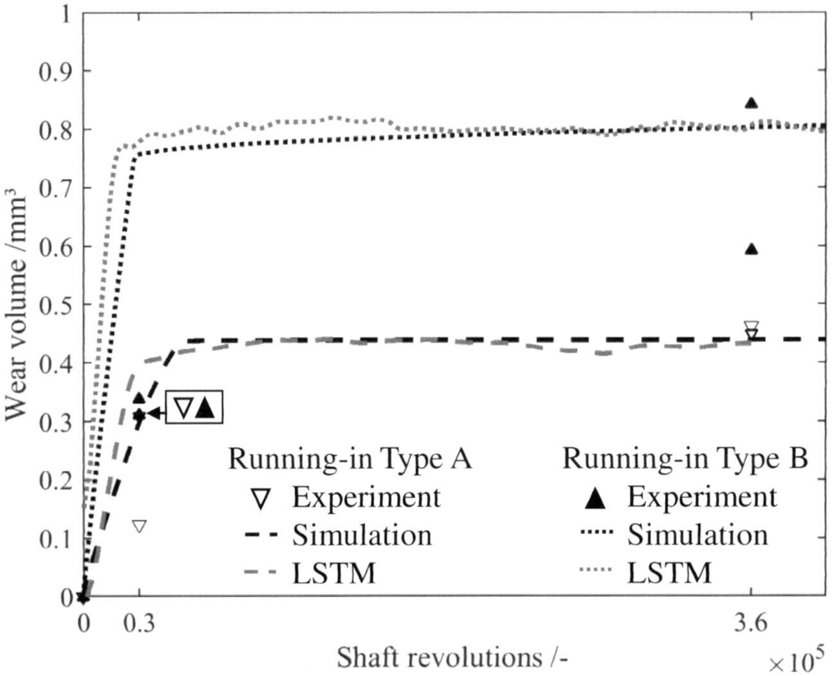

Figure 12. Results from the experiments, physical simulations and LSTM-based wear monitoring for running-in under mixed-friction conditions [25]

In both cases, the wear volume shows a sharp increase within the first thousand shaft revolutions, which can be attributed to mixed friction conditions. With increasing cycle count, wearing-in leads to reduced asperity contact pressures and reduced sensor activity. Thus, the LSTM predicts a lower wear-rate throughout the remaining revolutions. As an outlook, this method should be elaborated for transient operating conditions, such as start-stop operation [9].

4. Conclusions

In many sliding bearing applications, the wear and fatigue reliability of sliding bearings is a potential bottleneck for the systems' lifetime. To ensure a reliable operation, the degradation process up to the imminent bearing failure has to be monitored. For this purpose, the use of acoustic emission (AE) condition monitoring has proven to be advantageous due to its early warning capabilities. However, due to its complexity, the physical interpretability of AE is still a show-stopper.

In this study, an approach using Machine Learning (ML) algorithms for wear- and fatigue related failure diagnosis was presented. On the basis of these methods, the following conclusions can be drawn.

- Autoencoders can be used to detect anomalies from the normal, viz. hydrodynamic operational state, of a sliding bearing. The use of autoencoders allows to reduce that amount of data that is considered in further steps of the fault diagnosis.
- DNN are capable of multi-class classification with a limited amount of feature extraction and analysis. In context of AE signals, which show a strong locality in time and frequency domain, this leads to less systematic errors due to feature extraction.
- Wear regression modelling requires the knowledge of previous time steps, which is a typical task of sequence modelling. For this purpose, LSTM networks showed to be suitable.

Acknowledgements
Funded by the Excellence Initiative of the German federal and state governments. The present work is supported by funding from the Deutsche Forschungsgemeinschaft (DFG, projects: GRK 1856 „Integrated Energy Supply Modules for Roadbound E-Mobility", project no. 233292516 "Failure criteria for the prediction of compound journal bearing fatigue"). Furthermore, the authors wish to thank Mrs. Britt Westheim, Mr. Gerwin Pasch, Mr. Achraf Ouald Chaib and Mr. Matthias Kiesewetter and Dr. Christopher Sous for their support.

References

[1] Dong Q, Yin Z, Li H, Gao G, Zhong N, Chen Y. Simulation and experimental verification of fatigue strength evaluation of journal bearing bush. Engineering Failure Analysis 2020;109:104275. https://doi.org/10.1016/j.engfailanal.2019.104275.
[2] He Y, Zhao Z, Luo T, Lu X, Luo J. Failure analysis of journal bearing used in turboset of a power plant. Materials & Design (1980-2015) 2013;52:923–31. https://doi.org/10.1016/j.matdes.2013.06.027.
[3] Sous C. Methode zur dauerfesten Auslegung hydrodynamischer Gleitlager: = Method for the design of hydrodynamic sliding bearings in terms of fatigue strength. 1st ed. Aachen: Verlagsgruppe Mainz GmbH; 2018.
[4] Wünsch H, Sous C, Broeckmann C. Multiaxial life prediction of hydrodynamic plain bearings. MATEC Web Conf. 2018;165(1):16001. https://doi.org/10.1051/matecconf/201816516001.
[5] Sous C, Wünsch H, Jacobs G, Broeckmann C. Prediction of fatigue limit of journal bearings considering a multi-axial stress state. Ind Lubrication and Tribology 2016;68(3):430–8. https://doi.org/10.1108/ILT-08-2015-0119.
[6] Ligier J-L, Noel B. Friction Reduction and Reliability for Engines Bearings. Lubricants 2015;3(3):569–96. https://doi.org/10.3390/lubricants3030569.
[7] Meyer T. Journal bearings in wind turbine gear units – validation – track record – extended investigations for reliability. Proceedings to the Conference for Wind Power Drives, Aachen 2017.
[8] Vencl A, Rac A. Diesel engine crankshaft journal bearings failures: Case study. Engineering Failure Analysis 2014;44:217–28. https://doi.org/10.1016/j.engfailanal.2014.05.014.
[9] König F, Sous C, Jacobs G. Numerical prediction of the frictional losses in sliding bearings during start-stop operation. Friction 2020;71(12):22. https://doi.org/10.1007/s40544-020-0417-9.
[10] König F, Ouald Chaib A, Jacobs G, Sous C. A multiscale-approach for wear prediction in journal bearing systems – from wearing-in towards steady-state wear. Wear 2019;426-427:1203–11. https://doi.org/10.1016/j.wear.2019.01.036.
[11] Poddar S, Tandon, N. Study of Oil Starvation in Journal Bearing Using Acoustic Emission and Vibration Measurement Techniques. Journal of Tribology 2020;142(12):154. https://doi.org/10.1115/1.4047455.
[12] Hernández-Peña A, Farfan-Cabrera LI, Gallardo-Hernández EA. Investigation of the wear of engine journal bearings approaching severe lubrication conditions using a microabrasion tester. Proceedings of the Institution of Mechanical Engineers, Part J: Journal of Engineering Tribology 2019;233(12):1939–49. https://doi.org/10.1177/1350650119852225.
[13] Tandon, N. Detection of particle contamination in journal bearing using acoustic emission and vibration monitoring techniques. Tribology International 2019;134:154–64. https://doi.org/10.1016/j.triboint.2019.01.050.
[14] Vencl A, Rac A. Diesel engine crankshaft journal bearings failures: Case study. Engineering Failure Analysis 2014;44:217–28. https://doi.org/10.1016/j.engfailanal.2014.05.014.
[15] König F, Jacobs G, Ouald Chaib A, Sous C. Machine Learning based condition monitoring and anomaly detection for wear lifetime prediction of journal bearing drivetrains. TAE: 22nd International Colloquium Tribology - Industrial and Automotive Lubrication 2020.

[16] Nikas GK. A state-of-the-art review on the effects of particulate contamination and related topics in machine-element contacts. Proceedings of the Institution of Mechanical Engineers, Part J: Journal of Engineering Tribology 2010;224(5):453–79. https://doi.org/10.1243/13506501JET752.
[17] Qiao W, Lu D. A Survey on Wind Turbine Condition Monitoring and Fault Diagnosis—Part II: Signals and Signal Processing Methods. IEEE Trans. Ind. Electron. 2015;62(10):6546–57. https://doi.org/10.1109/TIE.2015.2422394.
[18] Saufi SR, Ahmad ZAB, Leong MS, Lim MH. Challenges and Opportunities of Deep Learning Models for Machinery Fault Detection and Diagnosis: A Review. IEEE Access 2019;7:122644–62. https://doi.org/10.1109/ACCESS.2019.2938227.
[19] Hase A, Mishina H, Wada M. Correlation between features of acoustic emission signals and mechanical wear mechanisms. Wear 2012;292-293:144–50. https://doi.org/10.1016/j.wear.2012.05.019.
[20] Hase A. Early Detection and Identification of Fatigue Damage in Thrust Ball Bearings by an Acoustic Emission Technique. Lubricants 2020;8(3):37. https://doi.org/10.3390/lubricants8030037.
[21] DIN ISO. Zustandsüberwachung und -diagnostik von Maschinen - Verfahren zur Dateninterpretation und Diagnostik –Teil 1: Allgemeine Anleitungen. 2012nd ed.(13379-1); 2012.
[22] Varshneya A, Pinto R. Machine Learning, Anomaly detection and the world of making Smarter, Faster and Better decisions. Tavant Technologies; 2019.
[23] Liu R, Yang B, Zio E, Chen X. Artificial intelligence for fault diagnosis of rotating machinery: A review. Mechanical Systems and Signal Processing 2018;108:33–47. https://doi.org/10.1016/j.ymssp.2018.02.016.
[24] König F, Sous C, Ouald Chaib A, Jacobs G. Machine Learning Based Anomaly Detection and Classification of Acoustic Emission Events for Wear Monitoring in Sliding Bearing Systems: accepted. Tribology International 2020:106811. https://doi.org/10.1016/j.triboint.2020.106811.
[25] König F, Marheineke J, Jacobs G, Sous C, Zuo MJ, Tian Z. Data-driven wear monitoring for sliding bearings using acoustic emission signals and long short-term memory neural networks. Wear 2021;3(3):203616. https://doi.org/10.1016/j.wear.2021.203616.
[26] König F. Prognose des Verschleißverhaltens ölgeschmierter Gleitlager im Mischreibungsbetrieb. Aachen: Mainz, G; 2020.

IOP Conf. Series: Materials Science and Engineering 1097 (2021) 012014 doi:10.1088/1757-899X/1097/1/012014

Tribological analysis of the plain bearings in an external gear pump

Lars Brinkmann[1], Stefan Kock[1], Jochen Lang[2] and Gunter Knoll[2]

[1]Bosch Rexroth AG, Dieselstraße 10, 90441 Nürnberg, Germany
[2]IST mbH, Schloss-Rahe-Str. 12, 52072 Aachen, Germany

lars.brinkmann@boschrexroth.de

Abstract. Tribological loads in the plain bearings of external gear pumps are influenced by system-specific parameters (e.g. elasticity of the components, micro-geometry and manufacturing tolerances) and the operating point (rotation speed, operating pressure and temperature of the pressure medium). Due to the pressure build-up in the tooth chambers, fluctuating radial forces are acting on the gear shafts and lead to complex deformations of the bearing bushings and the housing. These deformations have an influence on the tribological loads in the plain bearings. To decrease tribological loads and failure rates, it is possible to modify the plain bearings in terms of their micro-geometry (e.g. clearance and contours) or to take constructive measures on the overall system (e.g. on the bearing bushings). Therefore, it is necessary to understand the influences of external loads, elastic environment and micro-geometry on the tribological loads in the plain bearings. This paper describes the first steps to understand these interactions by building up a simulation model of a specific external gear pump using the MBS/EHD-tool FIRST developed by IST [1]. The focus is to evaluate the required modeling depth regarding the consideration of an elastic housing.

1. Motivation and Objective

External gear pumps are one of the most robust pump types for hydraulics in mobile machinery. Increasing demands for lower emissions and higher efficiency requires the continuous development of these pumps.

The most important components of these pumps are the bearing bushings. Based on their elastohydrodynamic characteristics in the mixed lubrication regime, these bearings define significant frictional losses and the width of important sealing gaps that limit the hydraulic pump efficiency. The detailed design of gear pump bushings did already mature over decades and thus further design optimization is not an easy task at all. To exploit additional potential, it is essential to develop a deeper understanding of the overall system. This means that the influences of external loads, elastic environment and micro-geometry on the tribological loads in the radial and axial plain bearings have to be well known.

Endurance tests cause high costs and are only useful to develop a deep system understanding to a limited extend, because they do not provide an insight into the effects in the pump. A simulation model in contrast makes it possible to investigate various influences on the tribological loads in the plain bearings at comparatively low costs.

The objective of this paper is to introduce the approach for the build-up of a simulation model which helps to investigate the stated influences qualitatively for a specific external gear pump manufactured by Bosch Rexroth. Previous work did not consider an elastic housing [2, 3]. Therefore, the focus is to investigate the influence of an elastic housing on the tribological loads in the plain bearings. Based on these investigations it is possible to evaluate the need for an elastic housing implemented in the model.

2. Loads in the analysed external gear pump

The investigations were made for an external gear pump manufactured by Bosch Rexroth with involute gear profile and single flank sealing. It is a pump with axial clearance compensation. This means that the axial sealing gap is regulated by the operating pressure. The investigated pump has a displacement of 16 cm^3 per rotation. A maximum permanent operating pressure of 250 bar and an intermittent, maximum 20 s lasting operating pressure of 280 bar is permitted. The permitted rotation speed depends on the viscosity of the pressure medium and on the operating pressure. It is between 500 rpm and 3000 rpm [4]. A commonly used pressure medium (e.g. in endurance tests) is the Shell Tellus MX46 [5].

External gear pumps with axial clearance compensation according to **Figure 1** mainly consist of two gear shafts, aluminium bearing bushings and an aluminium housing. One of the two gear shafts (the driving gear shaft) is directly connected to a power unit. The gear shafts are each supported by two radial and two axial plain bearings. The tooth chambers of the two meshing gears transport the fluid under increasing pressure (internal pressure field) from the suction side to the discharge side. Due to the pressure build-up in the tooth chambers, fluctuating radial forces are acting on the gear shafts [6]. These radial forces lead to the radial sealing between the teeth and the housing, see **figure 1 a)**. Even at high operating pressures, a sufficient axial sealing of the tooth chambers between the gear front side and the bearing bushings must be ensured. Therefore, the bearing bushings are pressurized with the operating pressure (external pressure compensation field). Due to this external pressure compensation field the bearing bushings are pressed against the gear wheels, see **figure 1 b)**. Because of the pulsation of the operating pressure, the resulting force of the external pressure compensation field pulsates too. The area of the external pressure compensation field is sealed against the suction side by a sealing ring.

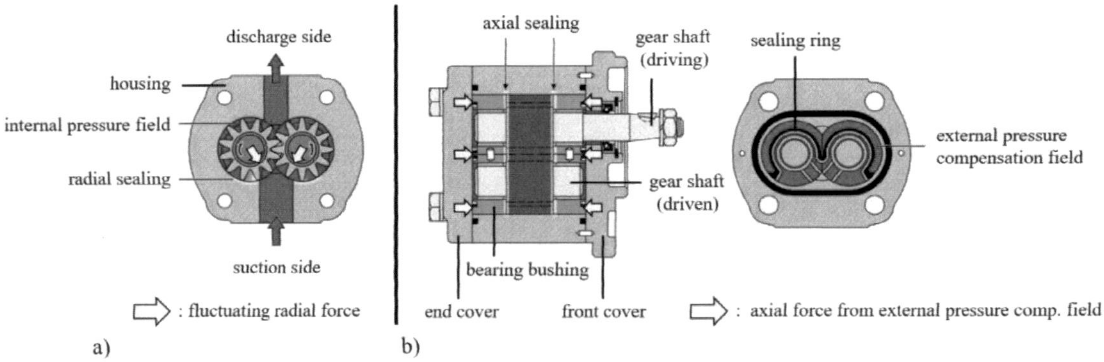

Figure 1. a) Radial forces and radial sealing b) Axial forces and axial sealing. [4]

3. Approach

As already stated, a simulation model in contrast to endurance tests offers the opportunity to develop a deeper system understanding. The focus of this paper is the investigation of the influence of an elastic housing. The MBS/EHD-tool FIRST provides a coupling between multi body dynamics and elastohydrodynamics and it allows to consider elastic components. Therefore it is a suitable tool to carry out this investigation.

The simulation model was built-up stepwise with increasing modeling depth and complexity. One advantage of this stepwise model build-up is the fact that the influence of different modeling approaches on the simulation results can be evaluated.

The following subsection gives an explanation of this stepwise model build-up. In further subsections the modeling approach of model stage 4 is described, because model stage 4 was used for the investigation related to this paper.

3.1. *Stepwise model build-up*

The simulation model was built-up stepwise using the MBS/EHD-tool FIRST. **Figure 2** shows this stepwise model build-up with an increasing modeling depth in four model stages. In model stage 1 the simulation model contains the driven gear shaft only. The driven gear shaft is supported in radial direction by two bearing bushings. All components are modeled as rigid or elastic. The radial forces from the tooth chamber pressures and from the meshing of the gears are applied to the gear shaft. Model stage 1 just considers radial forces and the radial bearings. In model stage 2 the internal and the external pressure compensation field was added. Therefore, modeling stage 2 also considers the axial bearings. In model stage 3 the driving gear shaft with its radial and axial bearings was added to the model. Model stage 4 contains additionally the housing. To evaluate the influence of an elastic housing, the housing is modeled as rigid and alternatively as elastic, while all other components are modeled as elastic. To reduce the complexity of the simulation model in this initial approach, both housing covers as well as their bolt connection were not modeled. The forces from the bolt connection between the housing and the covers are thereby applied to the housing. The effect of the housing covers (e.g. a stiffening effect) can be analyzed in subsequent investigations.

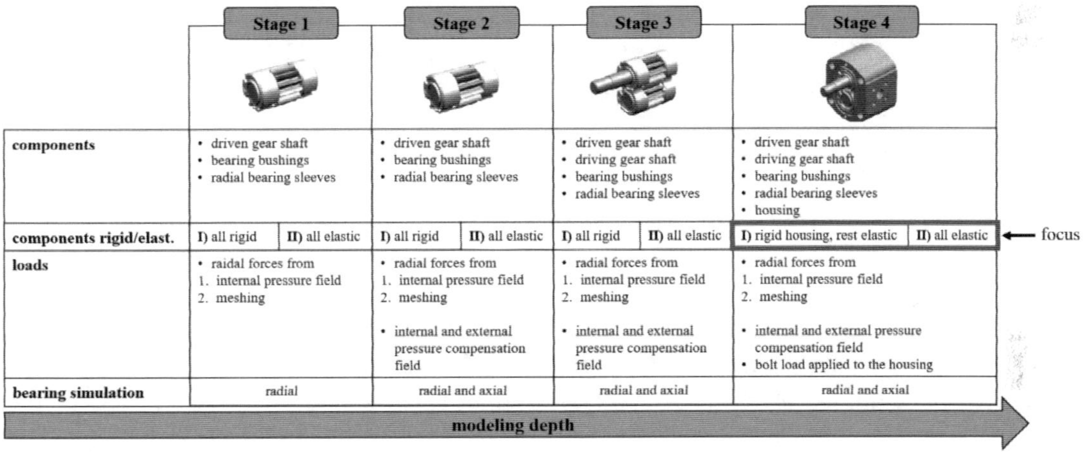

Figure 2. Stepwise model build-up in four stages.

3.2. *Forces on the gear shafts*

The radial forces on the gear shafts were determined with HYGESim [2, 3]. These radial forces are dependent on the rotation angle of the gear shafts and result from a superposition of two forces, see **figure 3**. One component results from pressure in the tooth chambers. The other component results from the meshing of the gears. The tooth contact is not in the focus of the investigations. Hence the meshing forces are applied to both gear shafts separately and the tooth contact is not modeled.

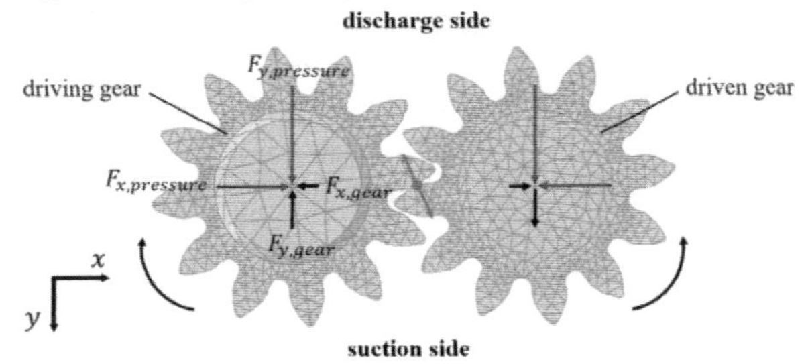

Figure 3. Radial forces acting on the gear shafts, qualitative.

The forces from the meshing and from the tooth chamber pressures add up in y-direction on the driven gear shaft. Consequently, the driven gear shaft takes a higher resulting radial force than the driving gear shaft.

3.3. Forces on the bearing bushings

Due to the bolting of the covers against the housing, the sealing of the external pressure compensation field is preloaded. The sealing ring and the covers are not modeled. The preload force is applied to the bearing bushing in the contact area of the sealing ring, see **figure 4 a)**. The operating pressure is applied to the area of the external pressure compensation field, see **figure 4 b)**. As a simplification, the external pressure compensation field is thereby assumed to be constant (no pulsation). The pulsation of the internal pressure field due to the pressure build-up in the tooth chambers is considered, see subsection 3.5. The contact area of the sealing ring belongs to the area of the external pressure compensation field because the sealing ring transfers the pressure force to the bushing. Areas of the bearing bushings on the gear side which are not part of the modeled axial bearing surface are pressurized with the operating pressure too, see **figure 4 c)**.

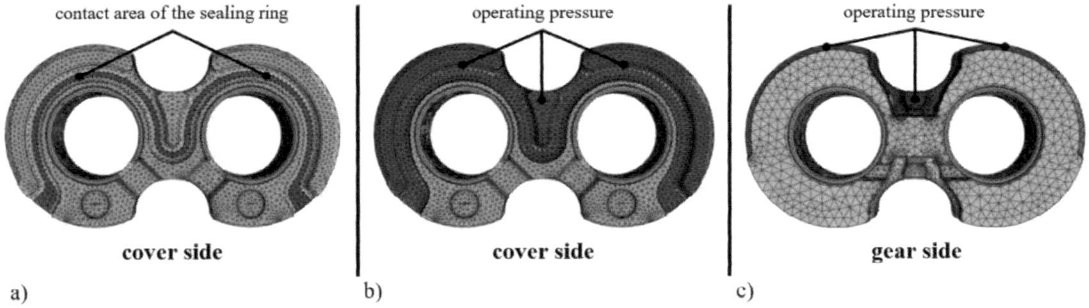

Figure 4. a) Contact area of the sealing ring b) Area of the external pressure compensation field c) Area with operating pressure on the gear side.

3.4. Forces on the housing

The operating pressure is applied to the marked surfaces of the housing, see **figure 5 a)**. Investigations have shown that contact stresses between the covers and the housing are greatest in the areas around the boreholes [7]. Therefore, the forces of the housing bolt connection are largely transferred from the covers to the housing around the boreholes. These forces were applied to the marked areas in **figure 5 b)**.

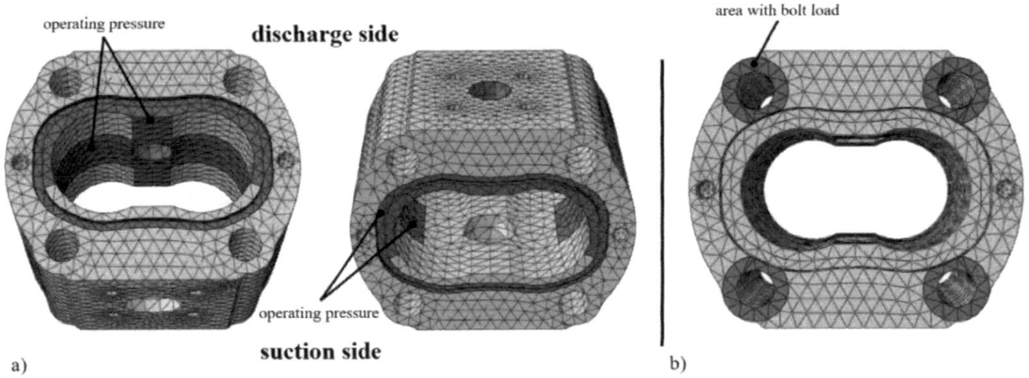

Figure 5. a) Surfaces of the housing with operating pressure b) Surfaces of the housing with forces from the bolt load.

The contact and the fit tolerance between the bearing bushings and the housing are considered. Furthermore, the flow of the pressure medium into the gap between bearing bushings and housing as well as hydrodynamic effects are considered. The bearing bushings and the housing are coupled by a calculation mesh on which the Reynolds equation is solved.

Figure 6 a) shows the contact surfaces on the bearing bushings and the housing. The radial forces from the pressure build-up in the tooth chambers are acting on the gear shafts in the direction of the suction side. The bearing bushings take these radial forces and consequently are pressed against the housing on the suction side. The size of the resulting contact area was evaluated in previous investigations at Bosch Rexroth. The pressure medium cannot flow into this contact area. Furthermore, there is suction pressure (p=0, overpressure above atmospheric pressure) on the suction side. Therefore, the contact area on the calculation mesh is bordered by a pressure boundary condition with p=0, see **figure 6 b)**. On the discharge side a gap between the housing and the bearing bushings arises (gap area). The pressure medium flows into this gap under operating pressure. Therefore, the gap area is bordered by a pressure boundary condition with the operating pressure. To allow hydrodynamic effects (e.g. squeeze effect), no pressure boundary conditions were set inside both areas.

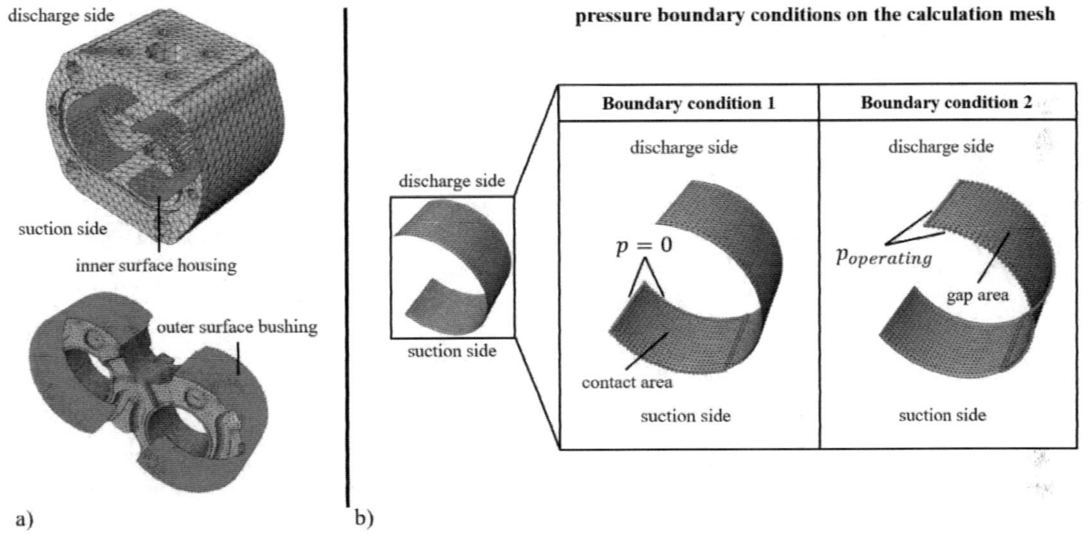

Figure 6. a) Surfaces on the housing and the bearing bushing b) Pressure boundary conditions on the calculation mesh.

3.5. Bearing simulation

The radial plain bearings consist of the bearing pins of the gear shafts and the bearing sleeves, see **figure 7 a)**. The solution of the Reynolds equation is carried out on a cylindrical calculation mesh. The pressure medium can flow into and out of the radial bearing due to grooves under atmospheric pressure. Therefore, a pressure boundary condition with p=0 is applied to the nodes of both edges of the calculation mesh.

The axial plain bearings consist of the gear front side and the axial sliding surface of the bearing bushings, see **figure 7 b)**. The pressures in the tooth chambers are dependent on the rotation angle ϑ of the gear shafts and were determined with HYGESim [2, 3]. The tooth chamber pressures are applied to the calculation mesh of the axial plain bearing in form of pressure boundary conditions. The node groups representing the tooth chambers are not variable. Additionally, all nodes of a certain tooth chamber receive the same pressure boundary condition at a certain rotation angle ϑ. Consequently, the sealing in the tooth contact cannot be considered.

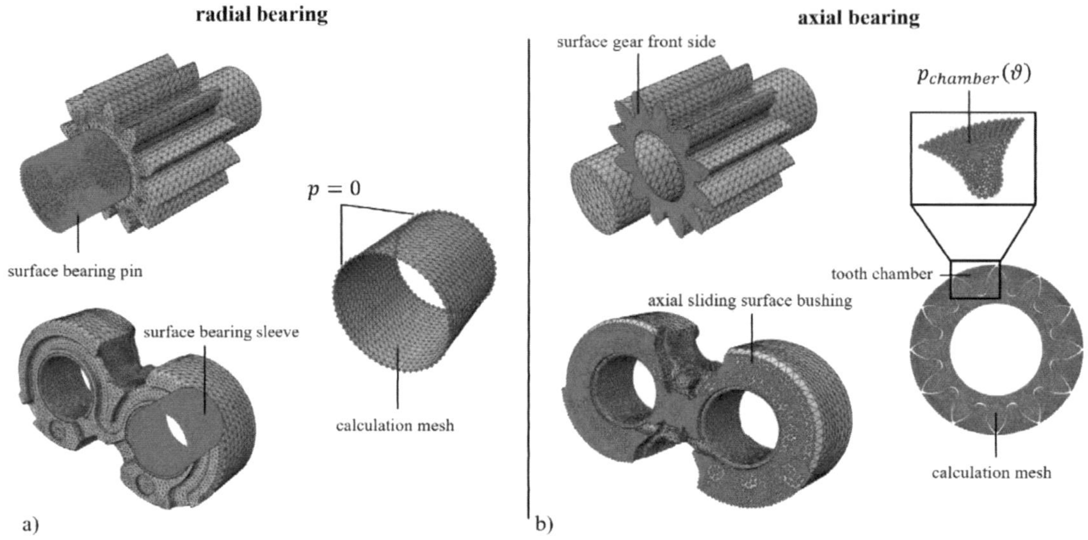

Figure 7. a) Modeling of the radial plain bearing b) Modeling of the axial plain bearing.

3.6. Assumptions and Simplifications

For the calculation of the hydrodynamic pressure build-up, the Reynolds equation is used. The viscosity of the pressure medium is assumed to be pressure independent. The viscosity of the pressure medium results from a temperature assumed to be constant in time and location independent. Furthermore, it is assumed that the pressure medium is free of particles and that there is no cavitation. Pressure and shear flow factors are not considered due to missing surface roughness measurements.

The rigid body contact is modeled according to Greenwood and Tripp. The roughness of the surfaces was selected within the manufacturing tolerances. Contours of the plain bearings (e.g. chamfers on the tooth edges) are not considered. However, these contours can be implemented in the model in subsequent studies. The calculation of wear is not included in the model.

The stated simplifications were made to reduce the complexity of the model in this initial approach. However, FIRST provides a huge amount of functions to overcome these simplifications if necessary.

4. Results

The focus of this paper is to evaluate the need for an elastic housing implemented in the model. Therefore, model stage 4 I (rigid housing, rest elastic) and model stage 4 II (all components elastic) are compared regarding two chosen comparison parameters.

One of this comparison parameters is the relative minimum gap width $h_{min,rel}$ in the radial and axial plain bearings. It is the minimum gap width h_{min} relative to the critical gap width h_{crit}, see **equation (1)**. The critical gap width is the gap width from which mixed friction occurs. The critical gap width is dependent on the roughness of the sliding surfaces. In FIRST, the gap width is defined as the distance between the base lines of the roughness profiles. When these base lines overlap, the gap width becomes negative. Therefore a negative gap width does not necessarily imply a penetration of the bodies.

$$h_{min,rel} = \frac{h_{min}}{h_{crit}} \cdot 100\% \tag{1}$$

The minimum gap width is directly related to wear in the bearing. Therefore it is a suitable parameter to investigate the tribological load in the bearing. However, the minimum gap width can occur at just one node of the calculation mesh. Consequently, the minimum gap width can be highly sensitive to variations on the model. Therefore, the resulting contact force in the plain bearings is chosen as a second comparison parameter.

The resulting contact force is the contact pressure integrated over the bearing surface. It is the proportion of the external force on the bearing that the bearing takes due to rigid body contact. Because the resulting contact force is an integral quantity, it is not as sensitive as the minimum gap width.

The investigations were carried out for the basic conditions in **table 1**. As already mentioned, the driven gear shaft takes a higher resulting radial force than the driving gear shaft. Therefore, the simulation results are shown for the driven gear shaft. In the investigated operating point, a radial force up to 13.5 kN acts on the driven gear shaft. The axial bearing takes an axial force of around 7 kN.

Table 1. Basic conditions

operating point	3000 rpm, 280 bar
viscosity of the pressure medium	25 mPa·s (at 50 °C)
radial bearing clearance	4 ‰

4.1. Results for the radial plain bearing

Figure 8 compares model stage 4 I and 4 II regarding the relative minimum gap width and the resulting contact force in a radial plain bearing of the driven gear shaft for one full rotation. For both model stages the relative minimum gap width is less than 100 %, see **figure 8 a)**. Consequently, mixed friction occurs for both model stages. Furthermore it is visible that the housing elasticity has no significant influence on the relative minimum gap width in the radial plain bearing. The same applies for the resulting contact force, see **figure 8 b)**. The deflection of the gear shafts shows a greater influence on the tribological load in the radial plain bearings than the deformation of the housing. As already mentioned, the calculation of wear is not included in the model. Due to a high roughness of the coating of the bearing sleeves in new condition, this simplification affects the simulation results. However, the simulation results allow to compare the effects in model stage 4 I and 4 II qualitatively.

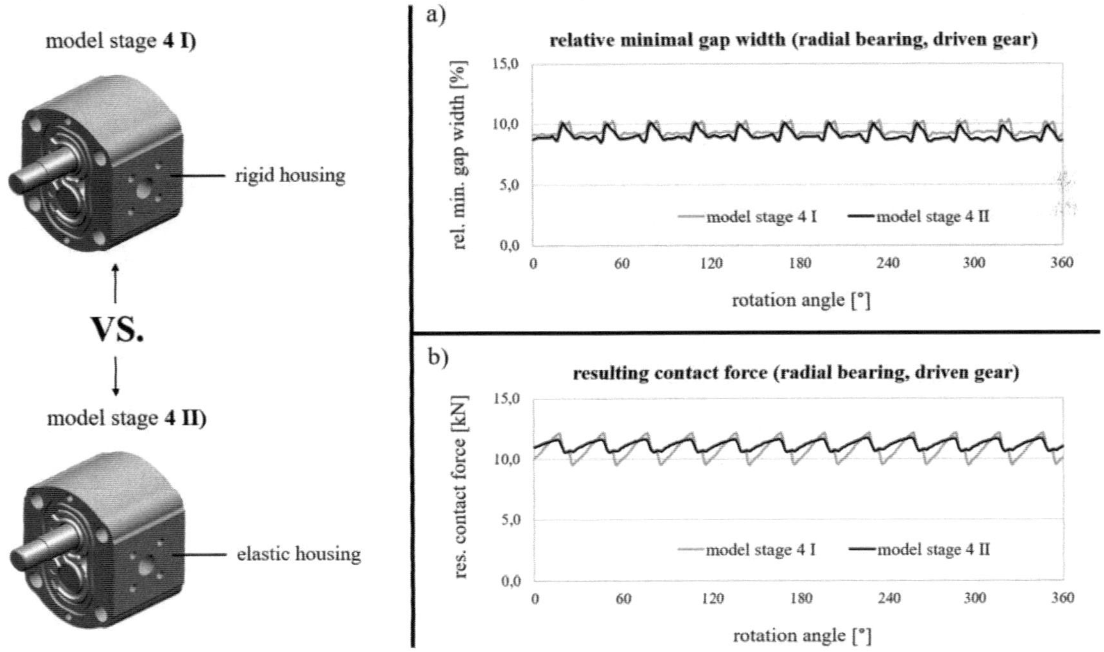

Figure 8. Comparison of model stage 4 I and 4 II for a radial plain bearing of the driven gear shaft a) relative minimum gap width b) resulting contact force

4.2. Results for the axial plain bearing

Figure 9 shows that in contrast to the radial plain bearing the housing elasticity has a significant influence on the relative minimum gap width and the resulting contact force in the axial plain bearing. For both model stages mixed friction occurs in the axial plain bearing (the relative minimum gap width is less than 100 %), see **figure 9 a).** In model stage 4 II a smaller relative minimum gap width occurs, because the deformation of the elastic housing causes a different displacement and deformation of the bearing bushings compared to a rigid housing. According to **figure 9 b)** the resulting contact force in the axial plain bearing increases under consideration of an elastic housing.

The relative minimum gap width and the resulting contact force oscillate with the frequency of the gear meshing, because over one full rotation all twelve teeth of the gear pass the area of the minimum gap width. **Figure 9 a)** and **b)** shows a phase shift between model stage 4 I and 4 II. This occurs because the angular location of the minimum gap width changes.

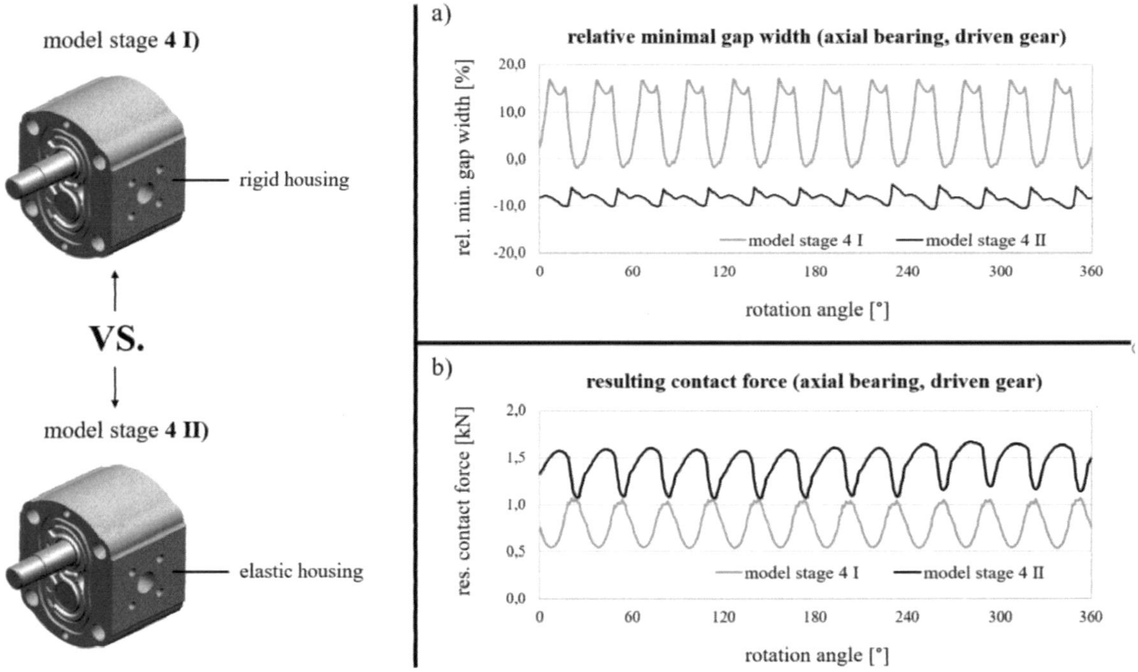

Figure 9. Comparison of model stage 4 I and 4 II for an axial plain bearing of the driven gear shaft a) relative minimum gap width b) resulting contact force

It is not apparent from **figure 9 a)** where the minimum gap width in the axial plain bearing occurs. For this reason **figure 10** shows the relative gap width in an axial plain bearing of the driven gear shaft for both model stages. The relative gap width h_{rel} is defined relative to the critical gap width h_{crit}, see **equation (2)**.

$$h_{rel} = \frac{h}{h_{crit}} \cdot 100\% \tag{2}$$

The grey areas in the color-plots in **figure 10** have a relative gap width greater than 100 % showing full fluid lubrication. With a rigid housing there is mixed friction on the discharge side and on the suction side with the minimum gap width appearing on the suction side. Under consideration of the elastic housing the areas of mixed friction mainly appear on the suction side. In addition the minimum gap width on the suction side is smaller in case of the elastic housing.

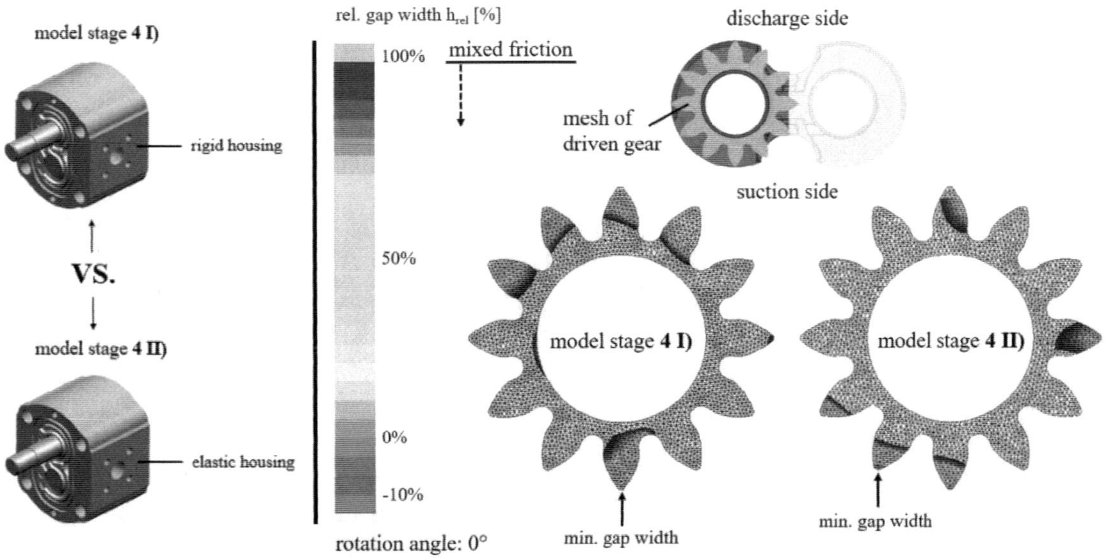

Figure 10. Comparison of the relative gap width in model stage 4 I and 4 II for an axial plain bearing of the driven gear shaft at a rotation angle of 0°

According to **figure 10** the elasticity of the housing leads to a shift of areas with mixed friction to the suction side and a reduction of the minimum gap width. **Figure 11** compares the deformations of the bushings for both model stages. Due to the pressure forces acting from the discharge side to the suction side, the bushings are pressed towards the suction side.

In case of a rigid housing the bushings flatten on the suction side in the contact area towards the housing. This moves the minimum gap to the suction side. At the same time the bushings tilt towards the gear on the discharge side because of the external pressure field and a moment of tilt due to the radial force acting on the gear. This leads to a large area of mixed friction on the discharge side.

In case of an elastic housing the contact surface between bushing and housing tilts, see **figure 11**. The bushings partially follow this tilt, resulting in a further reduction of the gap width on the suction side with a simultaneous increase in gap width on the discharge side.

In addition the investigated operating point with a rotation speed of 3000 rpm reduces leakage flows from tooth chambers with high pressure to tooth chambers with low pressure. Consequently, the pressure build-up takes place at a later rotation angle. Therefore, tooth chambers on the suction side have low pressure and cannot support the axial bearing hydrostatically. The influence of the elastic housing at other operating points has not been investigated yet.

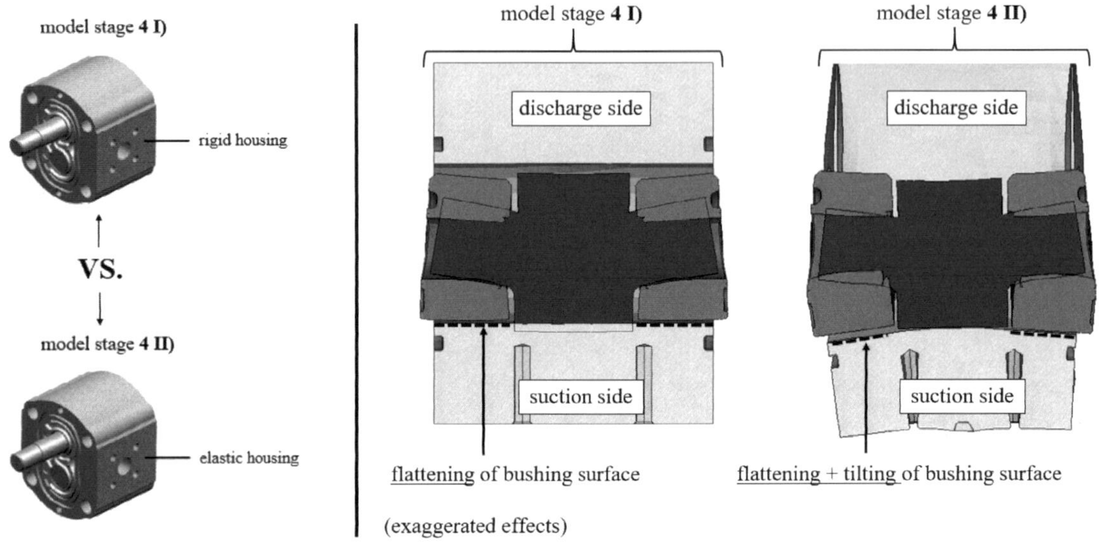

Figure 11. Comparison of deformations in model stage 4 I and 4 II

5. Summary

Electrification of mobile hydraulics and an increasing marked demand for higher operating pressures and rotation speeds make further development necessary. To ensure this, simulation-tools are intended to complement testing. This paper introduces an approach building up a simulation model of an external gear pump using the MBS/EHD-tool FIRST. The focus is to evaluate the required modeling depth regarding the consideration of an elastic housing. Therefore, the influence of the elastic housing on the tribological loads in the plain bearings was investigated for the driven gear shaft on the basis of the relative minimum gap width and the resulting contact force.

The investigation shows that the deformation of the housing has no significant influence on the tribological load in the radial plain bearing. However, a greater influence on the axial plain bearing was observed. The deformation of the elastic housing causes a different displacement and deformation of the bearing bushings compared to a rigid housing. This leads to a smaller relative minimum gap width and a higher resulting contact force and therefore a higher tribological load in the axial plain bearing.

A calculation of wear has not been included in the model yet. Therefore, it is not possible to make a statement regarding the influence of the elastic housing on the performance of the axial plain bearings. Assuming that the gap width is directly related to wear, the investigation indicates that the consideration of an elastic housing is essential to analyze the tribological load in the axial plain bearings. The stated effects are related to applications with high pressure and rotation speed. In low-pressure applications like lubrication they are negligible.

References

[1] IST: Simulationstechnische Auslegung, Optimierung und Schwachstellenanalyse thermo-elasto-hydrodynamisch gekoppelter Mehrkörpersysteme, URL: https://www.ist-aachen.de/ 18.11.2020
[2] Vacca A and Guidetti M 2011 Modelling and experimental validation of external spur gear machines for fluid power applications *Simul. Model. Pract. Theory* **19** 2007–31
[3] Thiagarajan D, Vacca A and Watkins S 2019 On the lubrication performance of external gear pumps for aerospace fuel delivery applications *Mech. Syst. Signal Process.* **129** 659–76
[4] Bosch Rexroth: External gear pump High Performance AZPF, URL: https://www.boschrexroth.com/en/xc/products/product-groups/mobile-hydraulics/pumps/external-gear-pumps/azpf 18.11.2020

[5] Shell: Tellus S2 MX 46, URL: https://www.shell-livedocs.com/data/published/de-AT/308ff126-e553-4cc6-b4be-d200538d7cf2.pdf 18.11.2020

[6] Ivantysyn J and Ivantysynova M 1993 *Hydrostatische Pumpen und Motoren – Konstruktion und Berechnung* (Würzburg: Vogel-Verlag) p 514

[7] Fiebig W and Heisel U 1994 Schwingungs- und Geräuschoptimierung von Hydropumpen durch Analyse des Schwingungsverhaltens von ihren Gehäusen *11. Aachener Fluidtechnisches Kolloquium* 171-88

IOP Conf. Series: Materials Science and Engineering 1097 (2021) 012015 doi:10.1088/1757-899X/1097/1/012015

Influences on the lubricant supply of grease lubricated gears

B J Siewerin, K J Winkler, T Tobie and K Stahl

Gear Research Centre (FZG), Technical University of Munich
Boltzmannstrasse 15, D-85748 Garching/Munich, Germany
siewerin@fzg.mw.tum.de

Abstract. Greases have a variety of advantages when special operating conditions apply. Mainly related to large, slow-running gear drives such as used in heavy industry applications, grease lubrication can also be the preferred solution for small, fast running gear drives. Consequently, the calculation of the wear service life within the gear design process is essential. Due to their flowing properties, there is a danger of losing the lubrication supply depending on the operating conditions, boundary conditions and grease properties. While a circulating lubricant ensures a continuous lubricant supply to the gear mesh, channeling includes the risk of starved lubrication and consequentially, even a discontinuation of the lubricating film that can lead to heavy damages of the gearbox. The experimental investigations on a modified FZG back-to-back test rig show a strong effect of operating conditions and grease properties on the lubricant supply: A higher amount of grease in the gearbox and a higher lubricant temperature support circulating, whereas, a higher consistency of the grease supports channeling. Based on the results, a first calculation approach is developed that approximates the lubricant supply of grease lubricated gears for the gear design process.

1. Introduction

Special operating conditions of a gearbox can require a grease lubrication. Especially in large, slow-running and open gear drives such as used in heavy industry applications (i. e. in the primary or mining sector), but also for small, fast running gear drives i. e. in power tools, greases are the preferred solution compared to oils. They have a variety of advantages that allow low-maintenance lifetime lubrication. In robot gearboxes they ensure a lubrication of the gear flanks, regardless of the spatial position the gearbox is in.

Continuous (or slow speed) wear (**Figure 1a**) is often the lifetime limiting factor when gears are running under grease lubrication. Wear appears continuously during the whole runtime and can be seen as scratches on the whole tooth flank. An exception is the pitch diameter, where almost no wear occurs, because the sliding speed is close to zero [1,2]. Consequently, the lubricating film thickness is higher than in other flank areas. For the gear applications mentioned prior, the calculation of the wear service life within the gear design process is essential. One common way to predict the wear lifetime of gears is the calculation method acc. to Plewe [3]. The calculation has originally been developed for oil lubrication and thus, assumes a continuous lubricant supply of the gear mesh. Greases can result in more critical lubrication conditions due to their flowing properties [4,5]. Different influences like operating conditions, boundary conditions and grease properties have strong effects on the lubrication supply and hence, on the lifetime of the gear pairing. While a circulating lubricant ensures a continuous lubricant supply to the gear mesh, channeling includes the risk of starved lubrication and accordingly, a discontinuation of the lubricating film. This discontinuation can lead to heavy damages or an immediate failure of the gearbox. As a consequence, continuous wear is no longer the critical gear failure mode, but scuffing (**Figure 1b**) or macro pitting (**Figure 1c**). Scuffing occurs typically in flank areas with high sliding speed and high contact temperatures. If a critical temperature is exceeded, the meshing tooth flank surfaces are welded instantaneous. The visual appearance of scuffing are deep scratches in sliding direction in corresponding flank areas of pinion and wheel. Macro pitting, in contrast, is a damage due to fatigue, but influenced by the lubricating conditions. It appears as larger, typically shell shaped, material breakout from the gear flank surface. Pittings are typically located on the tooth flank below the pitch diameter in areas of negative specific sliding [4,5].

a) continuous wear b) scuffing c) macro pitting

Figure 1. Typical gear failure modes

In order to prevent channeling by the gears, the knowledge of the expected lubricant supply, during the gear design process is essential. A first approach for the estimation of the lubrication supply mechanism of grease-lubricated gear pairings was proposed by Schultheiß [6] as a size-velocity-factor. Based on current experimental investigations, that first approach is extended within this course to include further influences into the model.

2. Lubricant supply mechanism

If a gear pairing operates under oil lubrication at a sufficient fill level, a lubricant supply of the tooth contact can be generally assumed. The oil is circulating with the rotation of the gears. With increasing circumferential speed, the lubricant film thickness increases and thus, causes better hydrodynamic conditions in the tooth contact (acc. to [7]). For grease lubrication, certain restrictions have to be considered. Although the wear behaviour can be related to the lubricant film thickness of the base oil [3], several more influences have an effect on the service life. The complex interactions between the grease components result in a different behaviour between a grease and its base oil in a sliding/rolling contact. The investigations of Poon [8] on a disc machine even showed a decrease of the lubricant film depending on the run time. Later, Fukunaga [9] could prove that the consistency and therefore, the thickener to have a greater effect on the grease characteristics than any base oil properties. The thickener determines the flowing properties that are essential to the lubrication of the tooth contact. If a grease is too viscous, there is a risk of starved lubrication. Consequently, the tackiness of a grease plays an important role for the ranking of greases acc. to Georgiou [10]. A lot of other investigations are dealing with grease lubrication, but the majority is based on investigations with the help of a ball-on-disc apparatus or a disc machine [8,11,12,13] and cannot directly be transferred to the conditions of a gear stage. Other experiments have been run with grease lubricated gears, but under non-critical conditions concerning a starved lubrication in the gear mesh [9,14,15]. First, Stemplinger and Schultheiß [4,5] observed the consequences of insufficient conditions on the gear service lifetime. They referred the results to two possible supply mechanisms grease lubricated gears can have. On the one hand, there is the so called "circulating" supply mechanism (**Figure 2a**) that mainly occurs at lower circumferential speeds and with low consistency greases; on the other hand so called "channeling" (**Figure 2b**) that mainly occurs at higher circumferential speeds and with high consistency greases. During circulation the grease builds a lubricating ring surrounding the gear and is transported into the tooth contact. The good lubricant supply enables a good mixing of the grease in the gearbox and thus, a good heat dissipation [4,5].

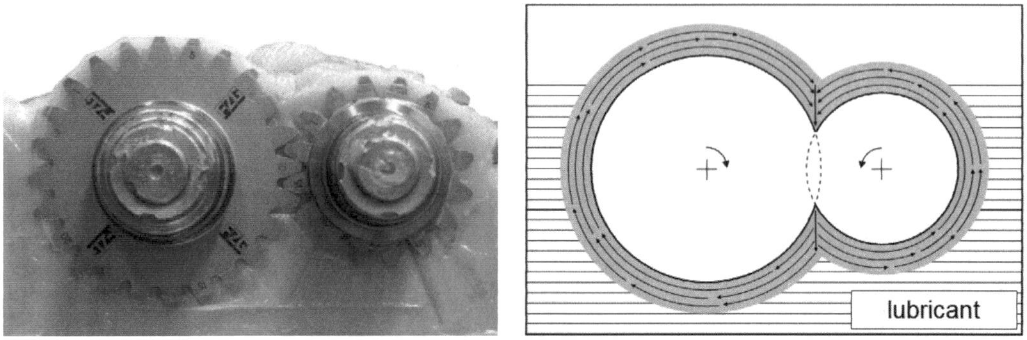

a) circulating

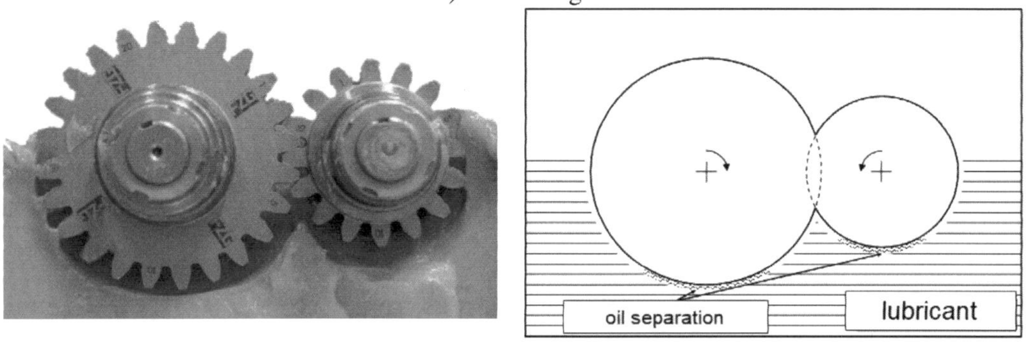

b) channeling

Figure 2. Lubrication supply mechanism acc. to [5,16]

During the supply mechanism channeling, no or only very little grease can flow back into the tooth gaps behind the gear mesh. Moreover, at higher speeds the lubricating grease is radially thrown off. Only a minimal amount of bleed oil can prevent the gear pairing from losing the separating lubricant film. Otherwise, the starved lubrication can result in an immediate failure of the gearbox. The insufficient mixing of the grease sump causes a poor heat dissipation and thus higher temperatures around the gears [4,5].

Based on the investigations of Schultheiß [6] a first calculation approach was developed in order to predict, whether a gear stage has a tendency towards circulating or channeling. The so called "size-velocity-factor" considers the rotational speed of a gear pairing and its size and therefore, the geometry of the gear teeth and their gaps in-between. Due to the large number of other influences on the lubrication mechanism and the complex interactions between them, this calculation approach needs to be validated and further developed. Especially, the transition area between the two supply mechanisms circulation and channeling is of particular interest.

3. Experimental investigation

The experimental investigations have been executed on a modified FZG back-to-back test rig. "It utilizes a recirculating power loop principle, also known as a four-square configuration, to provide a fixed torque (load) to a pair of precision test gears [...]. The slave gearbox and the test gearbox are connected through two torsional shafts [, one of which] [...] contains a load coupling used to apply the torque through the use of known weights [...] hung on the loading arm. The test gearbox contains heating elements to maintain and control the minimum temperature of the lubricant. A temperature sensor located in the side of the test gearbox is used to control the heating system as required by the test operating conditions" [17]. Additionally, within the herein described investigations an acrylic glass cover in the test gearbox allows to observe the lubricant supply by visual inspection as well as by a high-speed camera.

The test runs start at a slow rotational speed of the pinion of $n_1 = 9$ min^{-1} (acc. to the slow speed wear test [18]). Because the observation is difficult, if the speed increases constantly, 22 speed steps are defined up to a pinion speed of $n_1 = 3000$ min^{-1}. At each step, the lubricant supply mechanism is observed as soon as constant conditions can be seen. If not otherwise described, both gears are completely surrounded by the grease and the sump temperature is $\vartheta_S = 90$ °C. Only a minimal torque of $T_1 = 35.3$ Nm is applied to ensure, the gears are in contact. Since the lubricant supply is in the centre of the investigation, there is no further documentation of failure mechanisms within this course.

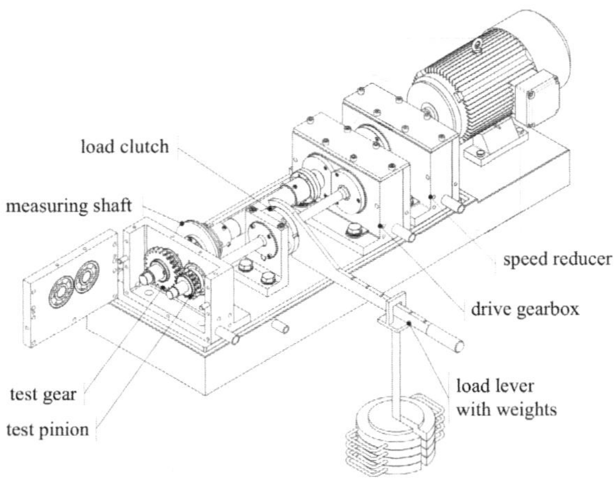

Figure 3. FZG back-to-back test rig [17,18].

The test gears are case-hardened standard gears of the type C-PT [18]. The main data of the gears are described in **Table 1**. To evaluate the influence of the grease consistency, three lubricants have been investigated within this course, which correspond to common greases for these applications (**Table 2**). They are based on a paraffin mineral oil with a nominal base oil viscosity $\nu_{40} = 110$ mm^2/s. The thickener is a lithium complex soap. The three lubricants only differ in their thickener concentrations in order to avoid further influences.

Table 1. Data of the C-PT gears.

Gear parameter	Symbol	Unit	Numerical value
Shaft center distance	a	mm	91.5
Module	m	mm	4.5
Number of teeth	z_1 / z_2	–	16 / 24
Face width	b	mm	14
Helix angle	β	°	0
Normal pressure angle	α	°	20
Profile modification factor	x_1 / x_2	–	0.1817 / 0.1715
Tip diameter	d_{a1} / d_{a2}	mm	82.46 / 118.36
pitch diameter	d_{w1} / d_{w2}	mm	73.2 / 109.8
Flank roughness	$Ra_{1,2}$	μm	0.2 – 0.4

Table 2. Data of the investigated greases.

Code		M110LiX00	M110LiX1	M110LiX2
Description	–	Grease	Grease	Grease
NLGI grade	–	00	1	2
Base Oil	–	Mineral oil	Mineral oil	Mineral oil
base oil viscosity ν_{40}	mm^2/s	113.1	113.1	113.1
base oil viscosity ν_{100}	mm^2/s	12.1	12.1	12.1
Thickener type	–	Lithium-complex	Lithium-complex	Lithium-complex
Percentage Thickener	wt%	4.3	9.2	11.5
Worked penetration (25 °C)	0.1 mm	412	318	285
Oil separation (90 °C)	%	5.2	5.3	3.1

4. Observations of the test runs

The results show a strong effect of the investigated operating conditions and grease properties: A higher amount of grease in the gearbox and a higher lubricant temperature support the supply mechanism circulating. In contrast, a higher torque and a higher consistency of the grease support channeling. **Figure 4** shows under which conditions and at which rotational speeds circulating or channeling of the grease sump is more probable. Between these two supply mechanisms, there is a range, which cannot be clearly assigned to any of the mechanisms. It is called "transition zone". In a first approximation the mean value of this zone has been selected as the critical rotational speed for the further analysis.

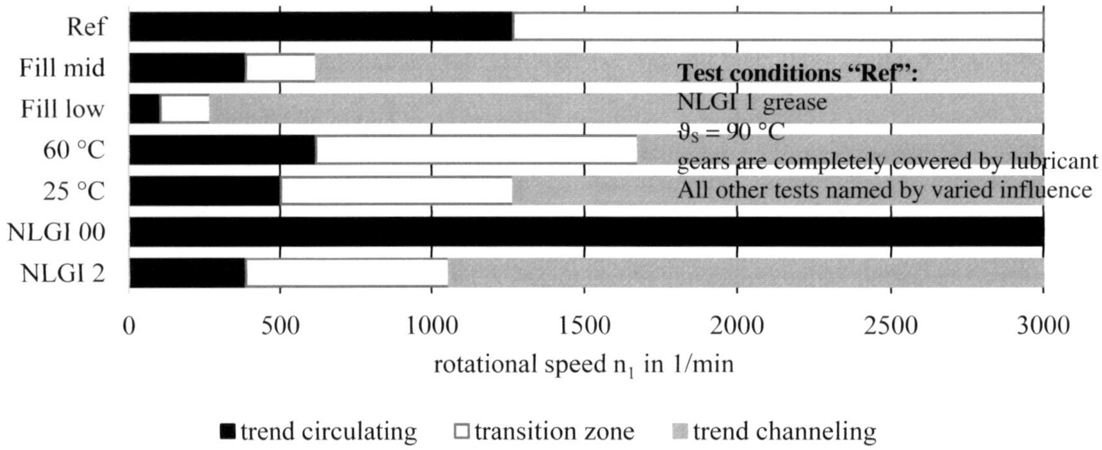

Figure 4. Lubricant supply mechanism in the tests.

In the beginning of the first test "Ref" under reference conditions, both the pinion and the wheel rotate within the lubrication sump of the grease M110LiX1. A ring of grease is rotating with the gears. Large parts of the grease sump remain stationary. The tooth spaces on the wheel are completely filled with the lubricant (comparable to **Figure 5a**). At a rotational speed of $n_1 = 845\ \text{min}^{-1}$, the grease starts to lift off from the gears and intermittently splashes against the gearbox housing. With increasing speed at $n_1 = 1250\ \text{min}^{-1}$, the thrown off grease jet becomes holey and inconsistent. Its flow is rather turbulent. This condition of the lubricant supply can no longer be described as pure circulating. The lubricant supply is in a transition zone. However, up to $n_1 = 3000\ \text{min}^{-1}$, there is still a certain amount of grease in the tooth gaps. Only at this rotational speed, the sump behaviour starts to match the criteria of channeling (comparable to **Figure 5b**).

4.1. Influence of the filling level

The tests "Fill mid" and "Fill low" are investigating the influence of the level of grease in the gear box. For "Fill mid", the grease level is up to the middle of the shaft centre (acc. to [18]). In relation to the immersion depth of the gear, "high" corresponds to 100 %, "medium" corresponds to 50 % and "low" corresponds to 25 %. It is suspected that the fill level of the gearbox has an effect on the lubricant supply, because the dipping time, the tooth gaps can be refilled with the lubricant, decreases with a smaller amount of grease inside the housing.

The comparison of the three tests shows a very clear influence of the fill level. In the beginning of the test "Fill mid", the grease forms a circulating ring around the gears. As the circumferential speed increases, a decrease of the lubricant in the tooth gaps can be observed immediately. Beyond a rotational speed of $n_1 = 386\ \text{min}^{-1}$, the first grease splashes are found on the gearbox housing. At a speed of $n_1 = 620\ \text{min}^{-1}$, large parts of the lubricating sump are thrown off. Because the tooth gaps are only a short time inside the lubricating sump, the grease can hardly flow back. The lubrication supply mechanism is clearly channeling. During the test "Fill low", the grease circulates at the first speed stages. However, even at those speeds, the tooth gaps are not completely filled. Already at a rotational speed of $n_1 = 105$ min^{-1}, the first grease splashes are visible on the housing. From a speed of $n_1 = 269$ min^{-1}, the supply mechanism can be described as channeling.

4.2. Influence of the temperature

The runs "60 °C" and "25 °C" are testing the influence of the sump temperature on the lubricant supply. Since the greases become more viscous at lower temperature, it is hypothesized that a reduction of the flowing properties (compared to the reference conditions) promotes a channeling of the test gears.

The test "60 °C" is operated with a sump temperature of $\vartheta_S = 60$ °C. In the beginning, at low speeds, the gears rotate within the lubrication sump in the area of circulating (**Figure 5a**). The tooth gaps are completely filled. The flowing behaviour is worse than under reference conditions. A large part of the lubricant sump is stationary. At a circumferential speed of $n_1 = 620$ min^{-1}, the parts of the grease are thrown off the gears. The fill level inside the tooth gaps is very low and the supply mechanism can no longer be described as circulating. At a rotational speed of $n_1 = 1677$ min^{-1}, the lubricant cannot flow back into the tooth gaps and the gear mesh is insufficiently supplied by the lubricant. At this point, the supply mechanism is clearly channeling (**Figure 5b**).

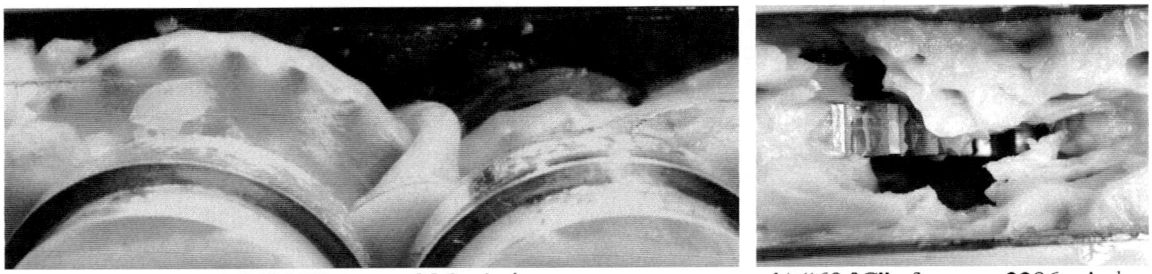

a) "60 °C" at $n_1 = 386$ min^{-1} b) "60 °C" after $n_1 = 2286$ min^{-1}

Figure 5. Visual inspection of the lubricant supply [16]

In the test "25 °C", the sump temperature is at the room temperature of $\vartheta_S = 25$ °C. At low rotational speeds, the gears are rotating within the lubrication sump in the area of circulating. The gaps between the teeth are completely filled, a circulating grease ring is visible. The surrounding lubrication sump does not take part in the lubrication, but rather gives the impression of a shrouding. The first splashes on the housing can be seen at $n_1 = 503$ min^{-1}. Due to the poor flowing behaviour at this low temperature, only a small part of the tooth spaces is filled with the grease. From the rotational speed of $n_1 = 1250$ min^{-1}, the lubricant supply can clearly be described as channeling.

4.3. Influence of the consistency

Two tests consider the effects of different grease consistencies. The low-consistency grease M110LiX00 ("NLGI00") and the high-consistency grease M110LiX2 ("NLGI2"). They cover a concentration range of the thickener soap from 4.3 % up to 11.5 %. Comparable to the prior tests runs at different temperatures, an influence on the lubricant supply due to the different flowing properties of the greases is expected.

The comparison of the runs shows a clear trend that a high-consistency grease promotes channeling. The grease M110LiX00 is gradually thrown off the gears against the gearbox cover at $n_1 = 620$ min^{-1}. It seems almost liquid during the whole test run and thus, the lubrication supply can be described as circulating up to the maximum speed $n_1 = 3000$ min^{-1}. Under identical test conditions, the grease M110LiX2 shows similar behaviour at low speeds and is initially comparable with the previous tests. Due to the low rotational speeds, the gears are circulating within the lubrication sump. As a result of the poor flowing properties, only a small part of the tooth gaps is filled. At a rotational speed of $n_l = 386\ min^{-1}$, the amount of grease in the gear mesh is reduced and the lubrication supply is in the transition zone. At $n_l = 1058\ min^{-1}$, parts of the grease are thrown off and spray onto the housing. The supply mechanism can be referred to as channeling.

5. Calculation approach

Based on the observations of the experimental investigations, a calculation approach is developed to determine the rotational speed n_{LSM}, when a grease lubricated gear pairing presumably changes its lubricant supply mechanism and starts channeling (**1**). The grease M110LiX1 is used as a reference, because its grease components are widely used in practical applications. The boundary rotational speed of 2135 min^{-1} is the mean value of the transit zone in this first approximation of the lubricant supply. The different factors F_K, F_F and F_ϑ consider the results of the experimental investigations and can change the limiting speed n_{LSM} depending on the running conditions, boundary conditions or grease properties. They result from different regression analyses of the test results.

$$n_{LSM} = 2135\ min^{-1} \cdot F_K \cdot F_F \cdot F_\vartheta \quad (1)$$

The grease consistency and therefore, the thickener concentration have a strong influence on the supply mechanism. A higher thickener concentration promotes the tendency of channeling (**2**). Due to the poorer flowing behaviour, the amount of lubricant in the tooth contact decreases significantly even at lower speeds (**Figure 6**).

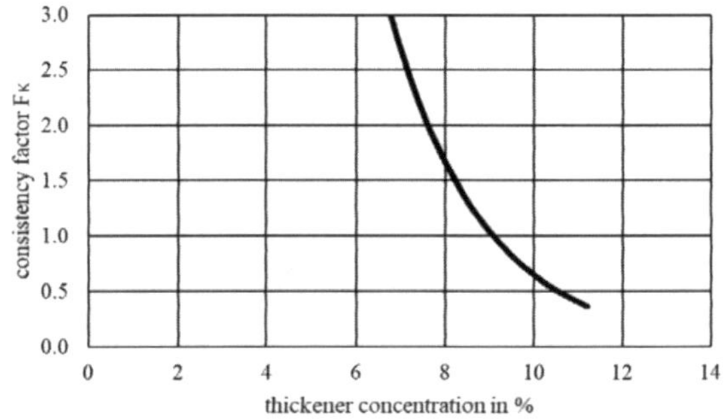

Figure 6. Influence of the consistency

The tests with different gear filling levels show a clear influence of the gear immersion depth on the supply mechanism. With a higher filling level, circulation of the lubricant is promoted (**Figure 7**). The longer a gear is inside the sump, the better is the possibility for the grease to flow back into the tooth gaps. Consequently, it enables better lubrication conditions and has a positive effect on the wear life of a gear pair (**3**).

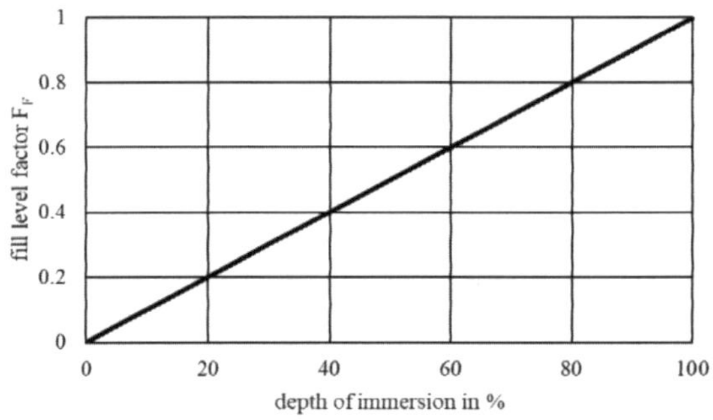

Figure 7. Influence of fill level

$$F_K = 76.64 \cdot 0.44^{\frac{V_\%}{\%}} \quad (2)$$

$$F_F = X_\% \quad (3)$$

$$F_\vartheta = 4.6 \cdot 10^{2.8 \cdot 1.002^{\left(\frac{\vartheta_S}{°C}\right)} - 4} \quad (4)$$

The temperature of a lubricating grease has an influence on the supply mechanism as well. At higher temperatures the grease becomes more fluid and have a better ability to flow back into the tooth gaps (**Figure 8**). This promotes circulation of the lubricant. When the sump temperature decreases, the grease is not able to refill the tooth gaps behind the gear mesh. The missing refill results in a tendency of channeling (**4**).

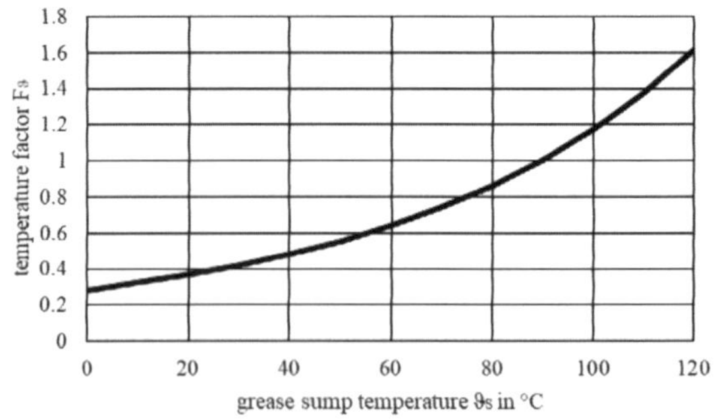

Figure 8. Influence of temperature

6. Conclusion

The results of the investigations show a strong effect of the investigated operating conditions and grease properties: A higher amount of grease in the gearbox and a higher lubricant temperature support the supply mechanism circulating. In contrast, a higher consistency of the grease supports channeling. Based on these results, a calculation approach was developed that offers the possibility to estimate in a first approximation the lubricant supply of grease lubricated gears for the gear design process. It calculates the rotational speed of the pinion n_{LSM}, at which a channeling of the gear pairing becomes probable. Therefore, it accounts the influences of the fill level, the temperature and the consistency. In order to reach a most possible validity for this approach, it is based on test results with standard test gears and a plain grease with a mineral base oil and a lithium complex thickener. These are common components in a lot of gearbox greases and have been used in a variety of research projects before. However, for a deeper understanding of the lubricant supply mechanism of greases and to consider other influences, it is necessary to further develop the investigation. Besides the running conditions and grease properties investigated in this course, several other aspects can have an effect on the lubrication supply of the gears. Thus, the geometry of the gear pairing (module resp. diametral pitch and tooth width) as well as grease components like the base oil, its viscosity and the type of thickener should be in the focus of additional investigations. It can be assumed that they have an influence of the flowing properties of a grease and its ability to get into the gear mesh. Consequently, these aspects can promote either circulating or channeling. Further tests can investigate these influences quantitatively and extend the validity of the calculation approach. This way, the relevant parameters as well as the interactions between the influences can be observed, analysed and taken into account in the calculation approach.

Acknowledgement

The research project the test runs is part of were conducted with the support of the DGMK (German Society of Petroleum and Coal Science and Technology) research association. The projects were sponsored by the German Federal Ministry of Economics and Technology (BMWi) through the AiF (Arbeitsgemeinschaft industrieller Forschungsvereinigungen e.V.), project number 19627 N [16].

References

[1] Siewerin B J, Dobler A, Tobie T and Stahl K 2019 Applicability of an Oil Based Calculation Approach for Wear Risk and Wear Lifetime to Grease Lubricated Gear Pairings *Proceedings of the ASME IDETC/CIE 2019*

[2] Linke H 2010 Stirnradverzahnung: Berechnung - Werkstoffe - Fertigung *Carl Hanser Verlag* Munich

[3] Winter H and Plewe H-J 1985 Calculation of Slow Speed Wear of Lubricated Gears *Gear Technology* **2**(6) pp. 8–16

[4] Stemplinger J-P, Stahl K, Höhn B-R, Tobie T and Michaelis K 2014 Analysis of lubrication supply of gears lubricated with greases NLGI 1 and 2 and the effects on load-carrying capacity and efficiency *NLGI Spokesman* **78**(1) pp. 18–22

[5] Schultheiss H, Stemplinger J-P, Tobie T and Stahl K 2016 Influences on failure modes and load carrying capacity of grease lubricated gears *Gear Technology* **33**(1) pp. 42–47

[6] Schultheiss H, Tobie T and Stahl K 2017 The Effect of Selected Operating Parameters, Material Pairing, and Gear Size on the Wear Behavior of Grease Lubricated, Small Module Gears *Journal of Tribology* **139**(6)

[7] Dowson D and Higginson G R 1966 Elastohydrodynamic Lubrication *Pergamon Press Oxford [u.a.]*

[8] Poon S 1972 An experimental study of grease in elastohydrodynamic lubrication *Journal of Lubrication Technology* **94**(1) pp. 24–37

[9] Fukunaga K 1990 Grease for gear lubrication *Lubrcation Engineering* **46**(9) pp. 557–564

[10] Georgiou E P, Drees D, De Bilde M and Anderson M 2018 Can We Put a Value on the Adhesion and Tackiness of Greases? *Tribology Letters* **66**(60)

[11] Kaneta M, Ogata T and Naka M 2000 Effects of a thickener structure on grease elastohydrodynamic lubrication films *Proc. Instn. Mech. Engrs*

[12] Cousseau T, Björling M, Graça B, Campos A, Seabra J and Larsson R 2012 Film thickness in a ball-on-disc contact lubricated with greases, bleed oils and base oils *Tribology International* **53** pp. 53–60

[13] DeVaal P, Möller V and Langenhoven J 2013 Investigating Seizure Load and Wear Characteristics *World Tribology Congress*

[14] Krantz T L and Handschuh R F 2004 A Study of Spur Gears Lubricated With Grease-Observations From Seven Experiments *58th Meeting of the Society for Machinery Failure Prevention Technology (MFPT)*

[15] Krantz T L, Oswald F and Handschuh R F 2007 Wear of Spur Gears Having a Dithering Motion and Lubricated With a Perfluorinated Polyether Grease *Proceedings of the ASME IDETC/CIE 2007*

[16] Siewerin B J, Tobie T and Stahl K 2021 Calculation Method and Boundary Criteria for the Continuous Wear Behaviour of Grease Lubricated Gears with Regard to Different Material Combinations *German Society of Petroleum and Coal Science and Technology (DGMK e.V.)* Hamburg

[17] DIN ISO 14635-1:2006-05 2006 Gears – FZG test method – part 1 FZG test method A/8.3/90 for relative scuffing load carrying-capacity of oils

[18] Bayerdörfer I, Michaelis K and Höhn B-R 1997 Slow Speed Wear Test in the FZG Gear Test Rig *5th CEC International Symposium on the Performance Evaluation of Automotive Fuels and Lubricants*

Investigation of the coating of hydrodynamic plain bearing contact surfaces by means of Extreme High-Speed Laser Material Deposition (EHLA)

Stephan Koß [1*], Achill Holzer [2][0000-0003-1190-18199], Sandra Megahed[1], Stephan Ziegler [1], Johannes Henrich Schleifenbaum [1], Katharina Schmitz [2][0000-0002-1454-8267]

1) Digital Additive Production DAP, RWTH Aachen University, Campus Boulevard 73, 52074 Aachen
2) Institute for Fluid Power Drives and Systems ifas, RWTH Aachen Univeristy, Campus Boulevard 30, 52074 Aachen

*corresponding author: stephan.koss@dap.rwth-aachen.de

Abstract: In the field of drive technology, plain bearings are used in almost all drives. During operation these bearings have a very low load resistance and almost no wear. However, increased mixed friction can occur during starts, stops, low speeds and/or high loads. As a result of the increasing demands on the drive systems, such as in the automotive sector with the automatic start-stop system to reduce exhaust gases or in the area of the angular force, a greater mixed friction occurs, which increases wear. Extreme High-Speed Laser Material Deposition (EHLA) is a coating process that allows the application of corrosion and wear protection coatings in a resource-saving and productive manner. Furthermore, alternative materials and material combinations can be created. In the framework of this study two different Metal Matrix Composites (MMC) are tribologically investigated and characterized in the area of mixed friction against three friction contacts each.

Keywords: Extreme High-Speed Laser Material Deposition (EHLA), Tribology, Coatings, Metal Matrix Composite (MMC)

1. Introduction

In the field of drive technology, plain bearings are an essential component for the support of axial and radial loads. These are used for example in wind power plants, pumps, electric drives and combustion engines. For applications requiring a long service life and low energy consumption, plain bearings with liquid friction are preferable (e.g. in drives, in turbines). Due to the hydrodynamic pressure of the relative movement, an oil film forms between the contact surfaces. Plain bearings are used to support moving components and to transfer loads. The most prominent representatives are hydrodynamic bearings in internal combustion engines. The functionality of such bearings relies on two fundamental aspects, sufficient lubrication and design aspects resulting in hydrodynamic film formation as well as suitable bearing materials [1]. The latter is addressed with unique solutions for specific applications and material selection has been found to be a compromise between contradicting demands such as fatigue strength

and tribological properties (i.e., a material needs to be hard and soft at the same time), as well as chemical compliance [2]. To meet these requirements, plain bearings are usually designed with a layered structure and heterogeneous materials. Each layer and material has to fulfil a specific task addressing each demand. The main layers are defined as steel back for mechanical support, a bearing lining and optional overlay systems [3,4].
At high loads and/or low rotational speeds, hydrodynamic pressure cannot be built up sufficiently and mixed lubrication occurs, which accelerates the wear of bearings. This mixed lubrication occurs particularly with a large number of start-stop processes. One example of this is the start-stop function in modern cars in city traffic. This can reduce fuel consumption by up to 8%, but this causes an increased occurrence of solid contact between crankshaft and plain bearing. This increases wear and reduces the service life of the motor [2]. In order to reduce wear, material pairs are selected in friction contact, which show the lowest possible wear in contact. Typically, the hydrodynamic plain bearings are made of bronze, white metal, aluminium alloys, polymers or ceramics. The contact surfaces of the shaft are usually hardened against wear [5]. However, the service lifetime depends on the following [6]:

- Load coefficient depending on pressure [N/mm^2]
- Rotation per minute depending on surface speed [m/s]
- Temperature [°C]
- Surface roughness
- Type of load (Point load/ Circumferential load)
- Factor depending on material/material combination
- Specific bearing load [N/mm^2]
- Sliding speed [m/s]

These factors can vary greatly depending on the field of application. With the increasing loads and challenges in drive technology, it is necessary to extend the service life in plain bearings; especially as the service life of the engine is often equivalent to the service life of the slide bearings used. The highest wear occurs in mixed friction, for example due to start-stop processes, such as in the automotive sector, and/or due to high loads, such as in wind power. According to Stickel et al., the tribological properties do strongly depend on the surface topography, in particular in the very first time of operation. If the surface is smoother, the frictional properties are not always better [18]. In order to achieve the longest possible service life, material combinations with minimum wear and greatest conceivable sliding properties must be used in these plain bearings. As contact materials, brass and bronze are often used in plain bearings. These materials have a hardness in the range of approx. 150 to 230 HV and demonstrates a coefficient of friction of approx. 0.15 in wear tests at a contact pressure of 1 MPa [23].
The additive process Extreme High-Speed Laser Material Deposition (EHLA) offers new possibilities to generate metallurgically bonded coatings with low heat input to the substrate. Due to the process characteristics it is possible to coat unconventional material pairings such as Metal Matrix Composites (MMC). This provides new possibilities to increase durability. First tests investigating the tribological characteristics of EHLA 316L coatings on 42CrMoV4 (1.7225)-substrates have already been completed. The coatings are tested on a tribometer at 100 rpm and a contact pressure of 100 N for 30 min. The results showed little amount wear (0.097 g for a sliding distance of 1 km) [6]. Currently MMC coatings manufactured with EHLA are investigated to improve wear resistance of brake rotors. During the DYNO brake test, the brake rotor demonstrated wear of 0 to <0.5 g in the simulated test over 1000 km [7]. In this paper, various material pairings to use in plain bearings are tested and characterised. Furthermore, Blau and Jolley investigated the abrasive wear behaviour of several titanium carbide reinforced matrix composites using 240 grit aluminum oxide band as counterpart [15].

2. Methods and materials

2.1. EHLA process

EHLA offers new possibilities due to its unique process properties to reduce wear in plain bearings, especially for the frequently occurring mixed friction case. Figure 1 illustrates a schematic representation of the EHLA process. The powder material is fed to the deposition process via a continuous powder feed nozzle positioned coaxially to the laser beam. During conventional manufacturing of material combinations such as MMC, for example, the individual materials are mixed in-situ before the nozzle and fed into the process via a powder gas stream. The focus of the powder gas stream is positioned above the substrate. In this process, the majority of the laser energy is absorbed by the powder particles before they hit the substrate, which enables new materials and material combinations to be processed (e.g. MMC, coating of aluminium/copper with Fe/Ni/Co-based alloys, coating of substrates with Al/Cu and Al-/Cu-based alloys, etc.). The residual energy forms a small melt pool on the substrate surface. Since the energy input is directed into the coating material, the heat influence for the substrate is small. Process speeds of up to 500 m/min can be achieved. In combination with the track displacement, coatings with layer thicknesses ranging between 20 to 350 µm are produced. EHLA is resource efficient, because it reaches a powder efficiency of up to 95% [8,9]. Process characteristics allow for a high productivity coating process in which the risk of layer detachment under high loading conditions is very low or even non-existent. Furthermore the option of repairing coatings is given with the EHLA process.

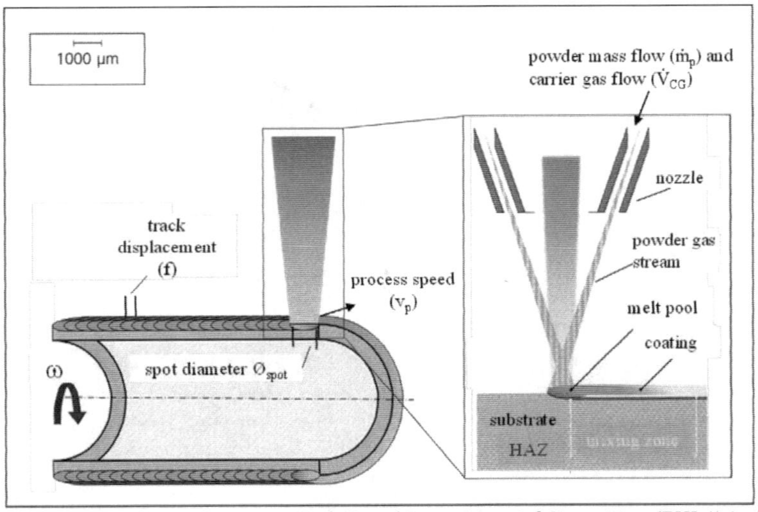

Figure 1: Schematic representation of Extreme High-Speed Laser Material Deposition (EHLA) including relevant process parameters

2.2. Coating materials

MMC consist of at least two materials: one soft and ductile matrix material, in this case 316L, and one harder material, e.g. silicon carbide (SiC) and titanium carbide (TiC). TiC and SiC have a high hardness of 3200 HV (TiC) and 2800 HV (SiC) as well as high melting temperatures of 3065°C (TiC) and 2797°C (SiC) [10,11]. Compared to pure hard metal coatings, the coating is more ductile due to the softer matrix material.

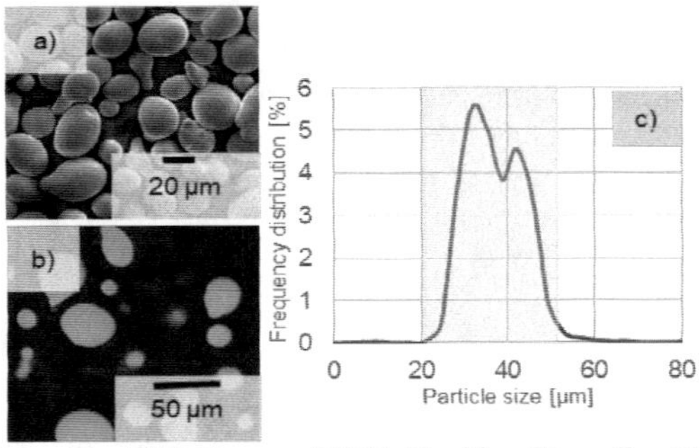

Figure 2: a) SEM image of 316L (Oerlikon Diamalloy 1003-1) with the magnification of 400x, b) Incident light microscope image of 316L cross sections with a magnification of 200x, c) measured particle size distribution of 316L particles

Figure 2 shows SEM images and the powder size distribution of the 316L powder from Oerlikon. The particles are oval and dense (see Figure 2 a), b)). The light blue area in the particle size distribution measurement highlights the manufacturer's specifications of 20 to 53 µm. The blue line in the diagram shows the measured particle size distribution. These lies within the range of approx. 30 to 45 µm and thus in the range specified by the manufacturer. The chemical composition of 316L is given in Table 1.

Table 1: Chemical composition of 316L (Oerlikon 1003-1)

Material	Fe	Cr	Ni	Mo	Si	C
Content [wt.-%]	Bal	17	12	2,5	2,3	0,03

The powder analysis of silicon carbides from GTV is shown in Figure 3 on the left-hand side. The particles are dense, no porosity can be identified in Figure 3b. The particle size distribution shows a size range of approx. 10 to 70 µm.

Figure 3: a) SEM image of SiC (GTV 70.85.8) and TiC (GTV 70.90.3) with the magnification of 400x, b) Incident light microscope image of SiC and TiC cross sections with a magnification of 200x, c) measured particle size distribution of SiC and TiC particles

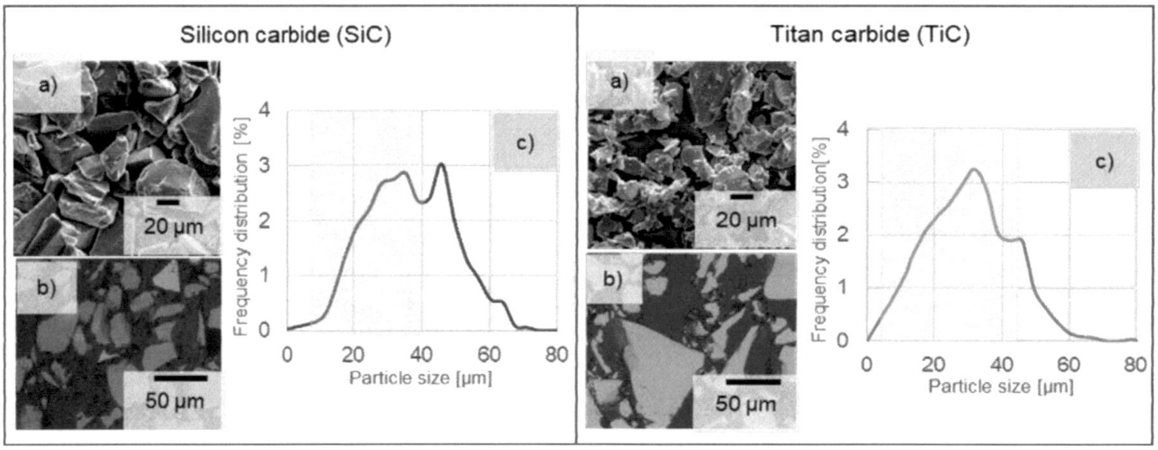

The titanium carbide (TiC) used from GTV is dense, no porosity can be identified in Figure 3b. The measured particle size distribution is in the range of approx. 5 to 60 µm. Compared to 316L and SiC, TiC has the highest percentage of fine particles (see Figure 3 on the right-hand side). The chemical composition of the materials is given in Table 2 for SiC (left) and for TiC (right).

Table 2: Chemical composition of SiC (GTV 70.85.8)

Material SiC	*Si*	*C*
Content [*wt.* −%]	70,2	29,8

Chemical composition of TiC (GTV 70.90.3)

Material TiC	*Ti*	*C*
Content [*wt.* −%]	80,8	19,2

2.3. Post processing
In order to obtain a smooth surface, all coatings are ground to a roughness of Ra 0.1 µm. For tribological examinations of lubricated contacts, Abbott-Firestone curves provide a summary of the roughness profile. High peaks increase the propensity of fretting, as the real pressure gets very high whereas deep valleys are expected to improve lubrication as valleys serve as fluid reservoir.

2.4. Tribometer
For the tribological investigation, a disc-on-disc tribometer test bench has been used. The tribometer test bench as shown in Figure 4, consists of two test discs, a stator and a rotor, which are pressed together by a hydraulic cylinder. The rotor is driven by an electric motor transmitting its torque to the stator. The stator is connected to the test bench by a torque sensor. Both test discs are placed in a temperature controlled oil basin. Air bearings make sure not to measure friction outside the disc-disc contact [14].

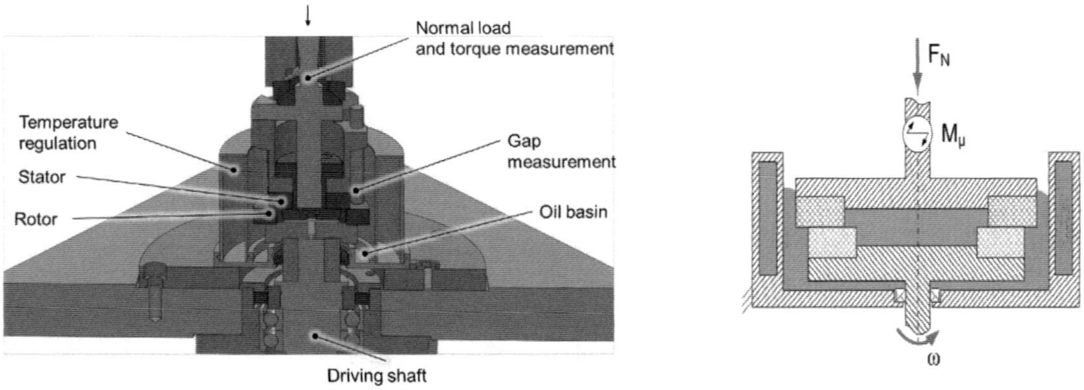

Figure 4 CAD model of the test bench (left) and schematic drawing of the disc-disc tribometer test with stator coloured in red, and rotor coloured in green (right) [14]

The tribological measurement took place at an oil temperature of 60°C which was maintained stable. For lubrication and cooling a fully additivated mineral-based hydraulic oil with the basis viscosity of 46 cSt has been used. Further technical parameters of the used test bench can be found in Table 3.

Table 3 Technical properties of the used test bench

Parameter	Value
Mean contact diameter	65 mm
Friction track width	5 mm
Area of contact	1021 mm^2
Surface pressure	1 to 30 MPa
Temperature range	0 to 120 °C
Gap measurement accuracy	<0,065 µm

2.5. Used counterparts
In order to cover a wide range of application, the MMCs have been tested with three different counterpart materials. As a bearing steel, 100Cr6 has been used. To cover the materials of lower strength, S355 material has been tested. As third material, a brass material MS58 has been evaluated.
All counterparts have been lapped to a roughness of Ra 0.1. Since lapping produces a stochastic surface, the specification of a one-dimensional Abbott curve is sufficient for the characterization (Figure 5) [22]. All surfaces have been lapped with the same parameter.

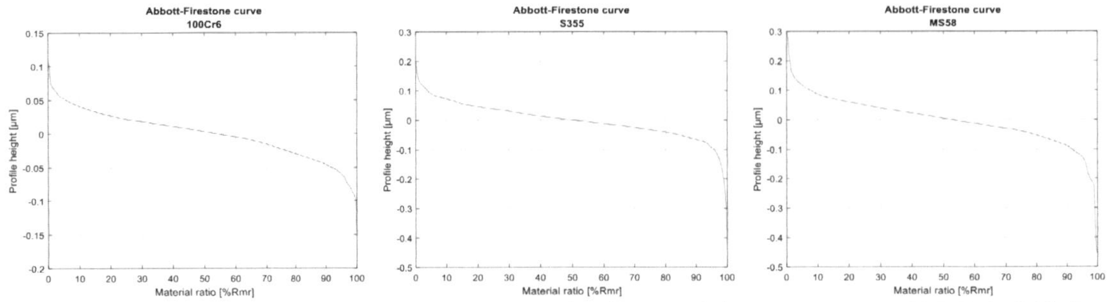

Figure 5 Abbott-Firestone curves of the counterpart discs made of 100Cr6 (left), S355 (centre) and MS58 (right) before the tribological tests

As shown in Figure 7 and Figure 5, the surfaces of the coated stator discs are slightly rougher than their counterparts. A comparison of the three Abbott curves shows that the 100Cr6 sample has slightly lower roughness peaks than the two other counterparts. It is assumed that this is due to the higher material hardness [22].

3. Results

3.1. Coating with MMCs

For the tribological tests, two MMC coatings were produced using EHLA, which are shown in Figure 6. The MMC produced from 316L + TiC contains no cracks or pores (see Figure 6 a)). Thereby, an average thickness of approx. 705 µm was produced with a mixing ratio of 1:1.

SiC is considered to be a difficult material to weld, as the material only has a thermal stability up to 1300 °C. A SiC content of 30 vol-% with an average thickness of approx. 445 µm was achieved (see Figure 6 b)). During the deposition process of the MMC, a large content of SiC is dissolved in the matrix material, which increases its hardness. The cross section in Figure 5b shows cracks, which is characteristic for brittle materials.

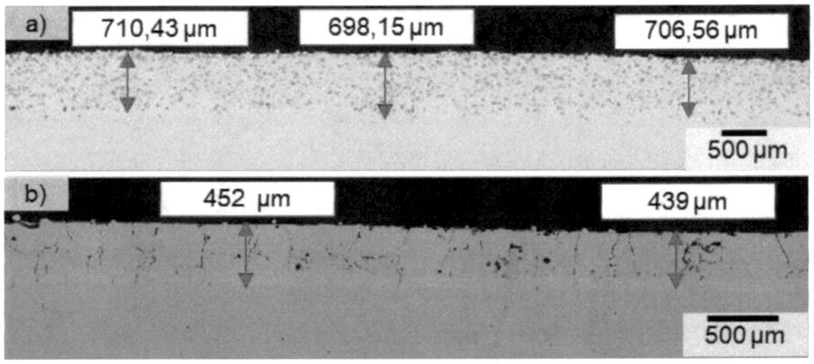

Figure 6: Cross sections of the two MMCs, a) 316L+TiC with a magnification of 50x, b) 316L+SiC with a magnification of 50x

By means of the energy input into the coating material and the high process speed, the generated melt pool is small, so that the boundary between substrate and coating is clear.

3.2. Post processing

Figure 7 shows the Abbott curves of both as built and ground stator discs. It can be seen, that the roughness peaks perpendicular to the grinding direction are more than double the height compared to grinding direction. It is expected, that fluid lubrications suffers due to the tunnel effect, preventing a fluid pressure build-up. Nevertheless, ground surfaces like this can be found in hydraulic manufactured parts.

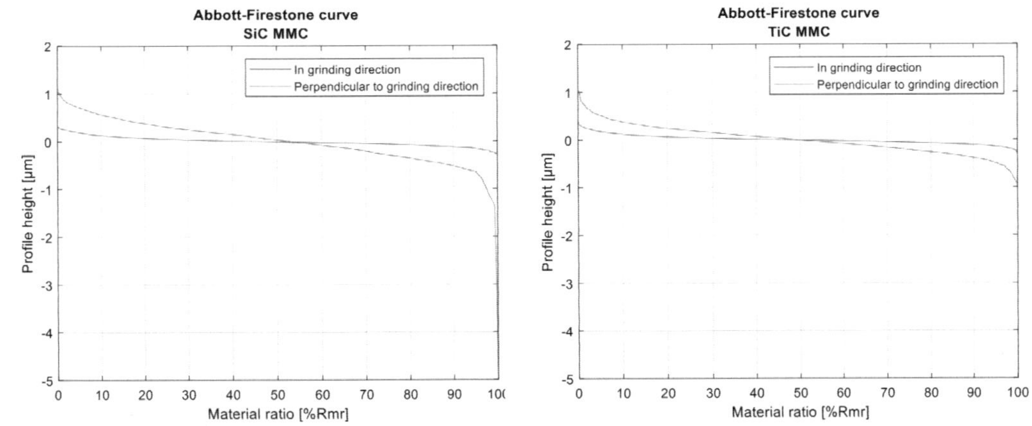

Figure 7 Abbott- Firestone curve of the coated and grinded SiC disc before tribological testing (left) as well as for the TiC coated disc (right)

It can be seen, that the SiC coated disc has, compared to the TiC coating, deep cavities perpendicular to the grinding direction, exceeding the depth about five times. Regarding the roughness peaks, both coatings are at the same level after the grinding process.

3.3. *Tribological investigations*

The aim of tribological testing is, one the one hand to analyse frictional behaviour. On the other hand, it is of special interest to investigate the bearing behaviour when insufficient oil pressure is maintained to support the bearing. This is the case, when the bearing is used at low speeds, without sufficient oil or at a very late state of life. The contact cannot be supported by the fluid, which results in damage, as mixed or even boundary lubrication occurs over a long time period.

A Stribeck curve has been plotted at a pressure of 1 MPa in a range from 0 to 14 m/s. As shown in Figure 8, the Stribeck curve can be seen as a superposition of solid state friction and fluid friction plotted over the sliding speed, where the solid friction is only applicable at low speeds until the contact gets fully supported by hydrodynamic pressure-build up. At higher speeds, fluid friction is an important parameter. Due to the temperature dependence of viscosity, the Hersey-Gümbel number is usually plotted on the y-axis [21]. In this case, temperature has been kept constant, allowing the sliding speed to be plotted.

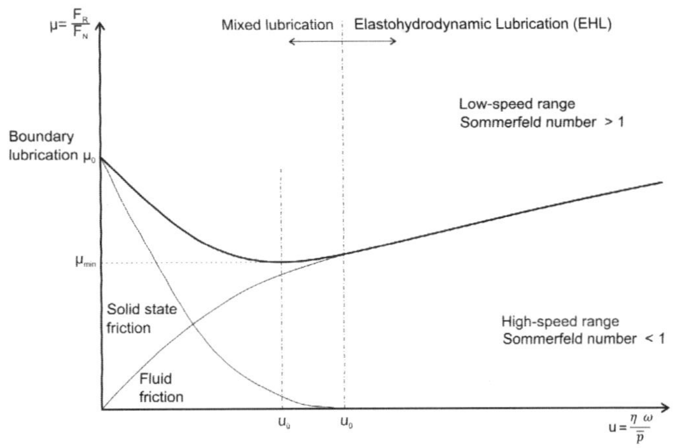

Figure 8 Stribeck test with coefficient of friction over Hersey Gümbel number

The pairing has been forced in the state of mixed lubrication by repeated acceleration from zero to 3 m/s under constant pressure (see Figure 9). This process is repeated 50 times to analyse the resistance capability regarding wear.

Due to a not ideally stiff bearing of the stator, a slight wobbling motion occurs, which we can detect by means of inductive distance sensors. This wobbling motion leads to the occurrence of convergent lubrication gaps between the rotor and stator. Even though we can get into the range of liquid friction under good conditions and the present specimen geometry, in this case, also due to the prevailing wear, it can be assumed that the increase in the coefficient of friction is rather due to the wear caused by the larger energy input than by fluid lubrication as showed in Figure 8.

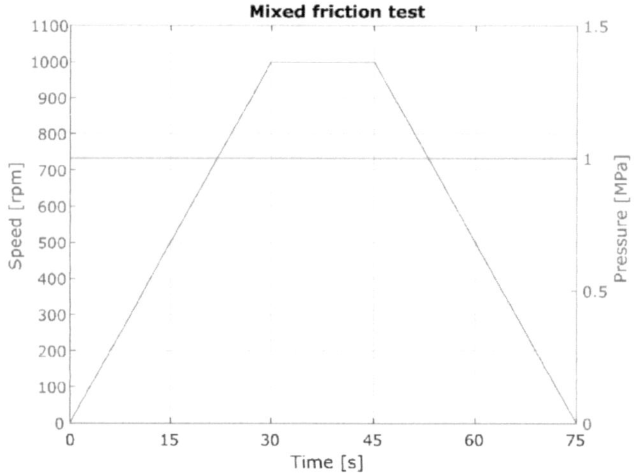

Figure 9 Mixed lubrication test, where the pairing is forced in mixed lubrication for a high number of times

3.4. *Results MMC SiC+316L coating*

In general, a very high level of friction has been observed for the SiC MMC against 100Cr6 discs. As shown in Figure 10, there is a noticeable drop of the friction coefficient at around 3 m/s and below. High friction coefficients on the right side of the curve (high speed range) may be explained with a run-in transition, as both discs were new, and the Stribeck curve has been captured from high-speed to low speed.

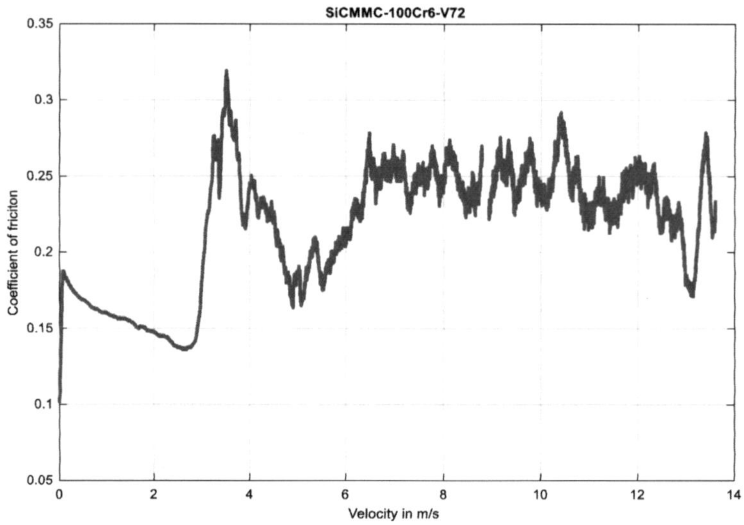

Figure 10 Stribeck curve, plotting coefficient of friction over sliding speed in m/s of SiC coating against 100Cr6 test disc

The 100Cr6 counterpart, as well as the coated stator disc showed a material loss of around 20 µm after performing Stribeck test and mixed lubrication test. In the Stribeck curve shown in Figure 11 a valley can be seen on the left side (coated disc). A possible explanation for this behaviour is edge contact leading to a non-uniform pressure distribution.

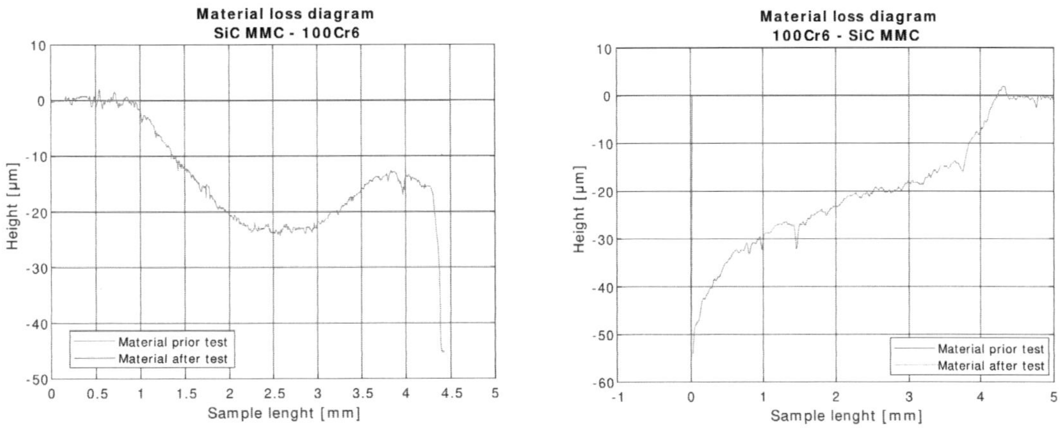

Figure 11 Material loss of coated stator disc (left) and counterpart (right)

In Figure 12, the Stribeck curve of the SiC MMC against the S355 steel disc is plotted. The behaviour at 3 m/s and less can be compared to the 100Cr6 material.

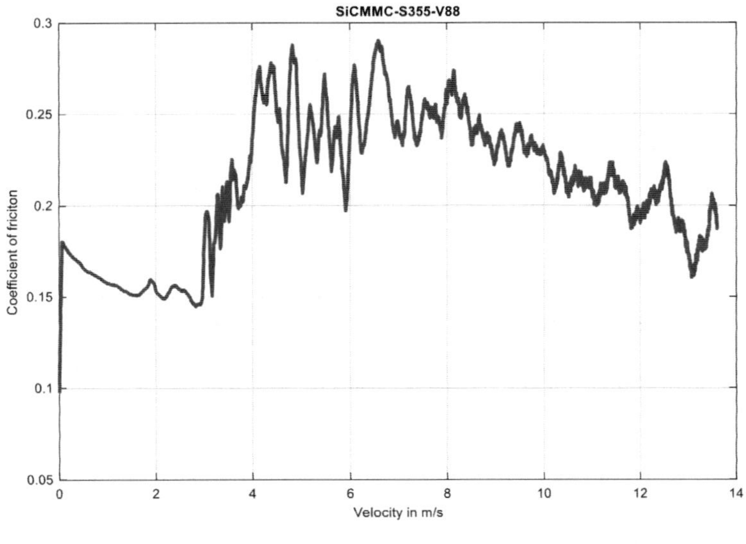

Figure 12 Stribeck curve, plotting coefficient of friction over sliding speed in m/s of SiC coating against S355 test disc

In contrast to the results of 100Cr6, the S355 disc showed less material loss than the SiC coated disc. Figure 13 shows the profile scan after performing both tribological tests.

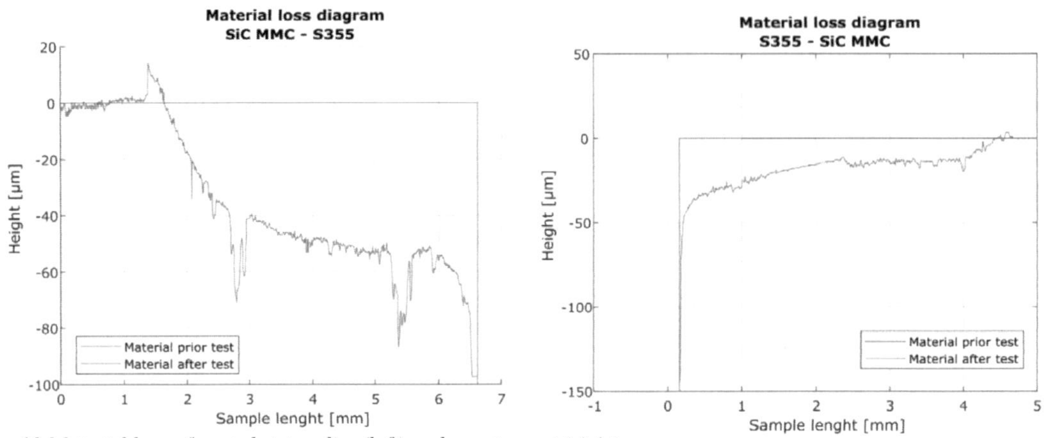

Figure 13 Material loss of coated stator disc (left) and counterpart (right)

Compared to the previous steel-based counterparts, brass has also been used which is much softer than the coated stator disc. The Stribeck curve (Figure 14) shows a drop at much higher velocity. The friction coefficient is slightly smaller, having its minimum at around 8 m/s with a value of 0.07.

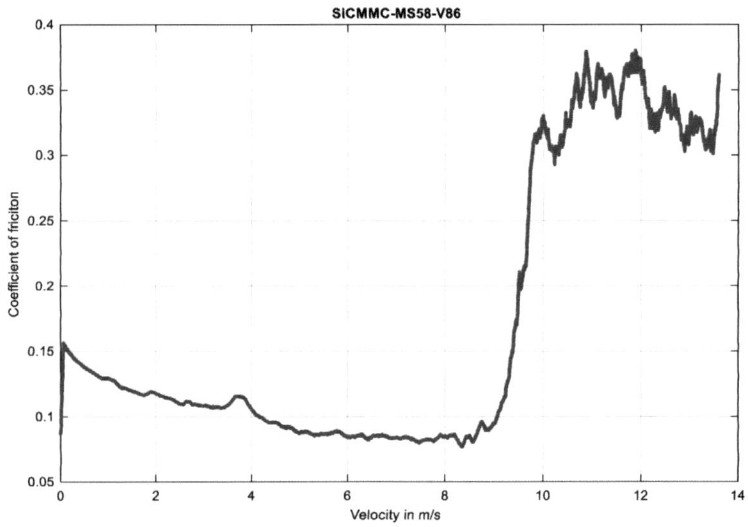

Figure 14 Stribeck curve, plotting coefficient of friction over sliding speed in m/s of SiC coating against MS58 test disc

Figure 15 depicts, that the SiC MMC coated stator (left) has almost no material loss, whereas the brass material (right) lost about 150 µm of its original height.

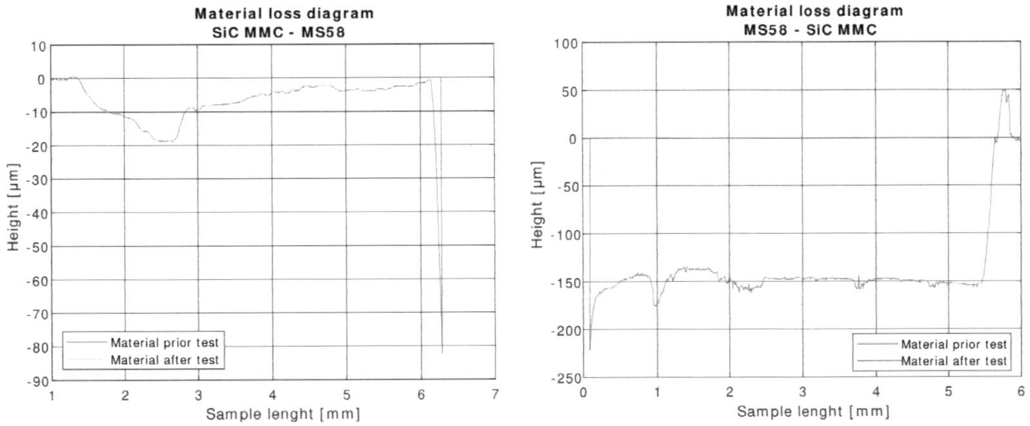

Figure 15 Material loss of coated stator disc (left) and counterpart (right)

The friction coefficients of the MMC 316L+SiC is fairly high (0.2-0.3) compared to the 100Cr6 and S355 samples (0.15-0.28 respectively). In the test against a brass contact, the friction coefficient lies within 0.08-0.15 at speeds up to 8 m/s. With a further increase in speed the friction coefficient increases to 0.3-0.36 is increased at speeds of ≥10 m/s against MS58. This behaviour is increased, as the mixed friction test passes through this area several times. The wear of MMC 316+SiC is in the range of 10 to 50 µm, whereby the greatest wear has occurred with S355. The wear of the opposite part is approx. 20 µm for 100Cr6 and S355 and approx. 150 µm for MS58. The wear data is listed in tTable 4 for the respective pairings.

Table 4 Material wear during mixed lubrication test of 316L+SiC

Coating material	Counterpart material	Coating loss [µm]	Counterpart loss [µm]
SiC MMC	100Cr6	20	20
	S355	50	20
	MS58	10	150

3.5. Results MMC TiC+316L coating

In Figure 16, the Stribeck curve of the TiC MMC against the 100Cr6 steel disc is shown. It can be seen, that the coefficient of friction is continuously increasing from 14 m/s to 5 m/s, where it reaches a value of over 0.4. As observed by previous contact partners, the curve lowers at low speeds at about 2 m/s.

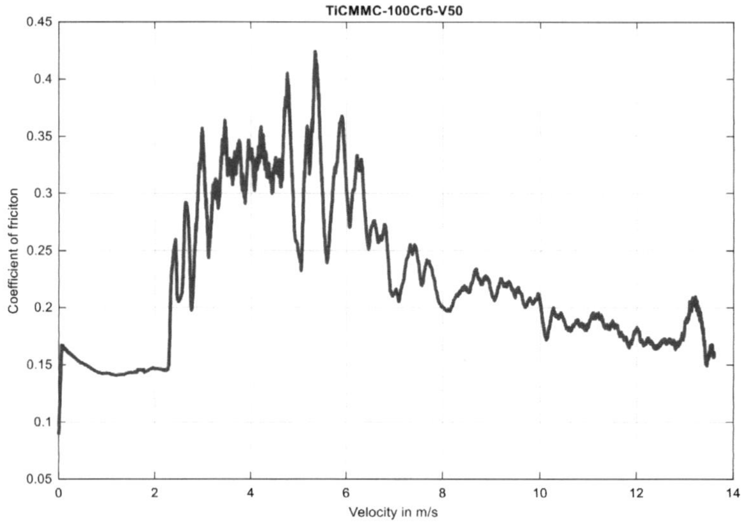

Figure 16 Stribeck curve, plotting coefficient of friction over sliding speed in m/s of TiC coating against 100Cr6 test disc

After testing, the TiC coated disc lost about 20 µm of material, shown in Figure 17 (left side), whereas the counterpart made of 100Cr6 lost 200 µm (right).

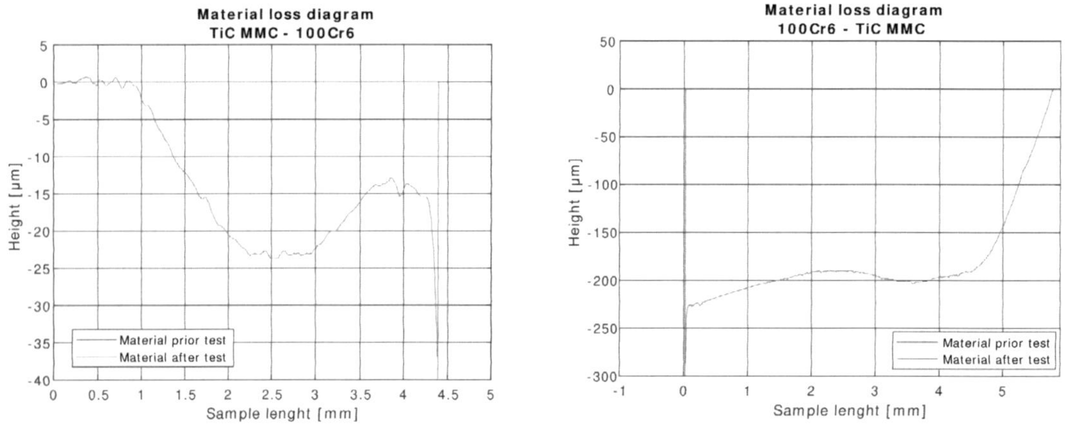

Figure 17 Material loss of coated stator disc (left) and counterpart (right)

Contrary to the behaviour of the SiC MMC coated disc, the measurement of the Stribeck curve started low, at a coefficient of friction around 0.2. From there on, the friction is continuously increasing until their peak of over 0.45 at around 3 m/s. At low speeds, under 2 m/s, the friction coefficient dropped to 0.2, and reached 0.3 in boundary lubrication (see Figure 18).

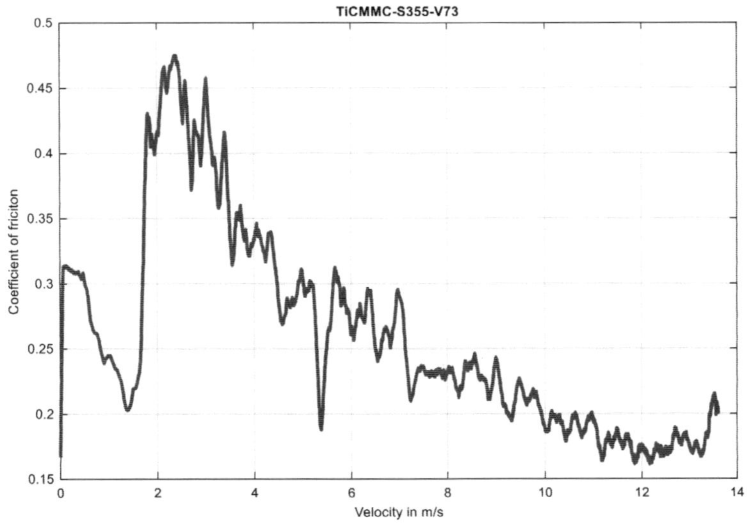

Figure 18 Stribeck curve, plotting coefficient of friction over sliding speed in m/s of TiC coating against S355 test disc

On the left side of Figure 19, the TiC coated disc is shown, having a material loss of 30 µm. The strain caused a bulge on both bodies.The results of the S355 disc showed a loss of almost 150 µm (Figure 19, right).

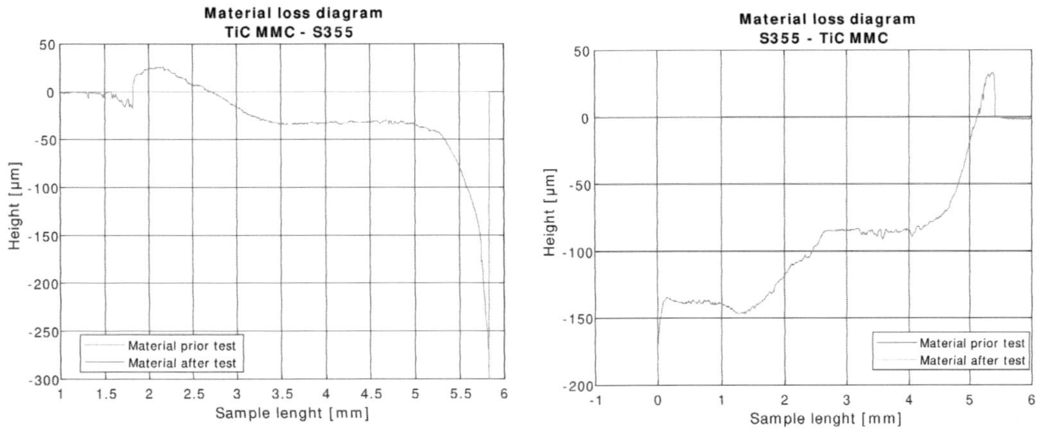

Figure 19 Material loss of coated stator disc (left) and counterpart (right)

When running the TiC MMC coated disc against brass, rather low friction coefficients can be achieved, especially in the range between 2 m/s and 9 m/s (see Figure 20). Higher COF, as they occur at speeds above 10 m/s, and also the two peaks located at 6 m/s and 4 m/s are suspected to be transitions while run-in.

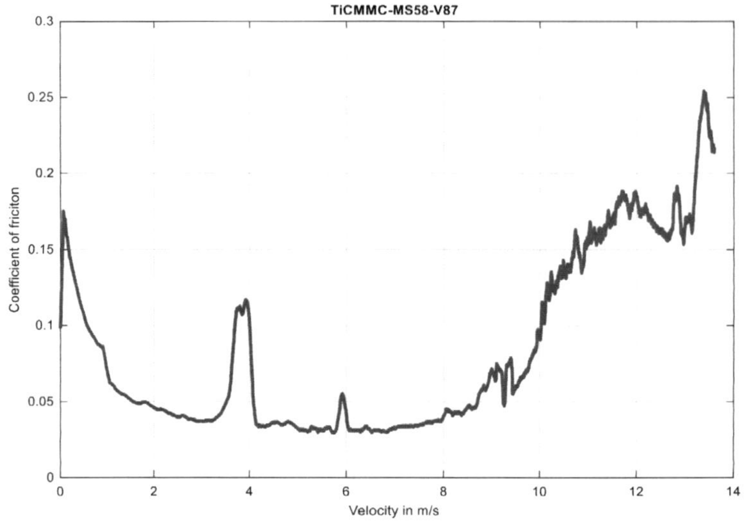

Figure 20 Stribeck curve, plotting coefficient of friction over sliding speed in m/s of TiC coating against MS58 test disc

Due to the fact, that the friction coefficient was the lowest measured of the whole test set, very few wear is expected. The corresponding material loss diagrams are shown in Figure 21, where the coated stator shows a material loss of around 10 µm (left) and the brass rotor suffered of 50 µm loss.

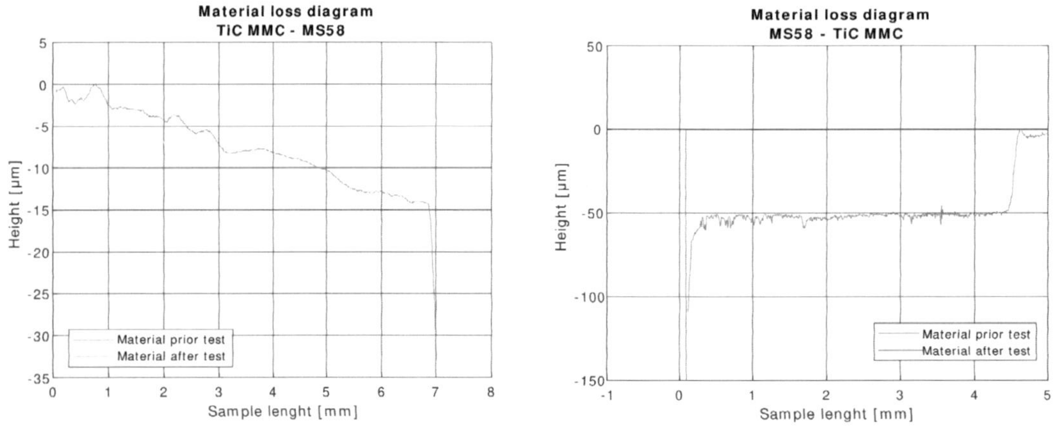

Figure 21 Material loss of coated stator disc (left) and counterpart (right)

For tribological contact against 100Cr6, the friction coefficient is 0.15 at 2 m/s and then climbs to over 0.4 in the range of 2 to 6 m/s. With a further speed increase, the friction coefficient drops to 0.15 to 0.2 (see Figure 16). Against the friction partner S355 (see Figure 18) the greatest coefficient of friction is at a velocity of 2 m/s. A further increase in velocity reduces the friction coefficient to less than 0.1 at 12 m/s. The friction coefficient profile for the MS58 reaches the smallest friction coefficient of <0.15 at a speed between 4 to 8 m/s. The highest friction coefficient with 0.25 is achieved at a velocity of 13 m/s (see Figure 20).

The wear of MMC 316+TiC is in the range of 10 to 30 µm, whereas the wear for Cr6 is 200 µm and for S355 100 µm. The smallest wear occurs in the tribological contact of the MMC 316L+TiC with MS58, MS58 only has a wear of 50 µm (see Table 5).

Table 5 Material wear during mixed lubrication test of the MMC 316L+TiC

Coating material	Counterpart material	Coating loss [µm]	Counterpart loss [µm]
TiC MMC	100Cr6	20	200
	S355	30	100
	MS58	10	50

4. Discussion

The coatings produced show very different characteristics. The MMC of 316L + TiC is characterised by a homogeneous TiC particle distribution and no cracks The TiC particles increase the strength of the MMC due to solid solution strengthening. MMCs with a hard material content of >30 vol.-% show cracks. The increased number of hard particles going into solution increases the stress concentration within the material. The matrix cannot withstand unlimited stress causing cracks to form due to stress relief. The dissolved TiC precipitates form at the grain boundaries. These precipitates pin the grain boundaries increasing wear resistance. However, grain boundaries act as weak spots and reduce ductility. This explains the brittle behaviour observed. The processing of SiC is possible due to the direct energy input into the powder in combination with the high cooling rate (10^6 °C/s). However, with larger volume percentages of SiC, the reaction with iron produces gases, which leads to strong pore formation.The SiC is almost completely dissolved in the matrix material during the deposition process. However, SiC is considered to be a difficult material to weld, as it reacts with iron at temperatures above approx. 1100 °C [12]. The reaction produces iron silicides and the carbon dissolves in the iron, which leads to an increase in hardness. However, if large volume percentages of SiC are added, carbon precipitation can also occur when the iron is saturated [12]. These carbides preferably form at grain boundaries weakening those. According to literature the structure of these precipitates is typical for increasing strength, however the structure itself is brittle, further impeding manufacturability and ductility. Saturation is a sign for segregation, leading to inhomogeneous material composition, an in turn mechanical properties. All test subjects showed a very high friction coefficient, where according to Blau et al., typical values are situated far below 0.1 for long time operation without severe wear [14]. Compared to the steel counterparts, both MMCs showed rather expectable friction coefficients. At higher speeds of more than 8 m per second, the measured coefficient of friction was too high. Peaks can be also seen at 4 m/s and 6 m/s, which may be explained as caused by transition and therefore a measurement error due to the not statically measurement method of Stribeck curve. Where the MMC with TiC showed very good results against the brass material (10 and 50 µm wear), SiC showed a significant loss at the brass counterpart (10 and 150 µm wear). Contrary to the results gained on brass, the SiC coated disc showed very few wear running against the 100Cr6 disc (20 and 20 µm). This pairing may be a very durable combination, enabling a high resistance against mixed friction and supporting emergency running properties for a longer time. Blau and Jolley investigated the abrasive wear behavior of several titanium carbide reinforced matrix composites like Ti64, 5TiB, 10TiC and 7.5W/TiC, using 240 grit aluminum oxide band as counterpart at a sliding speed of 10 m/s and an applied force of 9.6 N. However, the results cannot be compared in a direct manner, but at least a tendency can be seen. The TiC material achieved a wear volume of 1 mm^3 after a sliding distance of 900 m, measuring a linear correlation between sliding distance and wear volume (low-stress abrasion) [15]. While doing the measurements, sliding distance has not been recorded. A calculation based on the speed profile gives an overall distance of each pairing of around 9 km. Compared to the 240 grid aluminum oxide, the wear against steel or brass is negligible. Brink studied the run in behaviour of lubricated steel – steel pairings, explaining that lower velocities are generally more suitable for the run-in transition of the materials [16]. As no run-in took place before performing the Stribeck test, all examined material had to run-in at high speeds of 14 m/s. This could possibly be an explanation of high COF, in particular at high speeds of the Stribeck curve. Often, a drop has been observed at 3 m/s or lower. Kragelski investigated the wear behaviour of lubricated steel – steel contacts on a similar geometry, obtaining wear rates of 0.015 mm/ km applying 0.1 MPa of pressure and around 1 mm/km applying 2 MPa of pressure [20].

Feser et al. measured the coefficient of friction in a lubricated steel – brass (CuZn5) contact between 0.02 and 0.4, at a sliding speed of 10 to 20 mm/s and 1 to 4 MPa [17]. According to Blau, steel on steel typically reaches a coefficient of friction of 0.6 [18]. Compared to the first tribological test, also performed on a disc on disc tribometer by the authors, the use of hard carbides integrated to the 316l coating material improved the wear resistance clearly [7].

5. Conclusion

Within the scope of this study two different MMCs made of 316L+TiC and 316L+SiC have each been successfully tribologically tested against three different counterparts made of 100Cr6, S355 and brass (MS58). The produced MMCs have varying coating properties in the cross section, while the TiC particles have only been dissolved to a small percentage and are present in the MMC in elemental form, whereas the SiC particles are almost completely dissolved into the matrix material. In the tribological tests the coatings show very different results.
It was shown that MMCs, strengthened by Ti- or Si-Carbides, have a high tolerance against fretting and abrasive wear. Compared to their steel counterparts, the wear of the matrix material was just 10% of the steel wear. There is no connection between the hardness of the wear resistance and the hardness of the tested counterparts. It was not possible to abrade or destroy the coating in any way in the test. Even under high mechanical and thermal stress the weld between basis material (S355) and coating was firm. The MMCs, applied via EHLA are a promising way to strengthen the wear resistance of standard base steel.
The use of two hard contact partners compared to the classical system, the use of one hard contact partner against one soft contact partner, poses many new questions to tribology.
It is now no longer necessary to regularly replace wear parts, usually bushings or bearing shells, which are made of soft brass.
At the same time, of course, it can happen that housing parts that should actually be protected by the brass metal are now also subject to wear.
Despite this solution, it must be carefully examined whether a hard-hard pairing should really be used and thus the risk of wear also exists in unintended places, or whether one would rather use the advantage of inexpensive "sacrificial" components.
In further investigations, the results obtained must be examined metallurgically and tribologically in order to describe the wear behaviour. This includes an in-depth metallurgical analysis of the MMC to determine the microstructure and segregations within the alloy. In this way, suitable friction pairings with the lowest possible coefficient of friction as well as the lowest possible wear should be identified.

6. *References*

[1] Florian Summer, Florian Grün, Emma R. Ravenhill. *Friction and Wear Performance of Various Polymer Coatings for Journal Bearings under Stop Start Sliding*. Basel, Schweiz; 2019.
[2] David E. Sander, Mario Witt, Hannes Allmaier, Athanassios Skiadas. *Gleitlagerreibung und -verschleiß im Start-Stopp-Betrieb.* 78th ed. Wien; 2017.
[3] Rainer Aufischer, Gunther Hager, Kamal Hamdard, Martin Offenbecher. *Bearing Technology Combinations for Low Friction Cranktrains*; 2016.
[4] Rainer Aufischer. *Modular Bearing Designs to Cope With the New Engine Designs Demanding High Performance, Lead-Free Solutions, and Robustness*; 2014.
[5] GGB. *Aus welchem Metall sind Gleitlager und Buchsen.* Available at: https://www.ggbearings.com/de/faq/aus-welchem-metall-sind-gleitlager-und-buchsen; 2020.
[6] Achill Holzer, Stephan Koß, Gunnar Matthiesen, Daniel Merget, Stephan Ziegler,Johannes Henrich Schleifenbaum, Katharina Schmitz. *Tribological investigation on additively manufactured surfaceses using Extreme High-speed Laser Material Deposition (EHLA) and Laser Powder Bed Fusion (LPBF)*. Dresden; 2020.

[7] Marc Rettig, Jarolslaw Grochowicz, Klaus Käsgen, Rob Eaton, Andreas Wank, Alexander Hitzek, Christian Schmengler, Stephan Koß, Maximilian Voshage, Johannes Henrich Schleifenbaum, Clemens Verpoort, Thomas Weber. *Carbidic Brake Rotor Surface Coating Applied by High-Performance Laser-Cladding*; 2020.

[8] Thomas Schopphoven, Johannes Henrich Schleifenbaum, Sadagopan Tharmakulasingam and Oliver Schulte. *Setting Sights on a 3D Process: Extreme high-speed laser material deposition, or EHLA, is now being developed into an additive manufacturing process*. Additive Manufacturing; 2020.

[9] TWI. *What is Extreme High Speed Laser Application (EHLA)*. Available at: https://www.twi-global.com/technical-knowledge/faqs/what-is-ehla; 2020.

[10] Matweb. *Titanium Carbide, TiC.*

[11] Matweb. *Silicon carbide, SiC.*

[12] Joshua Pelleg. *Reactions in the matrix and interface of the Fe–SiC metal matrix composite system*; 1999.

[14] Achill Holzer. Optimization of a disc-disc tribometer to avoid hysteresis during friction measurements, 7th European Conference on Tribology ECOTRIB, Wien, Austria, 2019

[15] Blau, Peter J.; Jolly, Brian. Relationships Between Abrasive Wear, Hardness, and Grinding Characteristics of Titanium-Based Metal-Matrix Composites. In: *J. of Materi Eng and Perform* 18 (4), S. 424–432. DOI: 10.1007/s11665-008-9227-3, 2009

[16] Brink, Angelika. Einlaufverhalten von geschmierten Stahl-Stahl-Paarungen unter Berücksichtigung der Mikrostruktur. Dissertation. Karlsruher Institut für Technologie, Karlsruhe, 2015

[17] Feser, Tim; Stoyanov, Pantcho; Mohr, Felix; Dienwiebel, Martin. The running-in mechanisms of binary brass studied by in-situ topography measurements. In: Wear 303 (1-2), S. 465–472. DOI: 10.1016/j.wear.2013.03.047. 2013

[18] Stickel, D.; Fischer, A.; Bosman, R. Specific dissipated friction power distributions of machined carburized martensitic steel surfaces during running-in. In: Wear 330-331, S. 32–41. DOI: 10.1016/j.wear.2015.01.010. 2015

[19] Blau, Peter J. Friction science and technology. From concepts to applications. 2. ed. Boca Raton, Fla.: CRC Press / Taylor & Francis. 2009

[20] Kragelsky, I.V., Dobychin, M.N., Kombalov, V.S. Friction and wear calculation methods. Pergamon Press, 1977

[21] Matthies, Hans Jürgen; Renius, Karl Theodor. Einführung in die Ölhydraulik. Für Studium und Praxis. 8., überarb. und erw. Aufl. Wiesbaden: Springer Vieweg. 2014

[22] Klocke, Fritz. Fertigungsverfahren 2. Zerspanung mit geometrisch unbestimmter Schneide. 5. Auflage. Berlin, Heidelberg: Springer Vieweg (VDI-Buch). 2017

[23] Bouyer, J., Fillon, M. Experimental measurement of the friction torque on hydrodynamic plain journal bearings during start-up. Institut Pprime, CNCS-Université de Poitiers-ENSMA, Chasseneuil Cedex, France

IOP Conf. Series: Materials Science and Engineering 1097 (2021) 012017 doi:10.1088/1757-899X/1097/1/012017

Experimental Validation of a New Design Concept for the Increase of Efficiency of the Hydraulic Drive System in Mobile Working Machines

S. Geffroy[1], A. Brautlacht[1], S. Wegner[2], S. Gels[3], K. Schmitz[1]

[1] Institute for Fluid Power Drives and Systems (ifas), RWTH Aachen University, Campus-Boulevard 30, 52074 Aachen
[2] Stackpole Powertrain International GmbH, Campus-Boulevard 30, 52074 Aachen
[3] Hydraulische Antriebstechnik P&G GmbH, Nettestraße 39, 58762 Altena

Stefan.Geffroy@ifas.rwth-aachen.de

Abstract. In mobile working machines like excavators, hydraulic actuators and drives are indispensable due to their advantages of a high power to weight ratio, robustness and the good power transmission over medium distances by the hydraulic fluid. Axial piston machines can deliver a large volume flow at a high operation pressure with high efficiency and are therefore widely used as system pump or drive motor. The three essential tribological contacts piston / bushing, sliding shoe / swash plate and valve plate / cylinder block strongly influence the machine's efficiency and so the complete drive system. The paper on hand focuses on the experimental validation of a new design concept to increase the efficiency of the tribological contact between valve plate and cylinder block in an axial piston pump. Due to the different pressure forces at the high- and low-pressure side, the cylinder block tilts and holds this preferred position almost constant. Therefore, the gap height is not constant. In the area of minimum gap height, the danger of solid body contact increases. In addition, the heat dissipation is decreased, which can lead to local constant high temperature hot-spots. Both can destroy the surface structure or the coating. A new design concept has been developed to reduce these risks, thus increasing the efficiency, the lifetime and making it possible to dispense of leaded coating materials. In this design concept, pressure pockets are added in the area of minimum gap height. The simulation study shows a high potential of this concept. The experimental results of the prototype parts, which are measured on a special test rig and presented in this paper, confirm the simulations and validate the potential of the new design concept in this way.

1. Introduction

Mobile working machines are used worldwide in several areas, for example construction, mining, agriculture and forestry. Due to their advantages like a high power to weight ratio, robustness and the easy realization of linear actuators, hydraulic drives are the main drive technology in mobile working machines like excavators. The state of the art is that an internal combustion engine drives a hydraulic pump, which generates the hydraulic power and delivers it via the working fluid to the various actuators. Axial piston pumps are widely used because they can deliver high volume flows in high pressure operating points with an overall efficiency of about 90 % in modern units [1]. **Figure 1** shows an axial piston pump with its main components. The losses of the three essential tribological contacts piston / bushing, sliding shoe / swash plate and valve plate / cylinder block [1] determine the efficiency of these pumps in a high degree. Research to improve these contacts is therefore of great importance for the efficiency of mobile working machines in the context resource efficiency, climate protection and the emissions law.

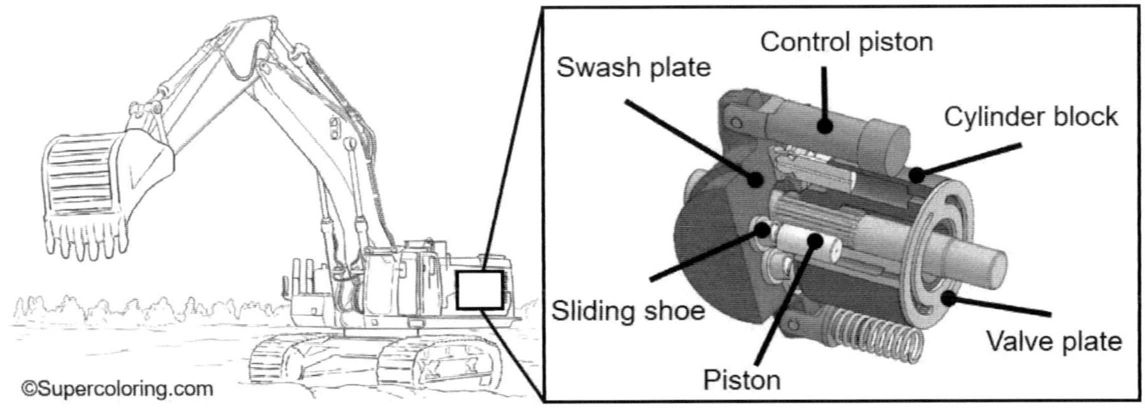

Figure 1: Main components of an axial piston machine [2]

Leaded alloys are used in the tribological contacts due to their emergency running properties and to compensate manufacturing errors. One part is therefore made of steel, while the other is coated with brass or bronze alloys. However, lead is a toxic heavy metal and its substitution is one of the main topics in hydraulics.

The topic of the here presented work is the tribological contact between the valve plate and the cylinder block. The drive shaft transmits the rotation movement to the cylinder block, which is pressed on the valve plate by the pressure forces and by springs to stabilize it and avoid lifting. The tribological system has two main tasks. The first is to minimize volumetric losses by sealing the high- and low-pressure side from each other and from the housing. The oil film in the gap fulfills the second task, to protect the contact surfaces against wear and to reduce the friction losses to a minimum. It lubricates the contact surfaces, dissipates heat and particles, and compensates the majority of the axial forces by hydrodynamic effects. However, this results in a conflict of objectives: The gap height needs to be small enough for low leakage but not too small, as this reduces heat and particle dissipation and increases wear. The perfect condition would be a small and even gap height over the complete contact area.

However, the wear pattern on the valve plate is not uniform. Early in the development of axial piston machines, it was assumed that the cylinder block tilts during its operation. This increases the load and thus the local wear. P. Achten investigates the tilting effect in [3]. He analyses and calculates the forces which are acting on the cylinder block in operation. By setting up and solving the equilibrium of these forces and torques he showed that they are unbalanced. On the one hand, the displacement chambers are filled with oil differently depending on the angle of rotation, so that the applied centrifugal forces are also uneven. On the other hand, the different compressive forces resulting from the narrowing of the cross-section are significantly higher. Here the difference is significant, especially in operating points with high load pressure, since a pressure of several hundred bar acts on the high pressure side, while the low pressure side is preloaded to reduce the tendency of cavitation. Because of this, a tilt torque to the high pressure side results. S. Wegner build up a special simulation model [4], which is briefly explained later in this paper, to investigate tribological contact of valve plate and cylinder block. In his simulation studies, he confirmed the results of the analytical investigation of P. Achten. The simulation results show that the cylinder block tilts rapidly to the high pressure side and consistently holds this preferred position steady. **Figure 2** shows the position of minimum gap height when 4 or 5 pistons are pressurized with high pressure. Due to the earlier explained negative effects of a gap height that is too small, the danger of solid body contact and the danger of a local constant high temperature hot spot is increased in this area [5]. Both can destroy the surface condition or coating.

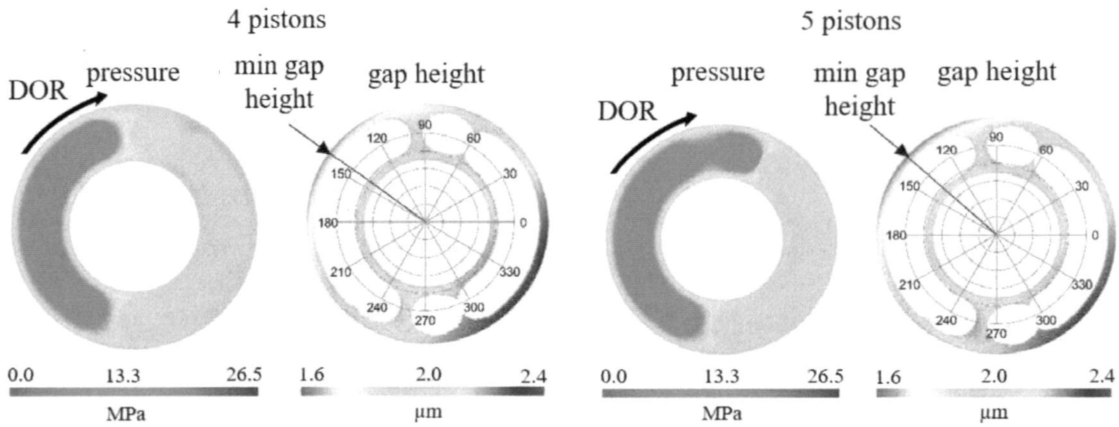

Figure 2: Position of minimum gap height [4]

The new design concept with additional pressure pockets in the valve plate's high pressure kidney, which is presented and validated experimentally in this paper, reduces the tilt angle of the cylinder block. This reduces the mechanical and thermal load significantly in the area of the minimum gap height. So, it could be the key for the implementation of new tribological optimized coating materials without lead.

2. State of the Art

For a long time, the development process focused mainly on the two problems, to avoid the lifting of the cylinder block and to improve its centering. In the context of the present study, different approaches have already been pursued. In [6], punctual pressure fields are positioned in the outer support ring and connected to the high-pressure kidney by means of holes. This supporting pressure field reduces the tilting of the cylinder block and thus increases the self-priming speed by 15 %. The disadvantage of this concept is the high additional production-related expenditure. The connecting bores must be drilled from the outside into a round surface. In addition, a thread must be produced for sealing with a plug. The actual support bores would also be difficult to produce in the case of a spherical valve plat, but they also represent an additional expense in the case of flat ones. In addition, the sealing surfaces would have to be designed accordingly wide. A completely different concept is patented in [7]. Here, an additional device prevents the piston drum from tilting. Compensation pistons, which are connected to the flow channel through a bore, generate a moment against the tilting. A new approach with an additional relief field in the control mirror is patented in [8]. This relief field is located in the area of the high pressure side. In this approach, the pressure field is externally pressurized. The supporting torque can thus be controlled independently of the operating point. Apart from the additional manufacturing costs, the main disadvantage of this approach is the additional components of the control system. For stable operation, the tilting over the entire characteristic diagram must also be known. S. Haug extends this approach and implements a further pressure field on the low-pressure side [9][10]. In his investigations, he also applies this concept to other tribological contacts and develops an operating point-dependent control. By means of this control, tilting of the cylinder block can be prevented almost completely over the entire characteristic diagram of the pump under investigation. The disadvantages of this concept are the additional components and the complex control design.

3. Concept Development

The research at ifas combines the experimental and the simulation investigation of the contact between the cylinder block and the valve plate in an axial piston machine. By combining simulation and experimental results, the effect of the new design concepts of optimized additional pressure pockets can

IOP Conf. Series: Materials Science and Engineering 1097 (2021) 012017 doi:10.1088/1757-899X/1097/1/012017

be thoroughly analyzed. The simulation tool, as well as the test rig, were specifically designed for the measurement of the cylinder block movement.

3.1. Simulation Model

To analyze the tribological contact between the valve plate and the cylinder block, a simulation model was set up at ifas. The simulation model is based on the model for the piston/bushing contact described in [11]. Its development a detailed description can be found in [4]. It will only be presented briefly in this paper. The simulation tool was designed with the aim of comparing several geometries at different operating points within an acceptable computation time and a sufficient accuracy. To reach this goal, the simulation tool includes the following physical effects:

- Hydrostatic and hydrodynamic pressure build up using the Reynolds equation
- Force and torque equilibrium on the cylinder block
- Squeeze film effect, viscous friction, microscopic part movement resulting from the first two points
- Macroscopic part movement, analytically derived
- Simple contact and part deformation model

The finite volume method is used to discretize the Reynolds equation. To calculate the gap height and the cylinder block movement, a linear system of equations using mass and momentum balances is set up. The Reynolds equation is added to the linear set of equations. Using the system parameters from the boundary conditions, from the geometry and the high- and low-pressure areas, the unknown variables and the number of equations are balanced, and the system can be solved numerically.

Solid friction is implemented by utilizing the contact pressure: The contact pressure at each finite area is multiplied with a constant friction coefficient (0.06), leading to the share of friction torque for each finite area. **Figure 3** shows a diagram of the basic steps of the simulation model.

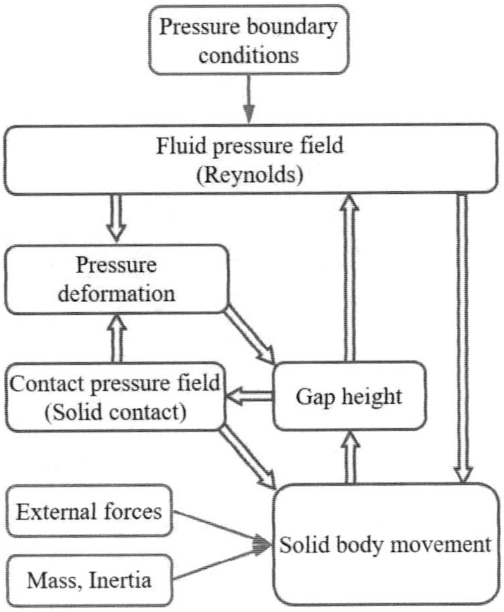

Figure 3: Simulation model [12]

3.2. Pre-study on additional forces to prevent cylinder block tilting

The pre-study had the purpose of determining the optimal force distribution to avoid the movement of the cylinder block into a preferred position. For this, four scenarios were simulated to show the effect of adding forces which create a torque that pushes the cylinder block out of the tilted position. The complete study can be found in [12]. The first option is to add one or more forces in z-direction. The

IOP Conf. Series: Materials Science and Engineering 1097 (2021) 012017 doi:10.1088/1757-899X/1097/1/012017

second is to add those forces in radial direction. The added forces can either be static or dynamic. This totals in four different scenarios: a static force in z-direction, a dynamic force in z-direction, a static radial force, or a dynamic radial force. In the pre-study the static force in z-direction showed the greatest potential. Thus, the results for an additional static force in z-direction will be discussed here. The added forces are described by their position through the radius, angle and size. **Table 1** shows the geometric parameters. The values are limited by the geometry of the valve plate's kidneys. The inner radius of the kidney is 29.45 mm, the outer radius is 44.55 mm and the medium radius is 37 mm as shown in **Figure 4**.

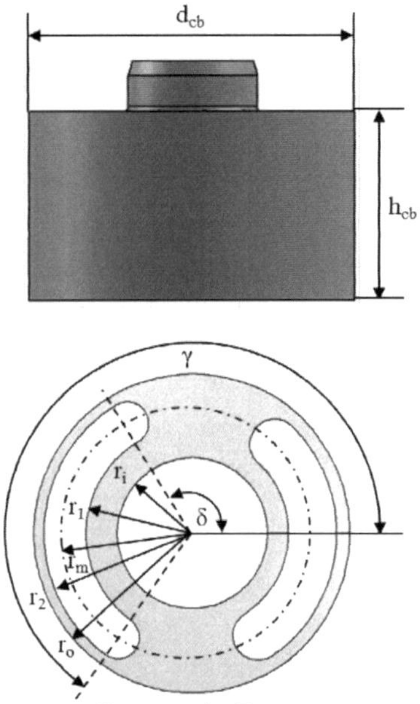

Figure 4: Geometric Parameters

Parameter	Value
d_{cb}	97.2 mm
h_{cb}	80 mm
r_1	29.45 mm
r_2	44.55 mm
r_i	23 mm
r_m	37 mm
r_o	48.6 mm
γ	225°
δ	120°

Table 1: Geometric Parameters

For the simulation, the added forces varied from 500 N to 2500 N in steps of 500 N. The change of the tilting behavior was simulated for the position at the inner, outer, and medium radii and for the angles of 120°, 135° and 150°. To portray the change of the tilt behavior, the relative change of the tilt angle was chosen as a comparative value. The tilting can be calculated with the gap height difference and the cylinder block's diameter by equation 1. The relative angle change is the ratio of the tilt angle with the additional forces and the tilt angle in the basic configuration.

$$\alpha = \sin^{-1}\frac{\Delta h}{d_{CB}/_2} \qquad (1)$$

The added force has two effects on the movement of the cylinder block movement. The first effect is a moment which reduces the tilt angle and the second is that the force lifts the cylinder block. Both effects reduce the gap height. The simulation results in **Figure 5** show that the minimum gap height rises with the value of the additional force. A larger radius increases this effect because the moment against the tilt angle increases.

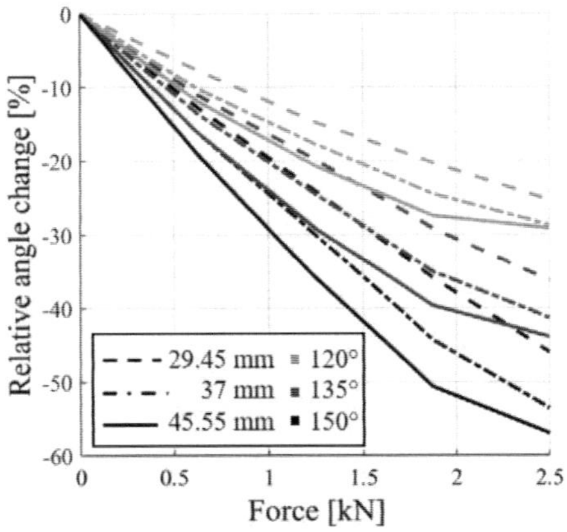

Figure 5: Change of relative tilt angle adding a static force in z-direction

3.3. Design Concept

To implement the concept of an additional force on the valve plate, a design concept is developed. Adding forces in hydraulic machines can be achieved by additional pressurized surfaces. For the concept presented here, pressure surfaces will be added to the sealing gaps of the valve plate's high-pressure kidney. **Figure 6** shows the geometry of the sealing gaps for the considered area as well as the additional pressure pockets as CAD and the prototype used in the experiments. The position of the pressure surfaces should be located around 150°. This position is chosen because it is the angle at which the minimum gap height was shown to be the smallest (Figure 2). The potential size of the pressure pocket results in 3.325 mm towards the shaft and 0.925 mm towards the housing, which results in maximum Surface of 145.72 mm². By assuming load pressure of 250 bar a force of 3.64 kN is added. For the prototype, the surface of 101.01 mm² was added, because a chamfer reduces the sealing width in the valve plate of the considered pump. In this way, the sealing gap width is reduced by 2 mm.

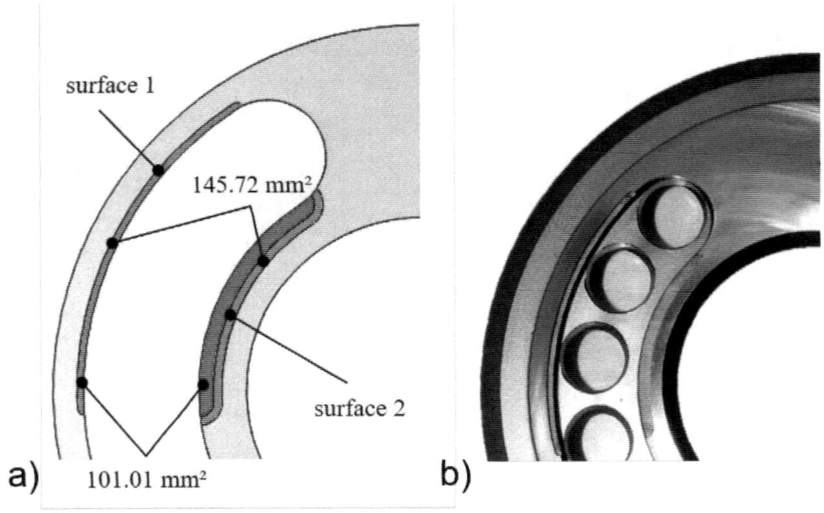

Figure 6: Geometry of pressure pockets a) CAD, b) manufactured

3.4. Simulation Results

To evaluate the impact of the pressure pockets on the axial piston pump, the previously described simulation model was used. **Figure 7** shows the simulation results for the minimum gap height, the contact pressure, the solid friction torque and the tilt angle. The change of the values depends on the speed and the pressure of the axial piston pump. Therefore, the simulation results are shown in a three-dimensional surface plot. The mesh grid is defined by the speed of the pump in rotations per minutes (rpm) and the pressure on the high-pressure side in bar. The simulations points of the pump speed are 500 rpm to 2500 rpm in steps of 500 rpm and the high-pressure goes from 50 bar to 250 bar in steps of 50 bar. Each simulation is run for the values with and without the pressure pockets. The values of the relative change graph were calculated with the ratio of the total values.

The minimum gap height increases when the pressure pockets are applied. Especially when the rotation speed is low, the relative change of the gap height is high. The minimum gap height increases up to 3.6 % with the added pressure pockets. The contact pressure and solid friction torque show a high decrease when adding the surfaces. Both decrease by up to 30%. The tilt angle also decreases. Especially in the area of low rotation speed and high pressure. When the rotation speed is low and the pressure high, the risk of solid friction is the highest. This is due to the lubricating fluid film between the valve plate and the cylinder block not being fully developed yet. According to the simulation results, the concept of added pressure pockets to the valve plate has great potential.

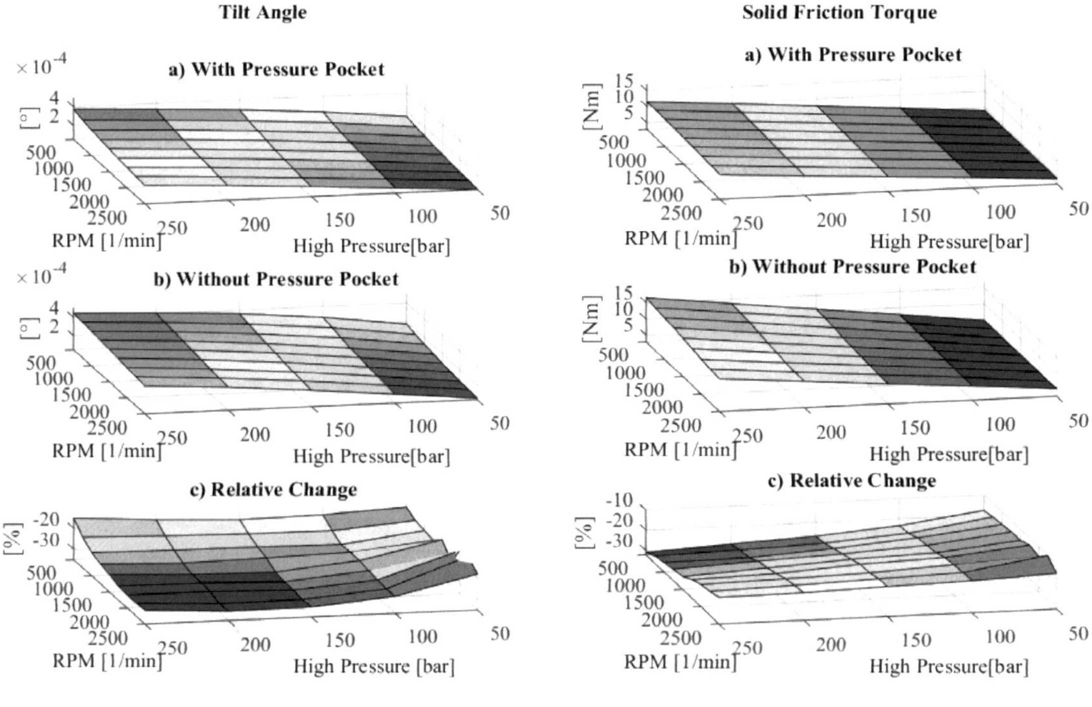

Figure 7: Simulation Results

4. Experimental Validation

4.1. Test Rig

For the experimental investigation of the tribological contact to validate the simulation model, a special test rig was designed at the institute. In **Figure 8**, a picture of the constructed test rig and a circuit sketch can be seen. The setup of the test rig is also described in [13] but is further explained in the following.

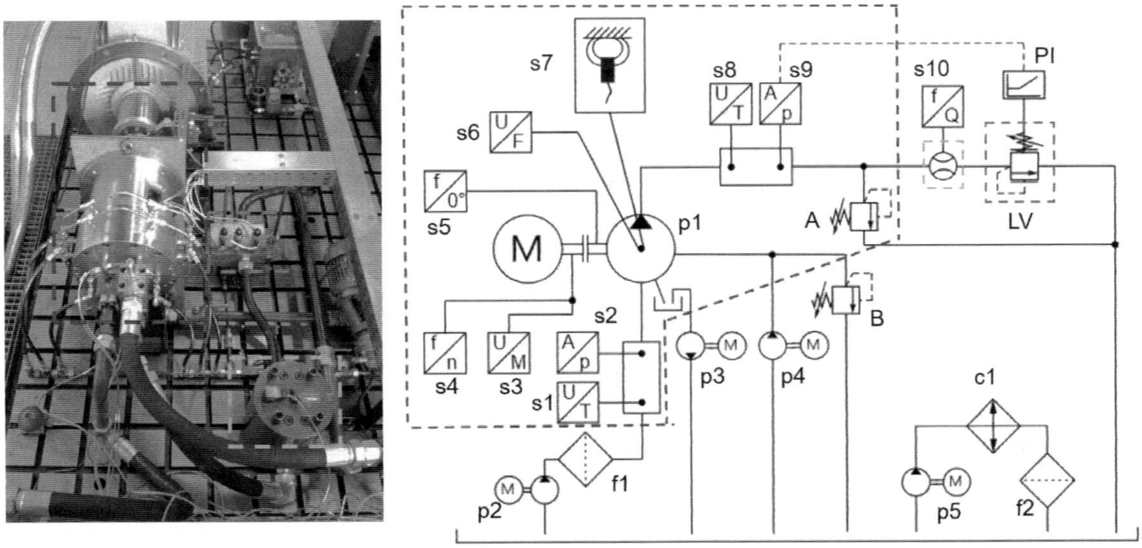

Figure 8: Test Rig and circuit sketch

The test rig is based on the test pump (p1) which consists of the engine of a serial pump with a displacement of 140 ccm, a maximum pressure of 340 bar and a maximum rotational speed of 2400 rpm as well as a hydrostatic bearing, which is described in more detail later. A force sensor measures the frictional force in the contact between the piston drum and the control mirror. The gap height as well as the movement of the piston drum is measured by 6 eddy current sensors. The pump is driven by an electric motor with a nominal power of 160 kW and a maximum rotational speed of 2600 rpm. The motor is connected to the test pump via a coupling. At the drive shaft, the rotational speed, the torque, and the position (angle of rotation) is measured by the sensors s3, s4 and s5 respectively. A centrifugal pump (p2) with a maximum volume flow of 250 l/min supplies the test pump with fluid. In addition, the pump preloads the low-pressure side to 10 bar. This ensures the reproducibility of the measuring points and reduces the risk of cavitation. In addition, the oil is filtered before entering the test pump (f2). The sensors s1 and s2 measure temperature and pressure on the low-pressure side. The pump p4 supplies the hydrostatic bearing with oil. The pressure can be adjusted manually via the pressure relief valve A. The leakage volume flow of test pump and hydrostatic bearing is discharged into the tank by pump p3. On the high-pressure side, the temperature and the pressure are measured by sensors s8 and s9. The pressure relief valve B is set to a value of 350 bar to protect the components behind it. The output volume flow is measured by means of the volume flow meter s10. The load valve (LV) consists of three proportional pressure relief valves with a maximum pressure of 350 bar and a maximum volume flow of 80 l/min. The operating pressure is set and controlled via a PI controller, which is integrated in the test bench control system. The oil conditioning is realized in an environmental circuit consisting of a pump p5, the cooler c1 and the filter f2.

In **Figure 9**, the test pump (p1) with the hydrostatic bearing is shown. The angle of the swash plate is set manually using the fixation bolts. The drive mechanism of the pump, which consists of the cylinder block, the pistons and piston slippers, is connected to the drive shaft through a gearing in the piston drum. During operation, the valve plate lies frictionless in the hydrostatic bearing. The hydrostatic bearing consists of the housing, the hydrostatically relieved internal rotor and the sensor holder. A force sensor and 6 eddy current sensors are integrated in the sensor holder.

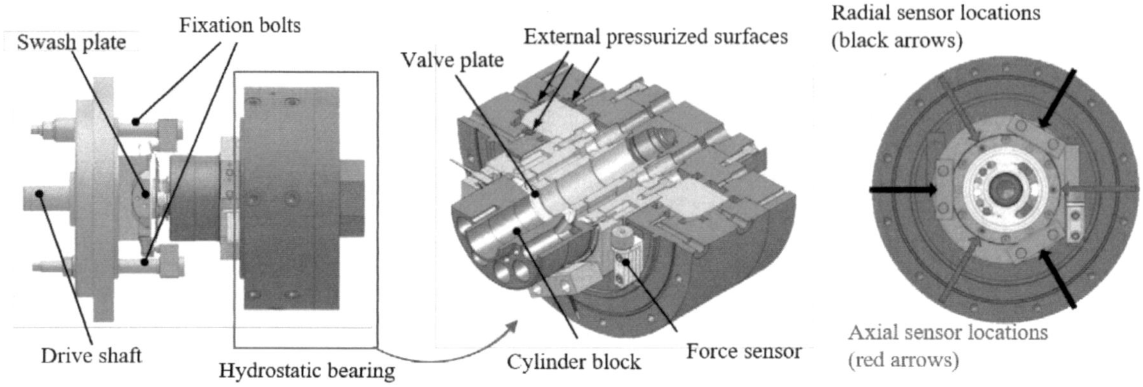

Figure 9: Test pump and sensor location

4.2. Experimental Results

For the validation of the simulation results, the standard valve plate and the prototype are measured at the test bench to analyse the friction torque for the different configurations. It is measured by the force sensor (s6). The hydrostatic bearing is pressurized constantly with 92 bar. The measurements from the run-in-process showed that the inner runner is completely relieved even when the load pressure is more than 300 bar. The low-pressure side is pre-loaded with 8 bar constantly for all considered operation points. The measurement points are as follows, four speed points (250 rpm, 500 rpm, 750 rpm, 1000 rpm) are approached and then the pressure is increased from 100 bar to 250 bar. The oil, a standard HLP46, is regulated to have a constant temperature of 40°C, as it was also used in the simulations. The results of the temperature measurements show that the temperature on the low-pressure side is 46°C. This can be explained with the heat dissipation of the pump p2 and the filter f1. The temperature on the high-pressure side is higher but less than 1 °C and increases less than 2°C during operation. **Figure 10** shows an exemplary measurement of the temperature for the measurement at 1000 rpm with the standard valve plate.

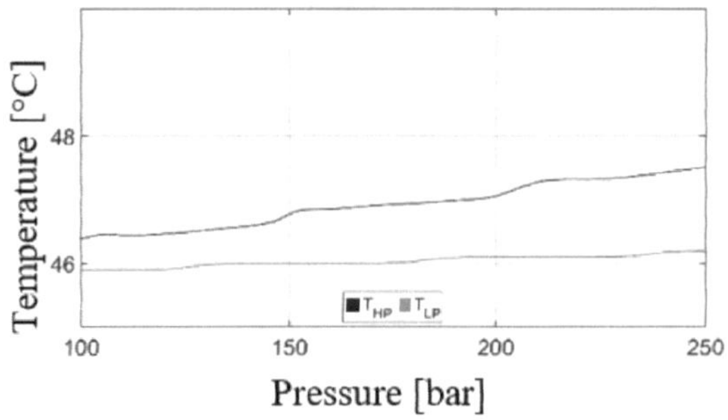

Figure 10: Temperature during measurement at 1000 rpm

Figure 11 shows the measurement results for both configurations. The friction torque increases almost linearly with the pressure increase. With the increase of the drive speed the friction torque also increases. These measurements are influenced by both the solid friction and shear friction, which increases with the drive speed.

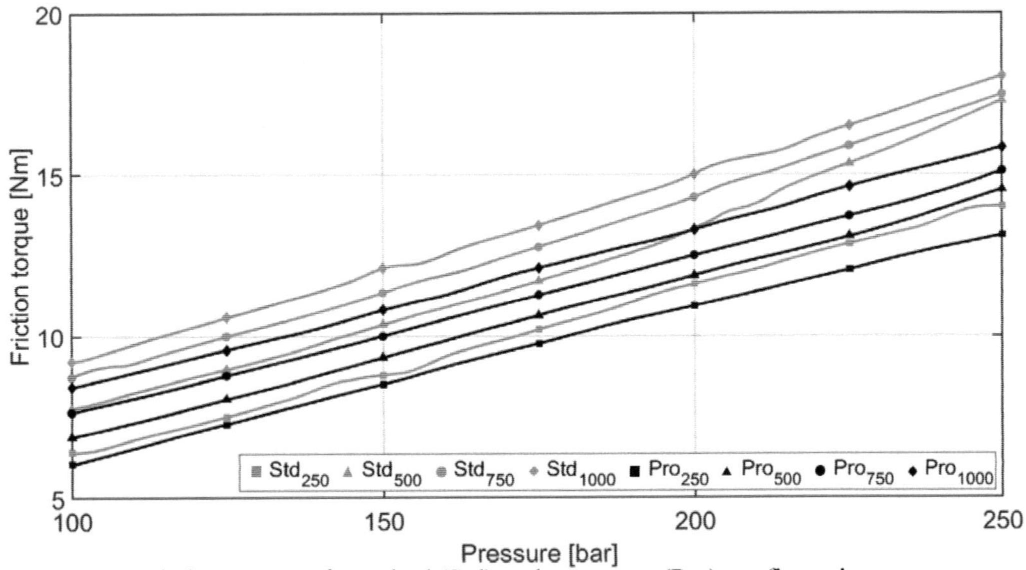

Figure 11: Friction torque of standard (Std) and prototype (Pro) configuration

The relative change of the friction torque when comparing the measurement results of both configurations is shown in **Figure 12**. A reduction of the friction torque can be noted for all operation points. The reduction is the highest at 250 bar and range from -6.3 % up to a maximum of -15.9 %. The results show that the effect of the pressure pockets is influenced by hydrodynamic effects and the increase of leakage. This can be seen in the measurement results of 250 rpm and 500 rpm. One advantage of the tilting towards the high pressure side is, that the decreased gap height increases the sealing, which is one of the main tasks of this tribological contact. The aim of the new design concept is to reduce the tilt angle, but this increases the leakage due to the higher gap height in this region. The experimental results show that the additional pressure pockets reduce the friction torque and support the simulation study even though the simulation showed a greater effect.

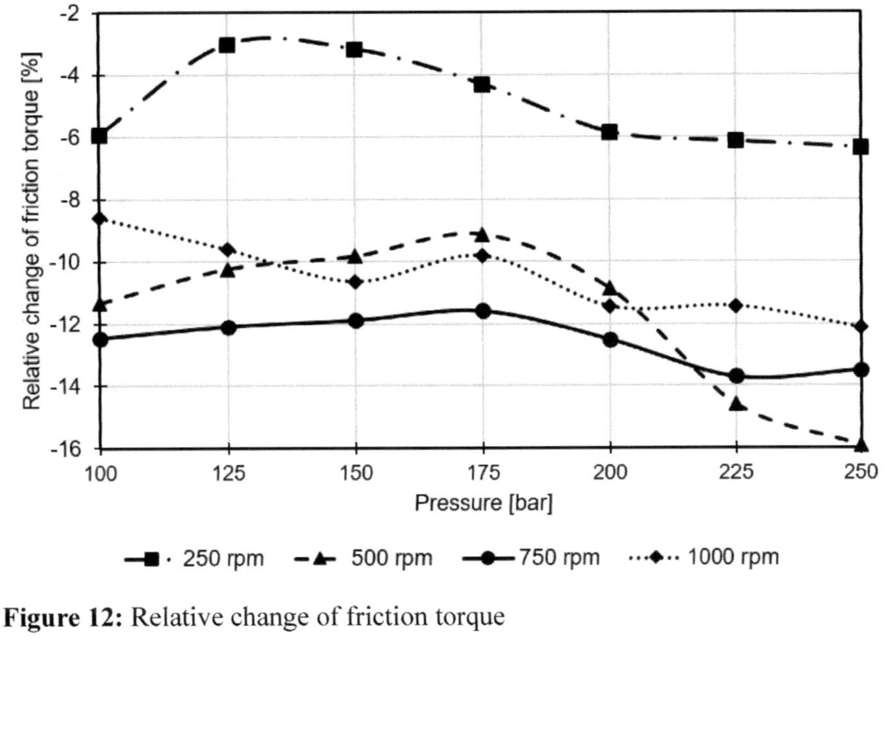

Figure 12: Relative change of friction torque

Due to the reduced sealing gap caused by the additional pressure pockets, leakage increases as expected. The relative change of the volumetric efficiency is shown in **figure 13**. On average it is less than 1 %. At low speed, leakage is comparatively high, especially at high load pressure. The measurements at 250 rpm show an increase in volumetric efficiency. However, this is probably due to the viscosity, which is strongly influenced even by small temperature differences.

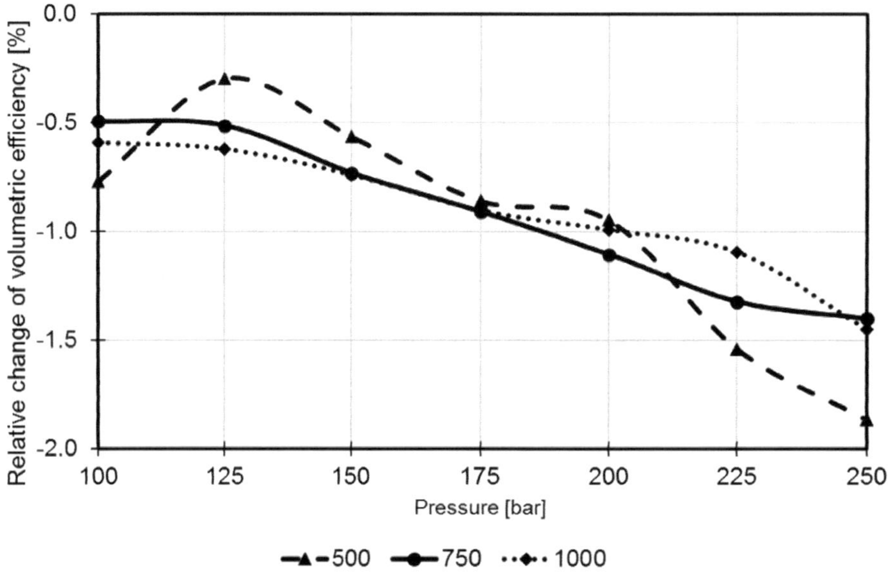

Figure 13: Relative change of volumetric efficiency

5. Conclusion and Outlook

Based on a simulation study a new design concept with additional pressure pockets at high pressure side was developed to reduce the tilting of the cylinder block. The simulations of the configuration with an additional surface of 101.01 mm² show great potential. The implementation of these pressure pockets reduces the tilting and the solid friction in all considered operation points. In this work it is shown that the simulation results could also be validate in experimental investigation of the new design concept. The friction torque for the configurations with a standard and a prototype valve plate is measured for several operation points. The experimental results show that the friction torque increases with the load pressure as well as the rotational speed. In all investigated operation points the friction torque is reduced for the prototype configuration. However, the reduction is at maximum 15.9 % and therefore less than predicted by the simulations. Additionally, the increasing effect of the pressure pockets with an increasing load pressure could not confirmed with the experimental results. One possible cause can be fluid friction, which is increased by the changed geometry. Basically, the experimental results show that the new design concept has the potential to reduce the friction in the considered contact.

The next steps in the ongoing research project are the extension of the experimental investigation for this design concept as well as the design concept presented in [14]. The effect of both design concepts to the tilting has to be measured with the explained single contact test rig. After, the results can be used by hydraulic displacement machine manufacturers to increase the overall efficiency of the components.

6. Acknowledgement

This work was performed in the scope of the project "Increasing the efficiency of the cylinder block/valve plate contact in axial piston machines"/"Steigerung der Leistungsfähigkeit des Kolbentrommel-Steuerspiegel Kontaktes in Axialkolbenmaschinen" (MU41225/44-1), which is funded by the German Research Foundation (DFG). The authors would like to thank the German Research Foundation for funding the project

Nomenclature

c1	Cooler 1
DOR	Direction of rotation
d_{cb}	Outer diameter of the cylinder block
f1...2	Filter 1…2
h	Gap height
h_{cb}	Height of the cylinder block
Δh	Gap height difference
ifas	Institute for fluid power drives and systems
l	Gap length
LV	Load valve
p1...5	Pump 1…5
PI	PI controller
r_1	Inner radius of high pressure kidney
r_2	Outer radius of high pressure kidney
r_i	Inner radius of valve plate
r_m	Medium radius of high pressure kidney
r_o	Outer radius of valve plate
s1...10	Sensor 1…10
T_{LP}	Temperature low pressure
T_{HP}	Temperature high pressure
α	Tilt angle
γ	Start angle of high pressure kidney
δ	End angle of high pressure kidney

References

[1] Gärtner M.
Verlustanalyse am Kolben-Buchse-Kontakt von Axialkolbenpumpen in Schrägscheibenbauweise, Dissertation, RWTH Aachen University, Aachen, 2019

[2] Schmitz K., Murrenhoff H.
Grundlagen der Fluidtechnik Teil 1: Hydraulik, Shaker Verlag, Aachen, 2018

[3] Achten P., Eggenkamp S.
Barrel tipping in axial pumps and motors, 15th Scandinavian International Conference on Fluid Power, Linköping, Sweden, 2017

[4] Wegner S., Gels S, Murrenhoff H.
Simulation of the tribological contact cylinder block/valve plate and influence of geometry and operation points on the friction torque in axal piston machines, The 9th International Fluid Power Conference, Aachen, Germany, 2014

[5] Ivantysyn R., Shorbagy A., Weber J.
An approach to visualize limiting factors in the cylinder block/valve plate gap in Axial Piston Pumps, ASME Symposium on Fluid Power and Motion Control, Bath, Great Britain, 2017

[6] InLine Hydraulik GmbH, DE 198 23 353 A1, patent, Germany, 1999

[7] Linde AG, DE 198 55 899 A1, patent, Germany, 2000

[8] Robert Bosch GmbH, DE 10 2010 006 895 A1, patent, Germany, 2011
[9] Haug S., Geimer M.
Optimierung von Axialkolbenmaschinen durch bedarfsangepasste Entlastung tribologischer Kontakte, 9. Kolloquium Mobilhydraulik, Karlsruhe, Germany, 2016
[10] Haug S., Geimer M.
Optimization of Axial Piston Units Based on Demand-driven Relief of Tribological Contacts, The 10th International Fluid Power Conference, Dresden, Germany, 2016
[11] Gels S.
Einsatz konturierter und beschichteter Kolben-Buchse-Paare in Axialkolbenmaschinen in Schrägscheibenbauweise, Dissertation, RWTH Aachen University, Aachen, 2011
[12] Geffroy S., Bauer N.,Mielke T., Wegner S., Gels S., Murrenhoff H., Schmitz K.
Optimization of the tribological contact of valve plate and cylinder block within axial piston machines, The 12th international Fluid Power Conference, Dresden, Germany, 2020
[13] Wegner S., Gels S., Jang D.-S., Murrenhoff H.
Validation of the physical effect implementation in a simulation model for the cylinder block/valve plate contact supported by experimental investigations, 10th International Fluid Power Conference, Dresden, Germany, 2016
[14] Geffroy S., Wegner S., Gels S., Murrenhoff H., Schmitz K.
New Design Concepts for the Tribological Contact of Cylinder Block and Valve Plate, ASME Symposium on Fluid Power and Motion Control, Bath, Great Britain, 2020

IOP Conf. Series: Materials Science and Engineering 1097 (2021) 012018 doi:10.1088/1757-899X/1097/1/012018

Simulation-Based Data Sampling for Condition Monitoring of Fluid Power Drives

Faried Makansi[1] and Katharina Schmitz[1]

[1]Institute for Fluid Power Drives and Systems (ifas), RWTH Aachen University, Campus-Boulevard 30, 52074 Aachen, Germany

Faried.Makansi@ifas.rwth-aachen.de

Abstract. Machine learning techniques are continuously gaining attention and importance in several technical domains. In the field of engineering, they can potentially provide manifold advantages for condition monitoring. However, availability of extensive operation data is a limiting factor. In this contribution, a simulation-based approach is presented, which allows an efficient generation of training data. Based on a lumped parameter simulation, a database of time-series data is generated for a hydraulic reference system. In order to incorporate states of faulty machine operation in the database, means to model component faults in the simulation are assessed. Further, a procedure for an automated training data generation is presented.

1. Introduction

Reliability and operational safety are crucial requirements for production machines and automation plants. Mostly, a high machine utilization is demanded to fulfill a required throughput and guarantee a firm return of investment. Monitoring the health status of such machines allows detecting looming failures at an early stage and therefore helps to prevent unplanned downtimes and optimize maintenance strategies. Data-driven techniques, often summarized by the term machine learning (ML), potentially provide manifold advantages in the area of condition monitoring (CM) compared to established approaches. However, the practical implementation of ML-based CM systems can often be time-consuming and costly. Major challenges are the availability of extensive data on regular and faulty machine operation as well as the selection of suitable data features [1].

One approach to circumvent the issue of costly and safety critical data acquisition in real world applications, is to generate training data by means of a lumped parameter simulation. Such simulations are often implemented during the engineering process anyways and are used to investigate a system's behavior in a few specific design points. However, training data for CM has to represent irregular, faulty system behavior as well, which leads to a vastly increased number of variations of the simulation model compared to the conventional use of the simulation. This can be managed by means of a variational study, where the simulation model is run with multiple sets of design parameters in an automated manner. Hence, if parametric fault models are included into the system simulation, this can be leveraged to automate the data generation procedure.

In this contribution, the process of simulation-based data generation is investigated on a hydraulic reference system. Fault related model parameters are identified and reasonable upper and lower boundaries for their variation are defined. Moreover, a procedure for an automated data generation is presented.

The following section reveals the state of the art of CM in fluid power applications and simulation-based data sampling. Subsequently, the considered hydraulic system and its simulation model are presented in section 3. In section 4, options for the simulation of faulty system behavior are investigated while in section 5 the procedure for data generation is described. Finally, the results are summarized and a conclusion is given.

2. State of the Art

2.1. Condition Monitoring in Fluid Power Systems

In the field of fluid power systems, numerous approaches have been investigated and applied for CM. Solutions range from conventional limit checking, to the comparison of observed and expected system behavior based on a mathematical process-model [2], [3], [4]. However, these approaches either require high manual implementation effort and process knowledge, or have low applicability in cases of complex multi-signal non-linear systems. Concurrently, data-driven methods have also frequently been subject of investigation for CM in fluid power applications. As in [5], [6] and [7], numerous publications in this field focus on the basic proof of applicability of ML-methods for CM of fluid power systems and mainly consider situations of single or few concurrent faults occurring at the same time. Accordingly, only small data sets are used in these studies and questions on the sampling strategy and the distribution of the sampled data are not focused. In [8], datasets in the scale of 3000 measurements are conducted on a test bench for data-based CM of an axial piston pump. Six different isolated faults are investigated which are either varied on binary levels or equally spaced within a defined design region. However, the high number of conducted measurements is a result of repetitions of individual fault conditions. A similar setup is found in [9]. Furthermore, Helwig studies data-based CM techniques on a system level, where sensors are distributed over different locations in the system. If sensor data is combined, higher level information can be provided from which component faults can be inferred as well [10]. Similar to other referenced publications, the sampling strategy for the variation of fault levels and the consideration of concurrent faults is defined heuristically and is practically limited as a consequence of the data collection on a physical test bench.

2.2. Simulation-based data sampling

An alternative to the data collection on a physical test bench is the generation of operation data from computer simulations. Especially for the task of CM this can be beneficial, as safety critical fault settings can be emulated without the risk of damages of the test equipment or harm of the operator. Moreover, it allows to efficiently generate data on a larger scale, in continuous fault settings and in various operating conditions [11]. This links to the statements in [1] and [12], that a major reason for the slow adaptation of ML-based CM techniques in industrial applications is not the lack of in-service data in general, but more the lack of comprehensive data, which covers a diversity of operation scenarios.

In [13], the training of virtual robots is realized with data, which is generated by means of a computer simulation. High fidelity simulation models are used to account for the gap between simulated and real system behavior. Tercan et al. successfully implement a performance estimation model for an injection molding machine, by pretraining an estimation model with data from a simulation and retraining the obtained model with real world data [14]. In the field of condition monitoring, simulation-based data generation is used for the estimation of pump leakage in [15]. Although only a single fault is considered in this study, it is stated that a relatively large number of training points is required, in order to achieve a high accuracy with the simulation-based approach.

By conducting computer simulations instead of measurements on a physical test bench, also different approaches for the experimental design can be considered. A group of experimental designs that are suited for computer experiments are *space filling designs* (*SFDs*). In general, these designs aim to spread the sampling points evenly over the experimental region with minimal gaps between them [11]. One type of SFDs are *Quasi-Monte-Carlo Designs*, which uniformly distribute the sampling points across the experimental region, based on a randomization scheme derived from pseudo-random sequences. Other commonly used SFDs are *Latin Hypercube Designs (LHDs)*. The basic idea of a LHD is to divide each design factor into an equal number of levels and then distribute each factor level uniformly among the samples, such that each factor level is only sampled once. As by this rule, good space filling properties are not guaranteed, several extensions of this design can be found in the literature. Most of these extensions aim to improve the space filling properties by considering distance measures between samples or by ensuring orthogonality of the samples [16]. Furthermore, an advantage of LHD designs is that they allow to incorporate arbitrary distributions for different factors by transforming the initially uniform distribution to the desired one.

3. Simulation of the Hydraulic Reference System

In this publication, a hydraulic system consisting of a cylinder drive is considered, as it is commonly used in various applications in stationary hydraulics. The simulation model is built according to a physical demonstrator machine, which is a test bench of a hydraulic press. The main components are a hydraulic power supply, a proportional directional control valve and a single acting hydraulic cylinder. The system is modelled in the *Modelica*-based simulation tool *SimulationX*, which allows the 1D-simulation of hydraulic circuits. The simulation consists of coupled component models, each containing information on their input and output variables, as well as equations describing their physical behavior. In figure 1, the simulation model of the considered system is depicted, while figure 2 shows the demonstrator.

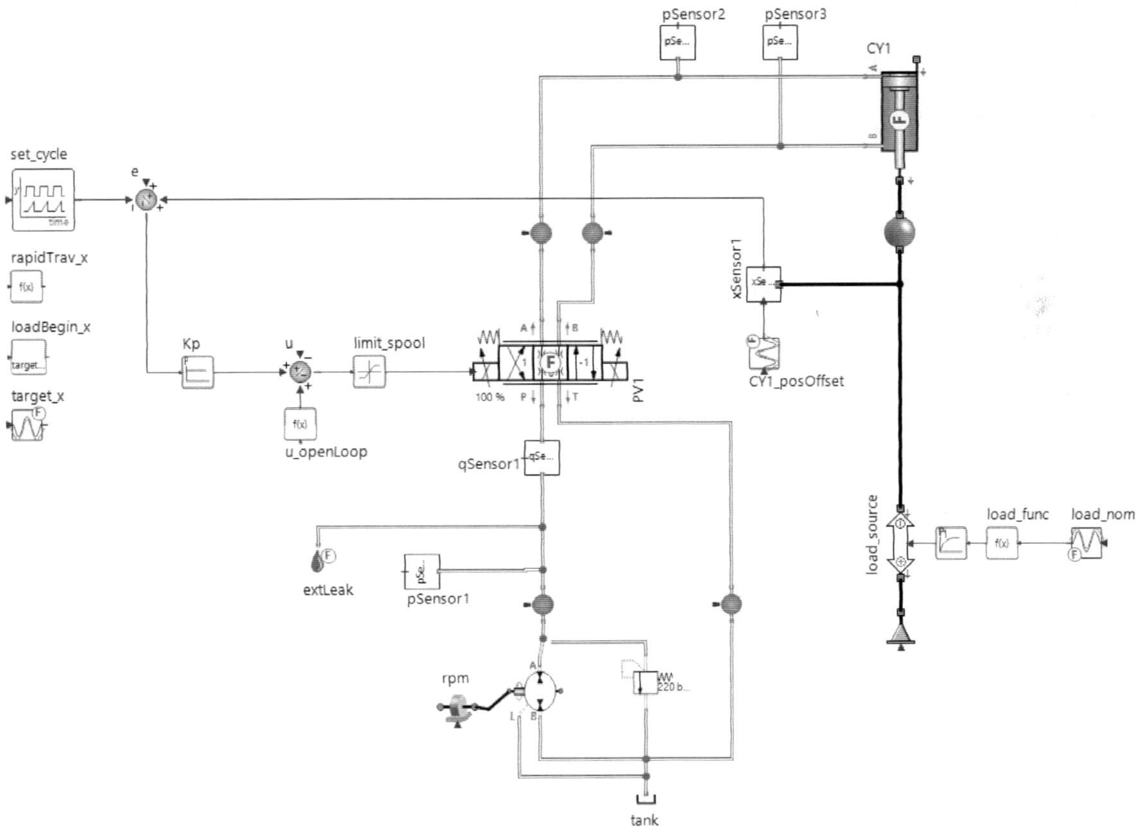

Figure 1. Simulation model of the considered hydraulic system.

The function of the press is to move a ram to a defined position. The ram is attached to the cylinder rod and acts against a load force, which represents the reaction force of a workpiece. In the simulation, the working cycle is modelled accordingly. To achieve a closed-loop position control of the cylinder, its actual measured position is fed back, compared to the target position and directed to the controller to finally manipulate the position of the valve spool. Figure 3 shows the progressions of the cylinder's position, velocity and load force over one working cycle. The working cycle starts in rapid traverse where the ram of the press moves towards the workpiece without external load and limitation of the velocity. As the ram reaches a defined distance to the workpiece, the velocity is restricted through limiting the control signal and the volumetric flow to the cylinder, respectively. After the ram is in contact with the workpiece, the load force increases linearly with the stroke of the cylinder. The cycle is defined such that the controller is given a fixed period to reach the set value, before the backstroke is initiated in open-loop control.

Figure 2. Demonstrator of the simulated hydraulic press.

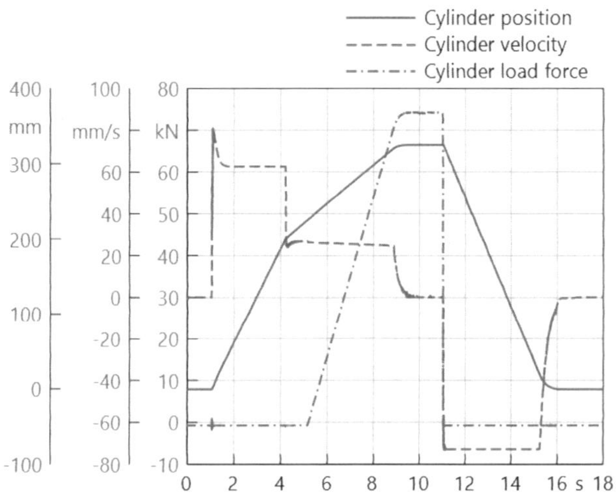

Figure 3. Working cycle of the hydraulic press.

Basic parameters of the component models, such as nominal dimensions, are set according to catalogue data of the components from the demonstrator. The hydraulic power supply provides around 18 l/min of volumetric flow at a pressure of up to 220 bar. Then, the fluid is controlled by a proportional 4/3-way valve with a nominal volumetric flow of 12 l/min. The piston diameter of the hydraulic cylinder is 80 mm with a piston rod diameter of 45 mm and a maximal stroke of 400 mm.

4. Fault Modelling and Fault Effects

In order to detect and diagnose a system's condition as part of a data-driven CM, the database used, has to include regular and faulty conditions of the system likewise. Therefore, faulty operations of the hydraulic system have to be emulated in the computer simulation as well. This can be done by varying parameters of the component models, which correspond to specific faults. The resulting system behavior of the hydraulic circuit then reflects the effects of these faults. Opposed to conventional process-model-based CM, a high fidelity model of the system behavior under fault conditions is not targeted. Hence, it should reveal general effects of component faults and fault combinations represented in a database to qualitatively identify root-cause dependencies.

4.1. Fault Modelling

In the following, considered faults are leakages and increased friction in the hydraulic cylinder, a faulty measurement of the cylinder position, worn control edges of the control valve, as well as external leakage at a connector in the hydraulic power supply.

In the hydraulic cylinder, leakages mainly occur at sealing spots between elements with relative motion. This is prevalent between piston and cylinder housing as well as between piston rod and cylinder housing, as depicted in figure 4. The clearance between piston and cylinder housing can cause a leakage flow between the two chambers of the cylinder and is therefore called internal leakage. Whereas a leakage flow between piston rod and housing is considered an external leakage, as it is emitted to the environment. In the simulation software used in this study, the internal leakage can be modelled as fluid flow through a gap. It is obtained from the superposition of a fluid flow driven by pressure difference across the gap and a kinematically induced transmission of fluid between the two sides of the gap. For a circular gap with concentric positioning of piston and cylinder, the internal leakage flow $Q_{\mathrm{C,Li}}$ can be described by equation (1).

$$Q_{\mathrm{C,Li}} = \frac{1}{2} \cdot \dot{x}_{\mathrm{P}} \cdot h_i \cdot D_{\mathrm{P}} \cdot \pi + \frac{h_i^3 \cdot D_{\mathrm{P}} \cdot \pi}{12 \cdot \eta \cdot l_i} \cdot \Delta p \tag{1}$$

Where $\dot{x}_{\mathrm{P}}$ denotes the relative velocity between piston and cylinder, h_i the gap height, D_{P} the piston diameter, η the dynamic viscosity of the fluid, l_i the length of the gap and Δp the pressure difference over the gap. In practice, an increase of the internal leakage can occur as a result of a worn or damaged piston seal which in turn is linked to a change in the gap geometry. Thus, the gap height h_i is considered the main geometric parameter impacting the internal leakage and therefore is used to emulate different levels of internal leakage in the model of the hydraulic cylinder.

Similarly, the external leakage at the hydraulic cylinder can be modelled as fluid flow through a circular gap. In contrast to the sealing system at the piston, the leakage flow is assumed to only be directed from the rod chamber to the environment as depicted in figure 4. Using the relation from equation (1) would therefore require a case separation, depending on the direction of the cylinder movement. In order to maintain model simplicity, and as the major part of the leakage is mostly induced by the pressure difference, the contribution of the relative motion is neglected here. Thus, the external leakage flow $Q_{\mathrm{C,Le}}$ at the hydraulic cylinder is modelled as described in equation (2) with the gap height h_e as fault related parameter.

$$Q_{\mathrm{C,Le}} = \frac{h_e^3 \cdot D_{\mathrm{R}} \cdot \pi}{12 \cdot \eta \cdot l_e} \cdot \Delta p \tag{2}$$

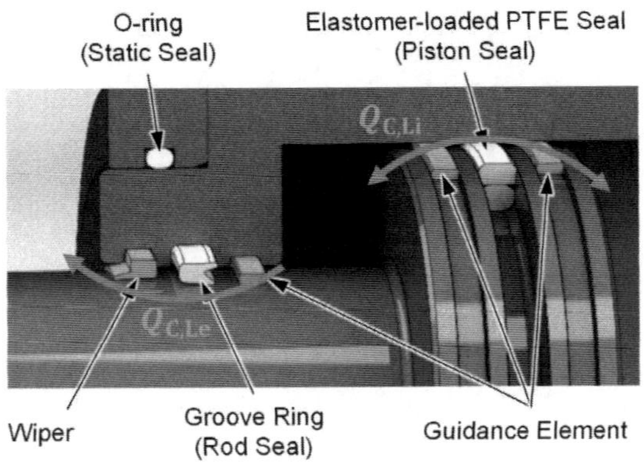

Figure 4. Sealing systems in a hydraulic cylinder.

Another fault that can result from damaged sealing elements or a bad alignment of piston rod and cylinder housing, is an increased friction between moving components of the hydraulic cylinder. Generally, friction is modelled by means of a Stribeck Curve, which can provide valid estimates of the friction force across different operating conditions and friction regimes. However, the parametrization of a Stribeck Curve requires detailed measurements of the considered system which raises issues of practicability. Alternatively, the friction in the hydraulic cylinder is modelled with constant static friction force and dynamic friction force, respectively. In order to reduce the number of fault combinations in the following data generation process, the static friction force is solely used for the indication and variation of the fault state while the ratio between the two friction forces is kept constant. Based on data from [2] and [18], the ratio is set to 2.5 as expressed in equation (3).

$$F_{\mathrm{Fr}} = \begin{cases} F_{\mathrm{static}}, & \text{for } \dot{x}_{\mathrm{P}} = 0 \\ \frac{F_{\mathrm{static}}}{2.5}, & \text{else} \end{cases} \quad (3)$$

Furthermore, worn control edges of the spool of a control valve are considered for the simulation of a faulty machine state. As a result of abrasion and erosion, the typically precisely manufactured control edges of a valve spool and sleeve can change their shape, especially over long-time usage. This can lead to a change in the relation between the spool position and flow characteristics and consequently the overall control behavior. However, without empirical investigation, this shift of characteristics is unknown and has to be modelled by other means. Consequently, the overlap of the valve is selected as fault defining parameter, such that a growing negative overlap correlates to an increased level of wear.

An external leakage in a connector in the hydraulic power supply is considered as well. It is modelled as a linear function of the system pressure and is varied through the proportionality factor of the leakage flow and system pressure, here denoted as leakage factor C_{L} in equation (4).

$$Q_{\mathrm{L}} = C_{\mathrm{L}} \cdot p_{\mathrm{S}} \quad (4)$$

Finally, a faulty position sensor at the cylinder is modelled by adding a static offset to the sensed cylinder position.

4.2. Fault levels and ranges

The considered faults, their corresponding parameters in the component models and their value ranges for regular and faulty operation are summarized in table 1.

Table 1. Nominal values and ranges of variation for fault related model parameters.

Fault	Model Parameter	Nominal Value	Range
Cylinder, Internal Leakage	Gap height h_i	1 µm	1 µm – 100 µm
Cylinder, External Leakage	Gap height h_e	0.5 µm	0.5 µm – 75 µm
Cylinder, Friction	Static Friction Force F_{static}	500 N	500 N – 2500 N
Cylinder, Position Sensor Offset	Offset	0 mm	0 mm – 410 mm
Proportional Valve, Worn Control Edges	Spool Overlap y_0	-2 %	-2 % – -4 %
Hydraulic Power Supply, External Leakage	Leakage Factor C_{L}	0 l/(min bar)	0 – 0.1 l/(min bar)

The nominal gap heights for the cylinder Leakages are selected according to data from [19] and [20]. The range for the leakage factor C_{L} is defined, such that states of no leakage, a leakage of a few drops per minute up to a broken connector can be emulated. The absolute leakage flows that result from the

IOP Conf. Series: Materials Science and Engineering	1097 (2021) 012018	doi:10.1088/1757-899X/1097/1/012018

selected ranges are listed in table 2 and refer to a peak differential pressure of 150 bar for the cylinder leakages and 200 bar for the leakage in the power supply.

Table 2. Resulting leakage flow rates for considered fault ranges

Fault	Parameter Range	Leakage Flow
Cylinder, Internal Leakage	1 µm – 100 µm	0 – 23 l/min
Cylinder, External Leakage	0.5 µm – 75 µm	0 – 5.5 l/min
Hydraulic Power Supply, External Leakage	0 – 0.1 l/(min·bar)	0 – 20 l/min

It is apparent, that the maximal leakage flows are not probable to be commonly observed in practice. However, these extreme cases are included in the database, to assure that the mapping of the system behavior to fault effects is valid for a broad range of operating conditions. This also applies to the other faults, where maximal values are heuristically defined, as no empirical base is available for estimation.

4.3. Fault Effects

A change of the previously derived fault parameters impacts the input-output-behavior of the hydraulic system. Hence, the fault effect is the change of operational characteristics which is caused by the change of a fault parameter. In order to outline options for an algorithmic mapping of faults and fault effects, exemplary cases are briefly illustrated in the following. For this, the pressure difference at the cylinder ports and the volumetric flow at the valve inlet are each compared at faulty and fault free settings.

Examples for the effect of internal leakage at the hydraulic cylinder are depicted in figure 5. The diagram on the left shows that with changing gap height h_i, the pressure difference between cylinder ports clearly changes at the beginning and end of the working cycle. Moreover, the internal leakage effects the volumetric flow throughout the working cycle, as shown in the diagram on the right. Especially in segments of the cycle where the volumetric flow is small, the relative change is apparent.

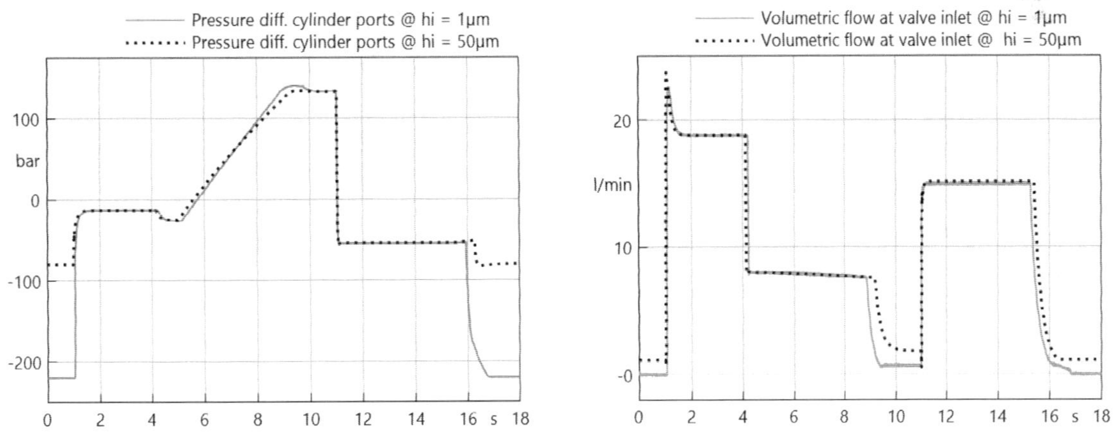

Figure 5. Effect of different settings of gap height h_i on the pressure difference over cylinder ports (left) and the volumetric flow at the valve inlet (right) over one working cycle.

Similarly, the severity of the effect of an increased friction varies over a working cycle. As the left diagram in figure 6 shows, an effect is mainly observable around the time stamp of 10 seconds, where the cylinder velocity is low and the position is finely controlled. Additionally, the diagram on the right hand side of figure 6 shows the effect of a change in valve spool overlap on the volumetric flow through the inlet port of the valve. In contrast to the fault effect of an internal leakage, the peak value is not affected by the change in overlap, nor is the volumetric flow at the beginning and end of the cycle.

As a conclusion, fault effects are reflected in system outputs that are commonly observable through sensor data. Providing a large number of training examples and describing signal characteristics with quantitative measures, moreover facilitates the use of ML methods for the mapping of system outputs to fault states.

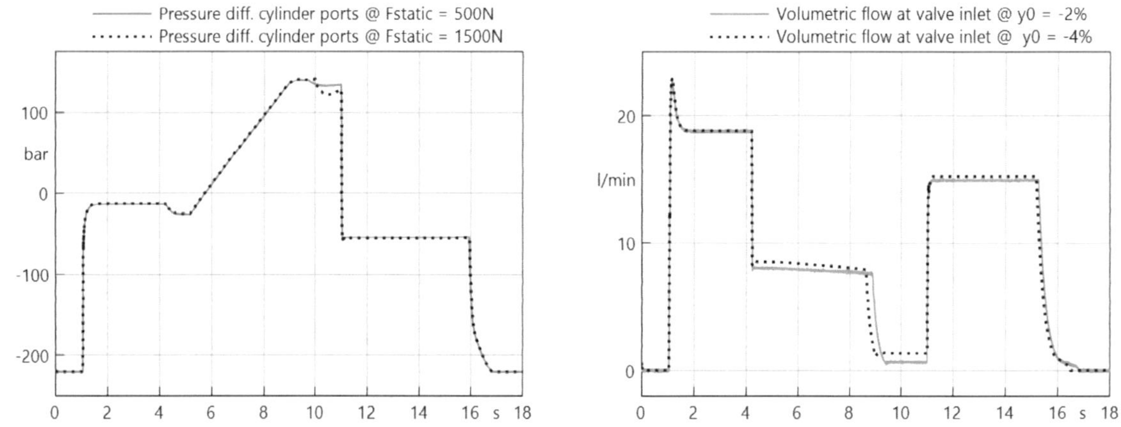

Figure 6. Effect of different settings of static friction force on the pressure difference over cylinder ports (left) and effect of different settings of valve spool overlap on the volumetric flow at the valve inlet (right) over one working cycle.

5. Automated Data Generation

In this section, the procedure of generating data from the simulation model is described. The general scheme is shown in figure 7.

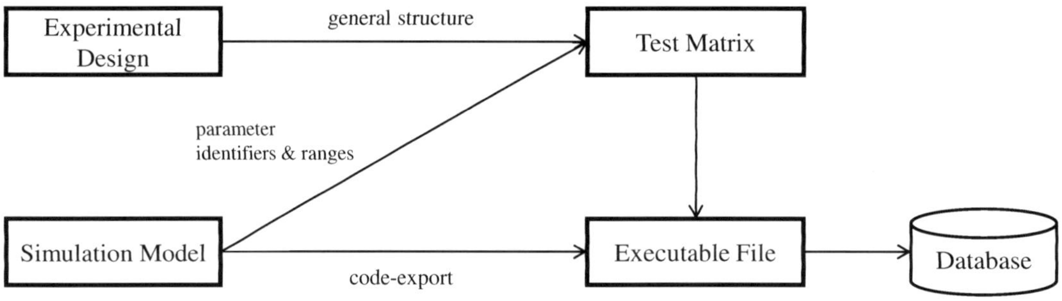

Figure 7. Procedure of automated simulation-based database generation.

The first step of this procedure is the setup of the experimental design, which is implemented in a *Python*-program based on the code-library *pyDOE2*. Here, a Latin Hypercube Design is selected as experimental design. Fault levels are uniformly distributed according the *maximin*-criterion and the number of samples is set to 10000, meaning that the same number of parameter sets is obtained, with which the simulation model is to be run. Provided the parameter identifiers and value ranges, the test matrix is generated and written to a simple text file. In addition to the fault related parameters

discussed in section 4.1, the load force acting on the system and the target position of the hydraulic cylinder are included as variable parameters.
Subsequently, the simulation model is exported as an executable file (exe-file) via the code export function of the simulation program. In this step, state variables to be output by the model are also selected. Sensor signals for different pressures, the volumetric flow entering the control valve, as well as cylinder position, valve spool position and control signals are chosen here. The signals are sampled with a rate of 1000 Hz. Using the *SRA*-add-in of *SimulationX*, the exported exe-file can additionally be linked to the previously defined experimental design. Thereby, the simulation is automatically run with all parameter sets listed in the test matrix once the exe-file is executed. Moreover, running the simulations from an exported exe-file allows parallel execution of simulation runs on multiple processor cores, which additionally increases the efficiency of the procedure.

The procedure is finally run on four processor cores of an *Intel Xeon Gold 6130* processor. Using the *CVODE* solver, the 10000 simulation runs are completed in approximately six minutes. This includes the writing and saving of the generated data to roughly 50 Gigabytes of text files.

6. Conclusion and Outlook

In this paper, a method for simulation-based data generation was presented, which allows to efficiently generate training data for a data-driven condition monitoring. The data was generated by means of a variational study in a lumped parameter simulation program. Options for parametric fault modelling were described for a hydraulic reference system. Furthermore, a procedure has been derived, which allows to generate large datasets of time-series data with a high variation of fault levels and operating conditions.

As a next step, the derived fault models are to be empirically assessed on the physical demonstrator of the reference system. Furthermore, the preliminary experimental design used in this study can be extended. The consideration of fault specific distributions in the Latin Hypercube Design, is expected to help focus the samples in design regions, which are of higher relevance regarding the separability of fault states and their probability of occurrence. Finally, the applicability of the obtained database is to be evaluated in the frame of a data-driven CM-setup.

Acknowledgements

The authors thank the Research Association for Fluid Power of the German Engineering Federation VDMA for its financial support. Special gratitude is expressed to the participating companies and their representatives in the accompanying industrial committee for their advisory and technical support.

References

[1] Döbel I, Leis M, Molina Vogelsang M, Neustroev D, Petzka H, Rüping S, Voss A, Wegele M and Welz J 2018 *Maschinelles Lernen –Kompetenzen, Anwendungen und Forschungsbedarf: Projektbericht*
[2] Stammen C 2005 *Condition Monitoring für intelligente hydraulische Linearantriebe* (Aachen: Shaker)
[3] Münchhof M 2007 *at – Automatisierungstechnik* **55**
[4] Isermann R 2011 *Fault-Diagnosis Applications: Model-Based Condition Monitoring: Actuators, Drives, Machinery, Plants, Sensors, and Fault-tolerant Systems* (Berlin, Heidelberg: Springer-Verlag Berlin Heidelberg)
[5] Ramdén T, Krus P and Palmberg J-O 1996 *Proceedings of the JFPS International Symposium on Fluid Power* **1996** 567–72
[6] Watton J and Freebody N 2000 *International Journal of Fluid Power* **1** 59–67
[7] Zachrison A 2008 *Fluid Power Applications Using Self-Organising Maps in Condition Monitoring*

[8] Fischer A 2013 *Prognosekonzepte zur Reduktion von Betriebskosten im Lufttransport (ProReB): Abschlussbericht*
[9] Adams S, Beling P, Farinholt K, Brown N, Polter S and Dong Q 2016 *Condition Based Monitoring for a Hydraulic Actuator*
[10] Helwig N J 2019 *Zustandsbewertung industrieller Prozesse mittels multivariater Sensordatenanalyse am Beispiel hydraulischer und elektromechanischer Antriebssysteme (Aktuelle Berichte aus der Mikrosystemtechnik* vol 30*)* (Aachen: Shaker)
[11] Siebertz K, van Bebber D and Hochkirchen T 2017 *Statistische Versuchsplanung: Design of Experiments (DoE) (VDI-Buch)* 2nd edn (Berlin, Heidelberg: Vieweg)
[12] Paulmann G and Mkadara G 2018 Condition monitoring of hydraulic pumps – lessons learnt *Fluid power networks 11th International Fluid Power Conference (Aachen, Germany, 19th - 21st March 2018)* ed H Murrenhoff (Aachen)
[13] El-Shamouty M, Kleeberger K, Lämmle A and Huber M 2019 *tm - Technisches Messen* **86** 673–84
[14] Tercan H, Guajardo A, Heinisch J, Thiele T, Hopmann C and Meisen T 2018 *Procedia CIRP* **72** 185–90
[15] Wiens T and Fernandes J 2020 Application of Data Reduction Techniques to Dynamic Condition Monitoring of an Axial Piston Pump *Proceedings of the ASME/BATH Symposium on Fluid Power and Motion Control -2019- ASME/BATH Symposium on Fluid Power and Motion Control (Longboat Key, Florida, USA, October 7-9, 2019)* (New York, N.Y.: The American Society of Mechanical Engineers)
[16] Gramacy R B 2020 *Surrogates: Gaussian process modeling, design, and optimization for the applied sciences (Chapman & Hall/CRC texts in statistical science series)* (Boca Raton, FL: CRC Press, Taylor & Francis Group)
[17] Schmitz K 2018 *Grundlagen der Fluidtechnik: Teil 1: Hydraulik (Reihe Fluidtechnik. U)* (Aachen: Shaker)
[18] Abbas Kazemi-Moghaddam 2000 *Fehlerfrühidentifikation und -diagnose eines elektrohydraulischen Linearantriebssystems* (Darmstadt: Technische Universität Darmstadt)
[19] Mielke T 2020 *Wassereinzug über ölhydraulische Stangendichtsysteme (Reihe Fluidtechnik D* vol 98*)* (Düren: Shaker)
[20] Müller H K and Nau B 2020 *Hydraulikdichtungen* http://www.fachwissen-dichtungstechnik.de/PDFs_01_2020_mit_PW/fw-dt_Kapitel05_01_20.pdf (accessed 30 Nov 2020)

Damage Diagnosis of Cardan Shafts in Mobile Mining Machines using Vibration Analysis

Reza Golafshan[1 *], Catalin Dascaliuc[1], Georg Jacobs[1], David Roth[1], Joerg Berroth[1], Stephan Neumann[1]

[1] RWTH Aachen University, Institute for Machine Elements and Systems Engineering (MSE), Schinkelstr. 10, 52062 Aachen, Germany

[*] reza.golafshan@imse.rwth-aachen.de

Abstract. The ability of prediction of machine condition may become even more crucial in mobile mining machines considering their instantly-varying conditions. The field of early damage diagnosis for mobile machinery is gaining increasing attention in modern and predictive maintenance strategies. In general, vibration-based machine condition monitoring methods offer the possibility to detect and localize the damages in mechanical systems at their early stages. In vibration monitoring, the vibration signals are captured via accelerometers, and are processed in the time and frequency domains in signal processing packages to extract diagnostic information. However, in real-world mobile applications, vibration analysis alone may not reveal the diagnostic information due to extreme transient and nonstationary operating conditions. In this regard, additional metadata available on the machine's CAN bus system, can improve the reliability of vibration analysis for mobile applications. The current study aims to investigate the usage of metadata as additional information to the vibration analysis on cardan shaft of a test wheel loader in field conditions. Two various damage types, namely, misalignment and bearing clearance, both on cardan shaft, are investigated in some detail. The experimental results show how vibration analysis together with metadata processing can identify the state of the machine even in harsh operating conditions.

1. Introduction

Predictive maintenance is of high importance in particular for mining industry. The main goal is to maximize the availability of a machine by minimizing the downtime [1]. In mining companies, the maintenance cost itself can be 30-60% of total operation costs [2].

The operating condition of a mobile machinery, such as a wheel loader, can vary instantly, as it is heavily influenced by how the vehicle is operated and by the environment around it [3]. As an example system for mobile work machines, wheel loaders are vehicles that are mainly used in construction and mining industries for handling and transporting various materials. According to results in [2], maintenance of wheel loaders plays a major role in total maintenance of mining machines. Therefore, prediction of a wheel loader's condition is crucial to reduce downtime and costs, and increase availability and production in mining industry.

The results in Figure 1, from [4], suggest that among various parts of a wheel loader, drivetrain-related failures occur most often. Drivetrains of heavy-duty vehicles are complex mechanical and dynamical systems [5-7] and can require advanced analyzing and monitoring tools [8, 9]. Due to the critical role of drivetrain in availability of a wheel loader, in particular in mining industry, vibration-based condition monitoring can be applied on drivetrain components such as cardan joint shaft between drive and driven sides of the vehicle.

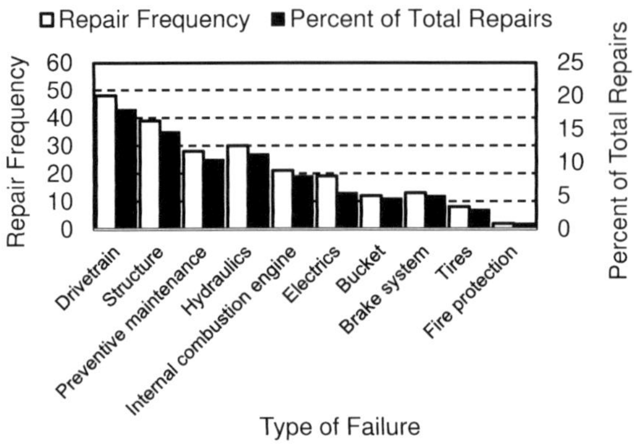

Figure 1. Repair frequency and percentage of total repairs versus type of failure, from [4].

The cardan joint shaft is amongst the most commonly used universal joints for torque transfer when the input drive and output driven shafts are not aligned, [10] such as the situation in wheel loaders, where front and rear sides of the vehicle may not be on the same axis of rotation. Cardan shafts, as one of the vehicle drivetrain components, can be heavily torque-loaded [11], and therefore, may experience misalignment between input and output sides, unbalances, and looseness or operating clearance of internal components such as rolling bearings [12].

The predictive or proactive maintenance strategies use primarily non-destructive Condition Monitoring (CM) testing tools to assess machinery drivetrain health status and performance [13]. The condition monitoring is mainly performed by external sensors, which are attached to the machine. These sensors can measure physical parameters related to the machine condition such as temperature, vibrations, and torque. By analysing sensor data, it is possible to enhance maintenance strategy and take required further repair actions. It is widely accepted that the vibration analysis is capable of finding machine element fault(s) and their location(s) in a reliable way compared to other monitoring methods [14]. The mechanical damage is usually referred to changes in geometric properties of the machine. These sudden changes may generate additional dynamic forces acting in the system, which can be diagnosed based on the system responses [9].

Although many studies have been conducted and published on the usage of vibration analysis for drivetrain components such as cardan shaft [12, 15], the lack of knowledge about the performance of available vibration analysis tools in real-world conditions motivates further experimental studies and investments in research works for CM in mobile applications. Mobile machinery are operated under harsh and non-stationary conditions. Therefore, in order to gain required diagnostic information, additional data on the operating status of the machine are required in a vibration-based CM tools.

The current study aims to show and investigate application of the metadata from the machine CAN bus in vibration-based cardan shaft condition monitoring. The remaining of the manuscript is organised as follows: in the next section a theoretical background of cardan shaft damage diagnosis is given. Then, the proposed fault diagnosis methodology together with the description of case considered studies are explained. Subsequently, vibration-based damage diagnosis of cardan shafts with the help of CAN bus data are studied experimentally using test rig and field data. Finally, some concluding remarks are highlighted.

2. Cardan shaft damage mechanism

The analysis of dynamic behavior of a joint for damage detection purposes requires appropriate understanding of its kinematics. The kinematics allow to analyze the characteristics of design and geometry parameters and also study their individual effects on the general motion and vibration of the joint. Cardan shafts or Hooke's joint [16] are considered as classic joints which are common in mobile machines. In the current study, two common types of faults are investigated in some detail.

According to [14], when a torque transmission shaft has three or more bearing supports, for example when front and rear axles of a wheel loader are coupled together, there is a potential for misalignment, which can be parallel misalignment, meaning that one of the two shafts is displaced laterally, but still parallel to the other, or angular misalignment, where the axis of one is at an angle to that of the other. The angular misalignment of cardan shafts can introduce bending deflections to both input and output shafts while rotating with respect to the shafts. The excited bending modes of shafts show themselves then as harmonics of the shaft's rotational speed in the frequency spectrum. This is considered as the basis for cardan shaft angular misalignment detection using vibration signals. Figure 2 shows a typical cardan shaft with an angular misalignment.

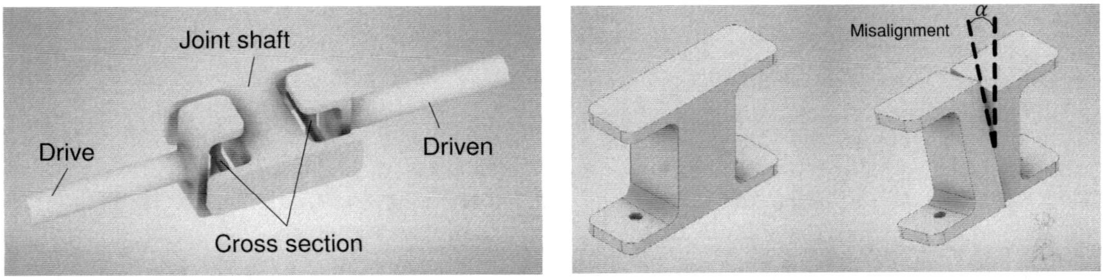

Figure 2. Simplified illustration of an angular misalignment at cardan shaft.

The angular misalignment can be studied mathematically using the equation (1), which shows the relative instantaneous angular velocity of output to input [17] as

$$\frac{\omega_2}{\omega_1} = \frac{\cos\alpha}{1 - \sin^2\alpha\,\sin^2\theta_i} \tag{1}$$

where $\frac{\omega_2}{\omega_1}$, α, and θ_i are output-to-input ratio of instantaneous angular velocity, the misalignment angle, and the instantaneous shaft rotating angle (depending on the vehicle steering condition), respectively. For small misalignment angles, say α up to about 10°, cos α and $\sin^2\alpha$ can be considered as 1 and α^2, respectively. Considering

$$\sin^2\theta_i = \frac{1}{2}[1 - \cos(2\theta_i)] \tag{2}$$

the following approximation of the output-to-input ratio of instantaneous angular velocity can be calculated:

$$\frac{\omega_2}{\omega_1} = 1 - \left(\frac{\alpha^2}{2}\right)\cos(2\theta_i) \tag{3}$$

Equation (3) shows that angular misalignment vibration may appear as second harmonic of shaft velocity in the spectrum. Moreover, according to Equation (3), the angular misalignment excites the system with second harmonic only when there is a steering angle between input and output.

Another common type of damages or faults that can appear in the cardan shafts is the axial rolling bearing looseness or bearing clearance of cardan shaft. In case of sudden change in speed or torque direction, the bearing clearance fault shows itself as impact-like excitations to structure as the rolling bearing itself oscillates and slides inside the cardan shaft. In practice, impulsive forces apply broadband excitations to the system over frequency range rather than producing a discrete frequency peak in the spectrum. However, if the direction of shaft velocity or torque remains unchanged, the rolling bearing inside the cardan shaft is expected to shift to a side and change the center of gravity from the center to aside, hence producing an unbalance dynamic force at each shaft revolution to the system. These periodic dynamic forces can then be picked by vibration sensors as increasing peaks at shaft rotational frequency harmonics.

Therefore, even though the vibration data can reveal the aforementioned fault types, without additional information about the operational conditions of the machine (e.g. steering and torque), it can be challenging to detect and diagnose misalignment or clearance of cardan shafts. Moreover, as vibration behaviour and trend of the machine also change with operational conditions such as speed and torque, vibration signals require to be analysed together with additional sensor data. The machine condition data, available on Controller Area Network (CAN) bus of wheel loaders, can significantly increase the reliability of vibration analysis for condition monitoring. The CAN bus protocols have been developed by Robert Bosch GmbH in 1986 at the Society of Automotive Engineers (SAE) congress in Detroit, Michigan [18]. In order to investigate and study the cardan shaft damage detection using vibration analysis with the help of additional metadata from the CAN bus, the current study focuses on basic time and frequency domain analyses applied on the cardan shaft signals.

3. Approach

In conventional vibration analysis approaches, vibration signals together with RPM data are processed and analysed to identify the damages without considering machine operating conditions [14]. However, using only vibration and velocity signals for detecting and localizing mechanical damages on mobile work machines may not reveal all required diagnosis information.

Under the light of previously discussed findings, in the current study, a damage detection procedure based on the combination of vibration, velocity, and machine condition signals is proposed and applied for cardan shafts in wheel loaders. Therefore, two typical damage types, namely, misalignment and bearing clearance, are considered and analysed by the proposed procedure. In what follows, detection of both damage types using vibration and metadata are discussed in some detail. The diagnosis of the misalignment fault is done based on the vibration analysis and expected behaviour of the system in the presence of the fault as described in the section 2. The so-called reference-based vibration analysis of the misalignment fault seeks to identify amplitude increase in the second order harmonic of the cardan shaft speed during a steering manoeuvre. Therefore, the synchronized steering data from CAN bus is used to select critical segments for further analysis.

The diagnosis of the bearing clearance fault is done by comparing statistical indicators, e.g. standard deviation, of the vibration signals for healthy and faulty states of the system at similar operational conditions when there is a change in torque/speed (e.g. non-linear acceleration manoeuvre). These operational conditions can be identified based on the conjunction of accelerator pedal, and vehicle speed all from CAN bus data. In addition, if the machine has a constant torque, the bearing clearance fault leads to an increase of the cardan shaft harmonics in the amplitude spectrum, as mentioned in section 2. In order to track speed fluctuations, order tracking-based amplitude spectrum is used to reveal diagnostic information in the frequency domain. Figure 3 shows the reference-based vibration analysis method for diagnosis of cardan shaft clearance fault. The CAN bus data is used for identifying the meaningful time periods in which the analysis has to be carried out and for selecting the appropriate vibration-based analysis routine.

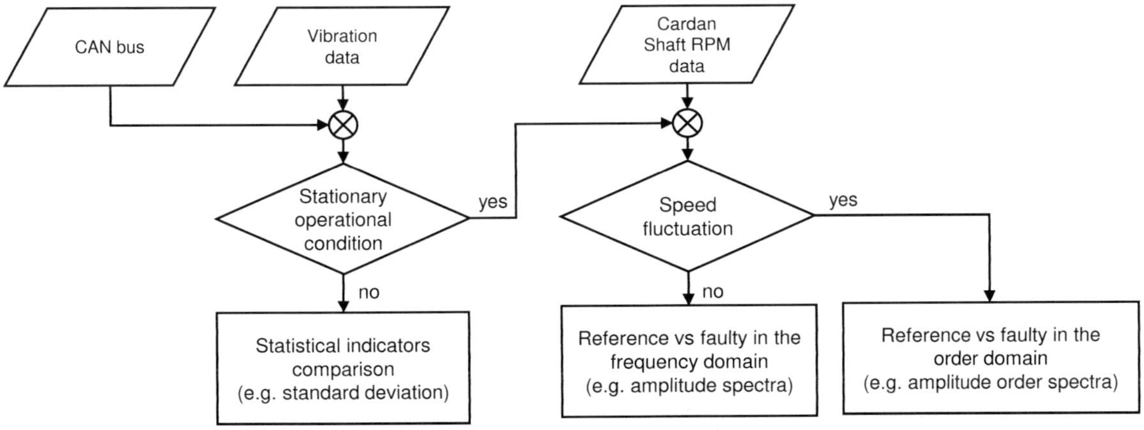

Figure 3. Vibration analysis for diagnosis of the cardan shaft bearing clearance fault methodology.

4. Case studies

In order to investigate the proposed methodology in section 3 using damage types discussed in section 2, two various experimental setups are used.

Firstly, to investigate the vibration signature of both damage types, the bearing clearance and angular misalignment, are examined using a test rig. For this purpose, a complete drivetrain of a wheel loader type GHH SLP3H was mounted on a test rig to apply required input excitation, i.e. torque, to capture vibration signals from two case studies using two different defective cardan shafts. Figure 4 shows a picture together with a schematic illustration of the complete test rig and the wheel loader drivetrain. A number of additional vibration sensors were used to capture vibration signals from the test rig mounted drivetrain.

For real-world applications, the field measurements were carried out in a salt mine, operated by company K+S in Werra, Germany. The measurements were conducted as part of a European Union-funded project called ProMaschinenDaten run by various partners from academia and industry including Institute for Machine Elements and Systems Engineering (MSE) of RWTH Aachen University. The mine is located approx. 850 m underground and accessible via an elevator. A total of 6 days (including preparation and measurements) were used to carry out the experimental investigations on the entire test vehicle, GHH SLP-3H.

Figure 4. Test rig setup for a GHH SLP-3H wheel loader drivetrain.

During the field tests, various signals were recorded via external sensors and CAN bus. The external sensor data such as vibrations, pressure and shaft velocity of the main shaft were recorded at a high sampling rate. The operational data from the CAN bus was recorded at a low sampling rate due to the limitations of the CAN bus. Figure 5 shows a picture of the vehicle at K+S underground facilities together with sensor positions used for measurements. Figure 6 shows schematically the location of cardan shaft in the wheel loader GHH SLP-3H. The cardan shaft in the middle of the vehicle can transfer the drive torque when vehicle's front and rear sides have a relative angle with respect to each other (i.e. when steering).

Figure 5. A picture of the test wheel loader type GHH SLP-3H together with additional sensors for vibration monitoring of the cardan shaft.

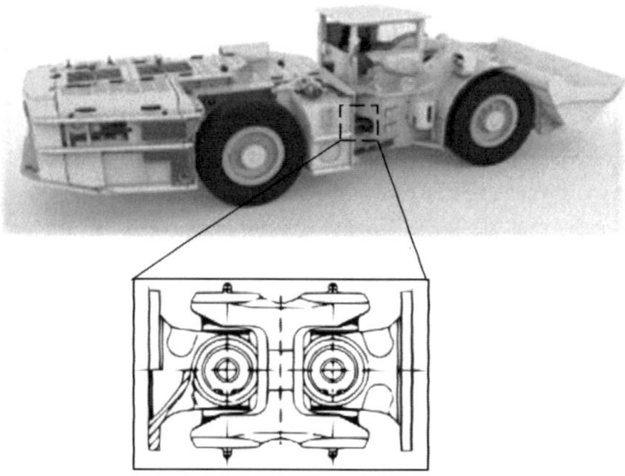

Figure 6. Schematical illustration of a cardan joint shaft in a GHH SLP-3H wheel loader.

In total, three sessions, healthy condition as reference, misaligned cardan shaft, and cardan shaft with bearing axial clearance, all using the same cardan shafts used for test rig investigations in Section 2, were measured. Each session has various parts ranging from wheel loader engine start to the end of measurement where vehicle arrives back at the mine workshop. In order to cover a wide gamut of operating conditions that may happen during the loading and unloading cycles of the vehicles, a few number of braking and accelerating manoeuvres together with steering conditions were carried out during the field measurements.

Figure 7 shows the time-frequency domain signal of an example session as the waterfall diagram alongside the vehicle's velocity. The different parts of the measurement are marked and corresponding time periods are illustrated on the plots.

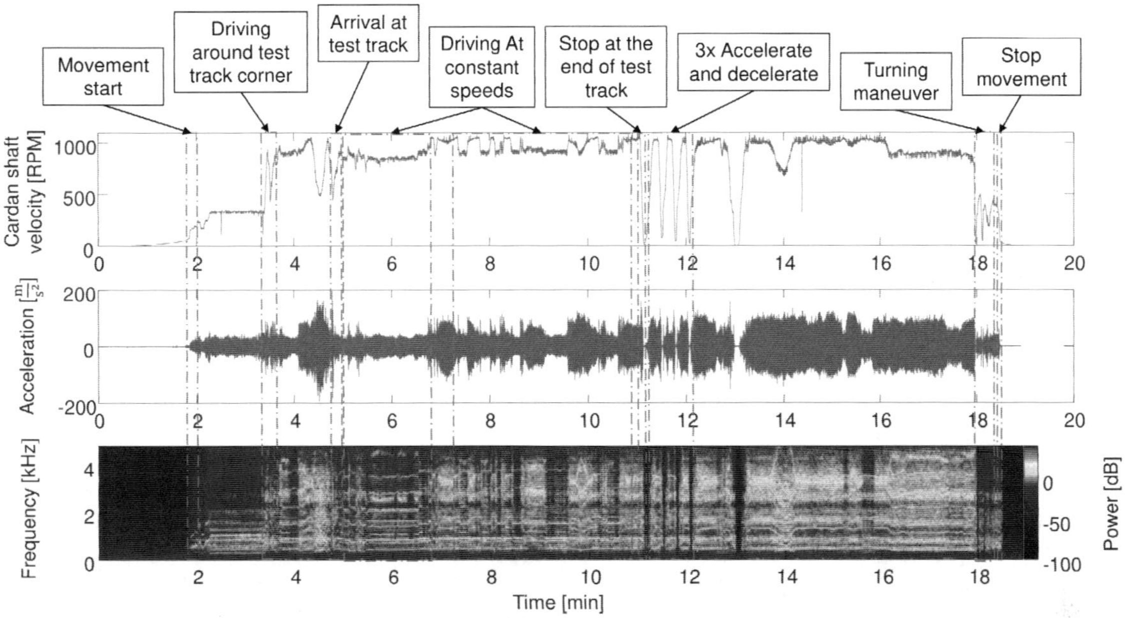

Figure 7. A complete operating cycles of the test wheel loader during the measurements.

5. Experimental results

The case studies described in Section 3 are used to perform the required analyses on the hypotheses for the damage types considered in this study.

5.1. Test rig experimental investigations

Figure 8 shows synchronised torque, vibration, and standard deviation signals for both bearing clearance and healthy states for comparison purposes. The vibration signal is captured by a sensor located near the cardan shaft on the test rig. As seen, the bearing clearance, as random impact-like excitation, gives a rise to energy level of the signal, hence an increase of the standard deviation in the time domain can be detected. Similarly Figure 9 shows the test rig experimental results, but for angular misalignment investigations. As seen, while the energy level of first shaft harmonic remains more or less same for misaligned cardan shaft, this fault produces higher energy at shaft higher rotational velocity harmonics. This result verifies the findings using Equation (3).

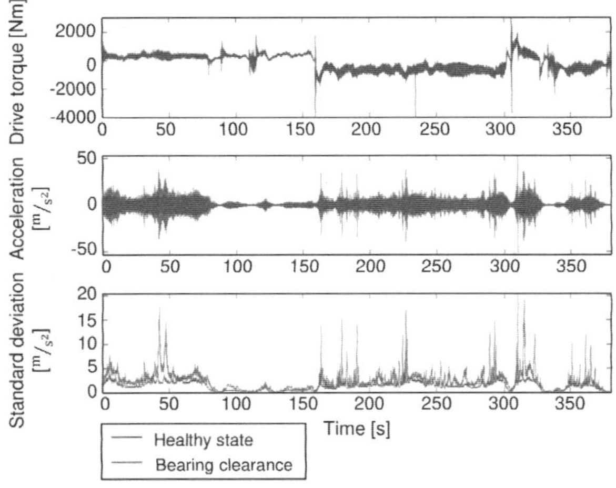

Figure 8. Test rig experimental diagnostic results of a cardan shaft with a bearing clearance fault.

IOP Conf. Series: Materials Science and Engineering 1097 (2021) 012019 doi:10.1088/1757-899X/1097/1/012019

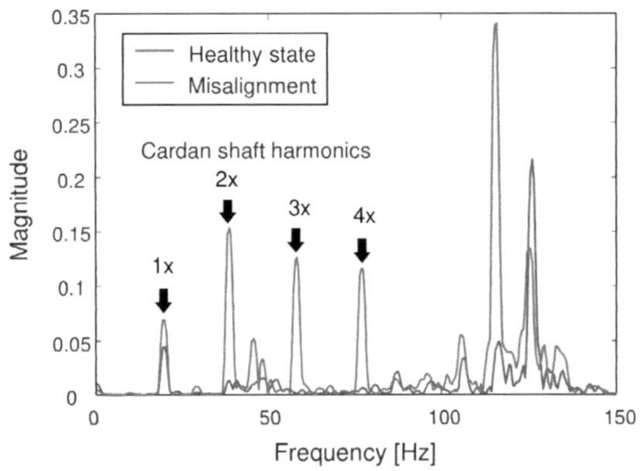

Figure 9. Test rig experimental diagnostic results of a cardan shaft with angular misalignment fault.

5.2. Field tests: cardan shaft with angular misalignment

As mentioned, while vibration signals were recorded using external accelerometers, the metadata, which includes the vehicle operational conditions, is recorded via CAN bus available on the machine. According to the hypotheses obtained in sections 2 and 3, the angular misalignment of a cardan shaft can be detected via vibration signals in the frequency domain when steering. Figure 10 shows 4 various spectrums obtained in different operating conditions and measuring directions.

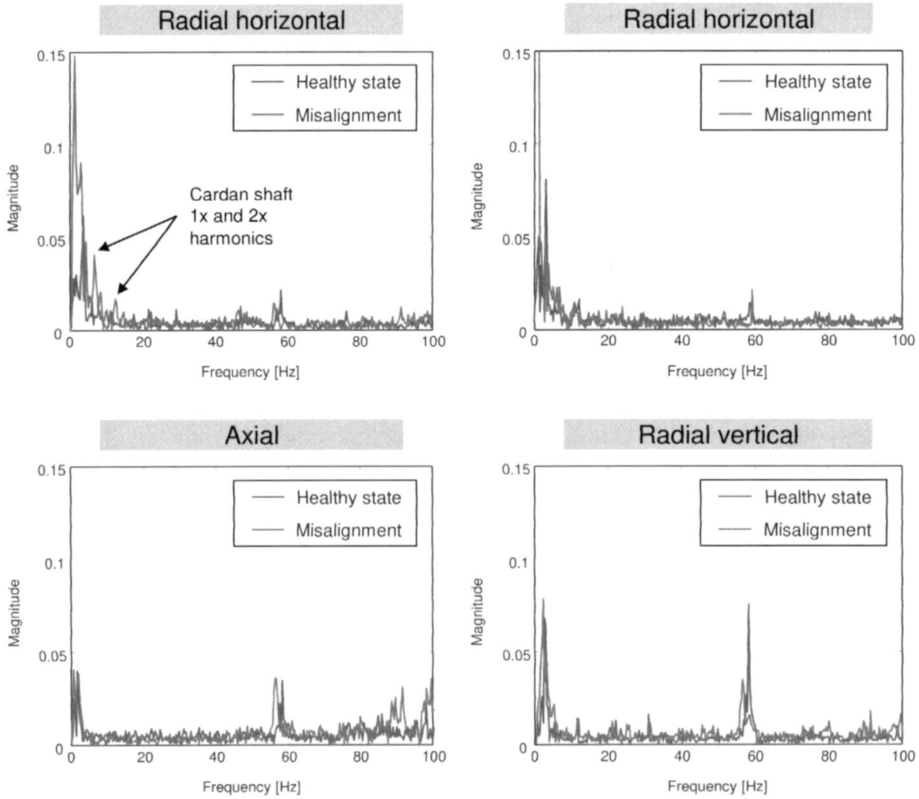

Figure 10. Amplitude spectrum of the vibration signal captured from a sensor nearby the cardan shaft shown in Figure 5, when steering (upper left) and without steering (upper right) both in radial horizontal direction, and in axial (bottom left) and radial vertical (bottom right) both when steering.

IOP Conf. Series: Materials Science and Engineering 1097 (2021) 012019 doi:10.1088/1757-899X/1097/1/012019

As seen, while the spectrum obtained in the radial horizontal direction when steering can reveal the misalignment successfully, other spectrums, obtained for other directions, do not show any peak at shaft velocity harmonics. In addition, the spectrum obtained in radial horizontal direction does not reveal fault characteristics when there is no steering. It is worth mentioning that all these spectrums are calculated for a short time period where the vehicle has constant and comparable velocity of 0.3 Hz as tire rotational speed for all these cases. In addition, by further analysing the upper left spectrum in Figure 10, despite the high external noise, the excitation caused by the misalignment is high enough to be detected in the spectra.

5.3. Field tests: cardan shaft with bearing clearance

In this subsection, the bearing clearance fault using the same sensor and CAN bus data discussed in section 4.2 is investigated. For comparison purposes, the measurement with defective cardan shaft is analysed and studied relative to the healthy condition of the vehicle as reference. According to hypotheses mentioned in sections 2 and 3, depending on torque and shaft velocity conditions, the bearing clearance fault can either be detected by parameter trend analysis in the time domain or by shaft harmonics in the frequency domain. Figure 11 shows the complete vibration measurements for healthy and defective cases. In order to compare healthy and defective states, two signal segments, each of which in one case with similar operating conditions, are selected for detailed investigations.

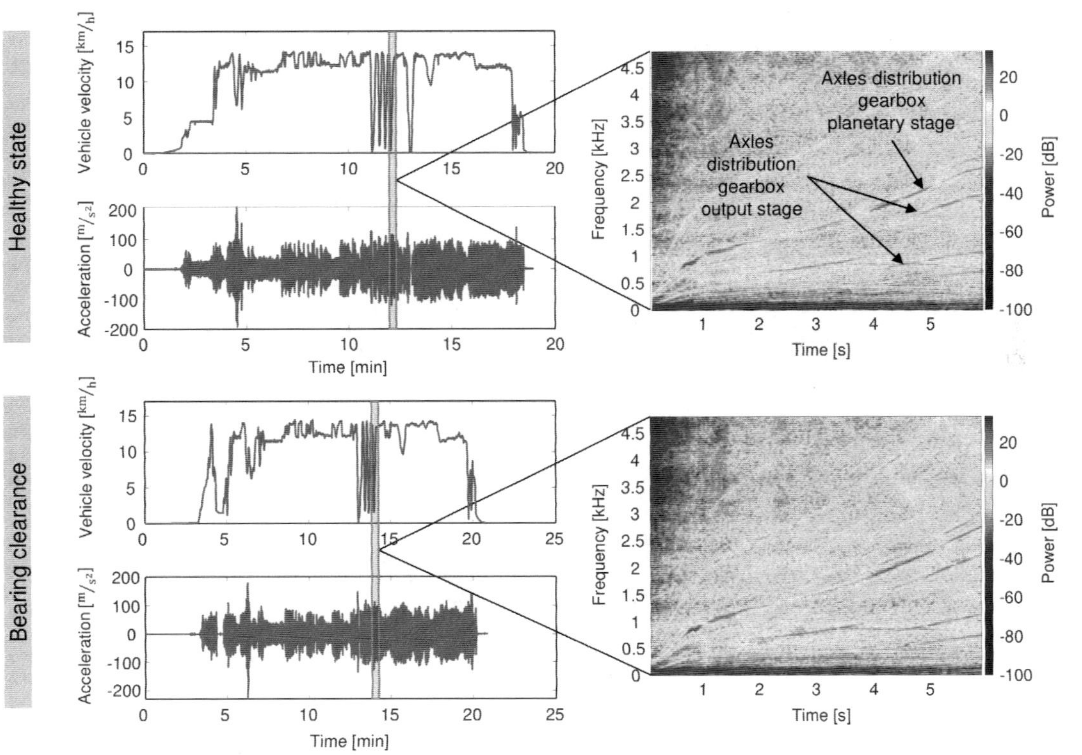

Figure 11. Signals corresponding to the complete cycles of measurements for healthy and defective cardan shafts (left), waterfall illustrations of short signal segments with similar torque and speed selected for further analyses (right).

Figure 12 (upper side) shows spectrums for the short signal segments selected and shown in Figure 11. According to CAN bus signals, since the velocity and torque directions of shaft remain unchanged, the bearing clearance should be detectable as a peak at shaft harmonics. Therefore, as seen in Figure

IOP Conf. Series: Materials Science and Engineering 1097 (2021) 012019 doi:10.1088/1757-899X/1097/1/012019

12, standard deviation trends corresponding to the short signal segments selected and shown in Figure 11, do not reveal any explicit energy level difference between healthy (reference) and defective cases. The frequency domain analyses results shown in Figure 12, however, indicate an increase at second harmonic of the shaft rotational frequency. The amplitude order spectra in Figure 12 suggest that the periodic dynamic force is applied in one direction only, which can be due to unbalance originating from bearing shift. Nevertheless, this may not be considered as a sharp and explicit indication of the bearing clearance fault due to uncertainties originating from field measurements.

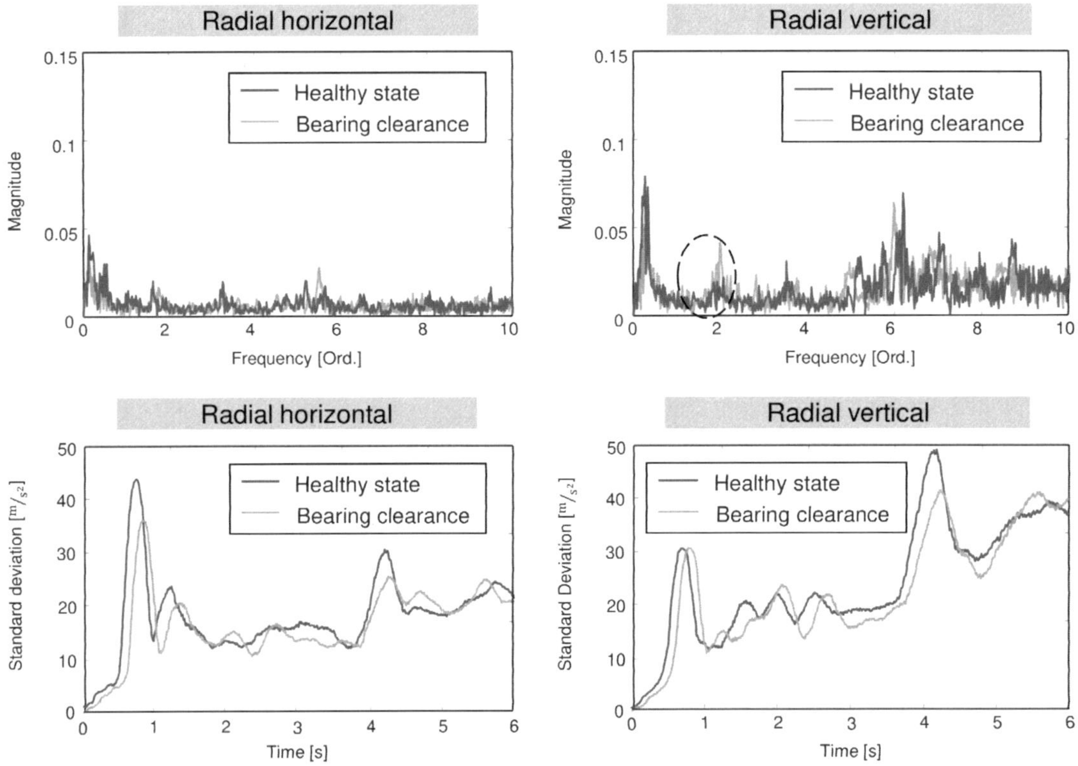

Figure 12. Amplitude order spectrum of the vibration signal captured from a sensor nearby the cardan shaft shown in Figure 8 in radial horizontal direction (upper left) and radial vertical direction (upper right) both for the short time signal segments in Figure 12, and corresponding standard deviations.

Using case studies in subsections 5.2 and 5.3, the great importance and role of metadata in CAN bus on the improvement of vibration analysis is highlighted. While analysis of vibration signals alone without information of vehicle driving condition may not be reliable way to detect the damage type considered here, a multi-dimension analysis using CAN bus and vibration signals can lead to a successful diagnosis of damage.

6. Conclusion

In this current study, cardan joint shaft, as one of the commonly failing components in the drivetrain of wheel loaders, is investigated in the context of vibration-based damage diagnosis.

In order to do that, two common cardan shaft failure types and their effects on the vibration signals are theoretically reviewed. Then, hypotheses for two various damage cases are derived. It is shown and discussed that how metadata, that is available in the machine CAN bus, can improve the fault detectability of a vibration-based diagnosis procedure. According to the derived hypotheses, a methodology based on combination of vibration, operational condition, and machine condition signals are proposed in the current study. The proposed methodology is then analysed and investigated using test rig measurements, as benchmark investigations, and field measurements under real-world operating conditions in an underground mine environment. The experimental results for the test wheel loader considered in this study confirm that CAN bus data can significantly improve the fault detectability of the on-board vibration-based diagnosis system, and are necessary for mobile work machine applications. It is shown that by merging the data from CAN bus and vibration signals, both angular misalignment and bearing clearance fault types on the cardan shaft are diagnosed even under harsh and non-stationary operating conditions of a wheel-loader in field scenario.

Finally, it can be concluded that in addition to stationary applications, vibration monitoring can also result in reliable diagnosis of non-stationary cases, providing that a deep insight into the target component and machine is available.

Acknowledgment

The research in this study was funded by the European Fund for Regional Development (EF-RE) program as part of the research project "ProMaschinenDaten" with the project No: EFRE-0800809. The authors would like to thank all the participating partners in the project including, but not limited to, K+S Aktiengesellschaft, GHH-Fahrzeuge GmbH, and ELBE Gelenkwellen-Service GmbH for their support in the field tests discussed in this study.

EFRE.NRW
Investitionen in Wachstum und Beschäftigung

EUROPÄISCHE UNION
Investition in unsere Zukunft
Europäischer Fonds für regionale Entwicklung

References

[1] US National Aeronautics and Space Administration, (2008) "NASA Reliability-Centered Maintenance Guide (RCM) for Facilities and Collateral Equipment."

[2] A. Laukka et al. (2016) "Condition based monitoring for underground mobile machines," *International Journal of Industrial and Systems Engineering,* 23(1): 74-89.

[3] E. Jakobsson, (2019) "Data-driven Condition Monitoring in Mining Vehicles," PhD Dissertation, Linköping University.

[4] N. Vagenas et al. (1997) "A methodology for maintenance analysis of mining equipment," *International Journal of Surface Mining, Reclamation and Environment,* 11(1): 33-40.

[5] G. Pasch et al. (2019) "Multi-Domain Simulation for the Assessment of the NVH Behaviour of a Tractor with Hydrostatic-Mechanical Power Split Transmission," *Land. Technik AgEng, VDI*, 2361: 19-27, Hannover, Germany, 8-9 Nov.

[6] C. Habermehl et al. (2020) "Influence of Drivetrain Hybridization on Transmission Lifetime," *Applied Sciences*, 10(20): 7086.

[7] C. Habermehl et al. (2020) "Availability Increase of Mobile Machinery through Condition Prognosis," *ATZheavy duty worldwide*, 13: 62-67.

[8] M. Wegerhoff et al. (2018) "Noise, vibration and harshness validation methodology for complex elastic multibody simulation models: With application to an electrified drive train," *Journal of vibration and control*, 25(2): 243-254.

[9] R. Golafshan et al. (2018) "Investigation on the Effects of Structural Dynamics on Rolling Bearing Fault Diagnosis by Means of Multibody Simulation," International Journal of Rotating Machinery, Article ID 51591.

[10] B. X. Tchomeni et al. (2019) "Modeling and vibration analysis of twin-rotor system interconnected by a Hooke's joint (Part A)," *Journal of Vibroengineering (Vibroengineering PROCEDIA),* 27: 1-6.

[11] C. M. Tanik et al. (2015) "Steel compliant Cardan universal joint," *Mechanism and Machine Theory*, 92: 171-183.

IOP Conf. Series: Materials Science and Engineering 1097 (2021) 012019 doi:10.1088/1757-899X/1097/1/012019

[12] M. A. Hili et al. (2006) "Failure analysis of a misaligned and unbalanced flexible rotor," *Journal of Failure Analysis and Prevention*, 6: 73-82.

[13] X. Zhou et al. (2007) "Reliability-centered Predictive Maintenance Scheduling for a Continuously Monitored System Subject to Degradation," Reliability Engineering & System Safety, 92(4): 230-534.

[14] R. B. Randall, (2011) "Vibration-based Condition Monitoring: Industrial, Aerospace and Automotive Applications," Chichester: John Wiley & Sons, Ltd.

[15] B. X. Tchomeni, et al. (2019) "Experimental setup for unbalance fault detection and vibration analysis in a cardan shaft (Part B)," *Journal of Vibroengineering (Vibroengineering PROCEDIA),* 27: 97-102.

[16] D. L. Dewell and L. D. Mitchell, (1984) "Detection of a Misaligned Disk Coupling Using Spectrum Analysis," *Journal of Vibration and Acoustics,* 106(1): 9-16.

[17] Y. Xia et al. (2019) "Nonlinear numerical and experimental study on the second-order torsional and lateral vibration of driveline system connected by cardan joint," *Journal of Vibration and Control*, 26(7-8): 540-551.

[18] W. Voss, (2005) "A Comprehensible Guide to Controller Area Network.Massachusetts," Copperhill Media Corporation.

IOP Conf. Series: Materials Science and Engineering 1097 (2021) 012020 doi:10.1088/1757-899X/1097/1/012020

Evaluation of the Additive Manufacturability of CAD-Parts for initial Data Labelling in AI-based Part Identification

Marcel Winkler, Stefan Stürmer, Christian Konrad

Institute for Machine Elements and Systems Engineering
Schinkelstr. 10, 52062 Aachen, Germany

marcel.winkler@imse.rwth-aachen.de

Abstract. The objective of this study is to develop and implement an analysis tool that identifies CAD geometries which impair Additive Manufacturing. A key performance indicator is generated to be used as data label representing the manufacturability for a future application of AI. Relevant geometric features are identified and algorithms to evaluate critical features are developed. The analyses include part orientation, build volume, wall thicknesses, gap widths, bore and cylinder diameters as well as the process-specific factors powder removal and need for support structures. The manufacturability of a part is calculated as Additive Manufacturing Feasibility Indicator (AMFI), depending on the identified critical features and a user-specific weighting. The AMFI successfully serves as data label which is suitable for application in AI methods. The developed GUI supports designers by highlighting critical features directly in the CAD environment and allowing the user in a purposeful part optimization for AM.

1. Introduction and Motivation

Additive Manufacturing (AM) processes are increasingly moving into the focus of product development nowadays. The potential application of AM processes can be found where products are to be manufactured with great freedom of design, individualised for customers or in a decentralised production to ensure independence from external supply chains. Conventional manufacturing processes often reach their limits regarding these factors. However, a high level of expertise regarding process parameters and design guidelines are required to manufacture parts using AM. Companies often lack the necessary expertise and methods to identify potential parts for AM, which is an obstacle to the successful introduction of AM processes. Overcoming these hurdles with the support of artificial intelligence (AI) methods is a promising approach. Generative design software already uses machine learning technology to propose solutions to a design problem that meet the technical requirements [1]. However, there is no assistance provided yet that offers a basic assessment regarding the manufacturability of a part using AM. An automated manufacturability assessment for AM could support the integration of AM processes into existing production landscapes to improve the value of technical products and provide an economic benefit for companies.

2. State of the Art

In the following, the state of the art regarding part identification for AM by AI methods as well as relevant parameters in AM will be discussed to disclose the need for research in this topic.

2.1. Artificial Intelligence in Part Identification for AM

A first study has shown that parts with high potential for AM can be automatically identified in a CAD database using AI algorithms that evaluate the geometric complexity of a part [2]. In this supervised learning approach, the relationship between given CAD-files as input data and pre-assigned output values is learned by an Artificial Neural Network (ANN) as AI algorithm. The derived classifier is suitable for assigning parts to the clusters "AM potential" and "no AM potential" with an accuracy of over 90%. The procedure and used toolchain are presented in figure 1.

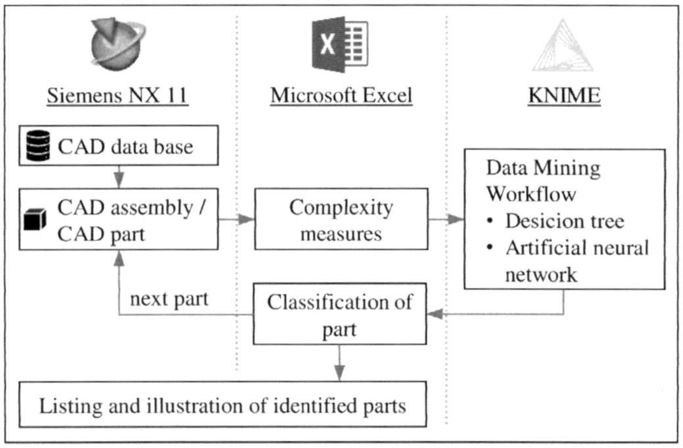

Figure 1. Procedure and Toolchain of an AI Part Identification for AM [2]

In order to train the AI algorithms for the part identification, a sufficient amount of data with known input and output information (labels) is required to obtain an optimal AI-model. The minimum amount of training data depends on the type and structure of the AI algorithm used, as well as on the quality of the training data. The use of a training data set of about 1500 parts as in the study mentioned above can already lead to valid results avoiding that the influencing variables do not describe the target variable sufficiently well (underfitting). In principle, the quality of the network increases with the number of different training records [3].

Literature research reveals that there are currently no methods or tools known to perform an automated labelling of CAD-data with regard to their manufacturability by AM. A manual labelling of all parts in the dataset as performed in the abovementioned study requires expertise in the field of AM and is not expedient with a required time of 1-2 minutes per manual label assignment.

2.2. Characteristic Indicators for AM

Geometric part complexity, as used in the above-mentioned study, can successfully be used as an indicator for AM potential. The literature research on additional characteristic indicators in Design for AM (DfAM) guidelines, reveals that the indicators can roughly be divided into two categories [4]. On the one hand, there are indicators that deal with the potentials of AM processes [5, 6], on the other hand there are indicators that deal with restrictions [7, 8, 9, 10]. Table 1 gives an overview of the most relevant characteristic indicators for AM and divides them in the two categories *opportunistic* and *restrictive*.

Table 1. Overview of Characteristic Indicators for AM [4]

Characteristic indicators	
Opportunistic indicators	*Restrictive indicators*
• Part complexity	• Suitability of material
• Degree of customizability	• Batch size of the part
• Functional integration and part consolidation	• Required design quality of the part

As a third category, some authors add the "combined" or "dual" DfAM approaches, which represent a combination of the two previous categories but will not be further considered in this study because they are considered subordinate. In some cases, the assessment of manufacturability is considered as separated indicator alongside of the restrictive indicator design quality, what further underlines the necessity of an assessment of the manufacturability of a part [4].

2.3. Manufacturability as restrictive indicator

Primarily, the limitations of a manufacturing process are crucial for successful additive manufacturing when considering the manufacturability as a restrictive indicator. A large number of different limitations and guidelines can be found in the literature, both with regard to the physical shape of the components and their quality.

Since the design quality of an additive manufactured product is always based on the design quality of the individual geometric elements of which it is composed, special features are often examined in literature for their manufacturability [5, 7, 8, 11, 12, 13]. The characteristic values mentioned apply equally to most of the relevant AM process. In addition to these generalisable factors (e.g. see table 2), process-specific limitations for the most industrially relevant powder-based and material extrusion processes have been researched in the literature [7, 11, 14]. Representatives of powder-based processes include the Selective Laser Melting (SLM), also known as Laser Powder Bed Fusion (LPBF), and Selective Laser Sintering (SLS). In the field of material extrusion, the Fused Filament Fabrication (FFF), often called Fused Layer Modelling (FLM), should be mentioned [15]. Table 2 presents a summary of the most important general and process-specific factors regarding the additive manufacturability of parts focusing on powder-based and material extrusion processes.

Table 2. Overview of the most important factors regarding the manufacturability of AM parts

General factors	**Process-specific factors**
• Minimum wall thickness	Powder-based processes
• Minimum gap width	• Critical hole diameter
• Minimum bore diameter	• Cavities
• Minimum cylinder diameter	Material extrusion processes
• Part orientation	• Min. cross-sectional area
• Build volume	• Critical overhangs
	• Islands

At this point it should be noted that the mentioned factors do not cover all existing process limits due to the complexity and variety of additive manufacturing processes. However, these factors represent the central general and process-specific factors that limit the most common AM processes. That is proven on the one hand by their frequent mention in the literature and on the other hand by the degree of their research. However, there is currently no known solution that automatically analyses these factors and makes an assessment in terms of additive manufacturability.

2.4. Automation in the AM Process Chain

The Mechanical Engineering Industry Association (VDMA), Europe's largest industrial association, founded the working group Additive Manufacturing (AG AM) already in 2014 [16]. The ambition of the AG AM is to increasingly integrate AM processes into industrial process chains. A detailed documentation of the entire process chain of AM has been prepared by the AG AM in order to capture the current state of development of the individual process steps on the path from manufactory to Smart Factory. This has resulted in the roadmap "Automation for Additive Manufacturing" (see figure 2), in which an extensive amount of know-how for the integration of AM technology in industrial production processes is compiled and potentials for future development towards automation are listed [17, 18].

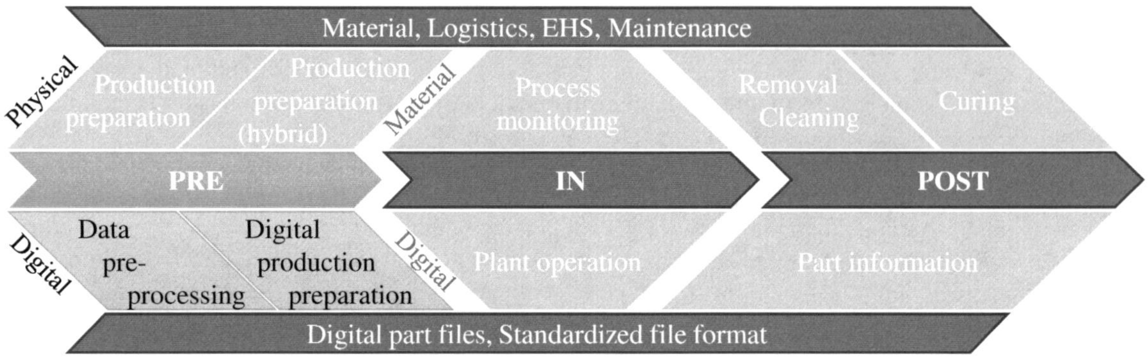

Figure 2. Roadmap "Automation for Additive Manufacturing" adapted from [18]

The necessary pre-processing steps regarding the digital part preparation are marked in the figure above. Prior to the manufacturing process, the CAD data from the design domain have to be prepared for the application in the layered manufacturing process. Information on component geometry, tolerances and functions are incorporated into the production data. These data are adapted according to the desired manufacturing process. In the subsequent digital production preparation process, this data is imported to arrange the parts in the build space, if necessary define support structures and simulate the build process.

When examining the individual process steps in this digital area of the manufacturing process chain, many of today's processes require manual inspection by a manufacturing employee. If deficits in the part design are found which impair production, the component design has to be adapted by the design domain and a new inspection needs to be carried out which is a time-consuming iteration. The simulation and optimization of the build process is carried out in a similar manual process. Current work in progress primarily focuses on a software-supported screening of part defects as errors in the mesh structure. However, tools are only available as individual solutions or at a demonstrator stage. To reach a full automation in the digital data preparation the desired future solution is described as an assistance system implemented in CAD software. It should include intelligent algorithms for automated data creation and manipulation, that consider economic efficiency, high quality and process safety already during the design phase of a part [17, 18]. The consideration of the additive manufacturability of a part in early design process stages is also identified as a goal by consultants and service providers in order to close the existing gaps in a broad industrial application of AM [19, 20].

3. Need for Research

The preceding remarks on the state of the art reveal the need for research in the field of an automated manufacturability assessment of parts for AM. To reach the goal of economic efficiency, there is a need to support the design domain regarding design for AM by early assessment of the manufacturability to reduce cost for design iterations and time to production. The application of AI for part identification for AM also needs be further researched to exploit its promising potential. The following research questions and hypothesis are derived to address the white spot in the research area.

Research Questions

How can the effort for data labelling of parts with potential for additive manufacturing be decreased by an assessment of the manufacturability?

- Which geometric elements of a part does the feasibility of additive manufacturing depend on?
- How can these geometric elements be evaluated automatically with regard to their additive manufacturability?
- How can the result of the AM feasibility analysis serve as a data label for an application of AI methods?

IOP Conf. Series: Materials Science and Engineering 1097 (2021) 012020 doi:10.1088/1757-899X/1097/1/012020

Research Hypothesis

Based on the part geometry depicted in CAD the manufacturability of the parts in specific AM processes can be evaluated to derive a suitable characteristic value as data label.

Objectives of the Study

The intention of this study is to leverage applications of AI and AM in product development. The overall objective is to develop and implement an analysis tool that evaluates restrictive part geometries in CAD files that impair AM and that generates a key figure that can be used as data label representing the manufacturability. Based on this overall objective, the following sub-objectives were defined for this study:

1. The most relevant geometric features that might impair AM are identified and graded
2. Methods to evaluate these features are developed and implemented in CAD software
3. The manufacturability is determined by a calculation method based on feature evaluations

4. Methodical Approach and Materials

The methodical approach for the assessment of manufacturability in AM first of all intends to determine the relevant restrictive factors that disqualify parts for manufacturing using AM. This step is followed by the evaluation of these factors in a general and process-specific analysis using algorithms in the CAD software. Finally, a key figure representing the manufacturability is calculated. The used methods for each step are described briefly in the following chapters.

Further, the given constraints for the research and development have to be determined to achieve the objectives of the study. The concepts and solutions developed within the framework of this study target the integration into the AI part identification toolchain for AM presented in the state of the art. Since the part analysis and the derivation of characteristic values in the AM toolchain is carried out in the CAD software Siemens NX, the NXopen interface is also intended for this study. NXopen offers a programming interface to access the part data, perform analyses and calculate characteristic values. To further process the calculated values in the AI toolchain, data transfer via an Excel sheet is used.

Regarding the process-specific analysis three AM processes are focused to cover a wide range of industrial relevant applications. The SLM process is considered as a representative of powder-based processes using metals. At the other end of the relevant range the FLM as a material extrusion process using polymers is acknowledged. The third process between the abovementioned is the SLS process. These AM processes are sufficiently widespread in industrial practice and guidelines for the design of components, as well as the potentials and restrictions of the respective processes are well researched in order to derive generally valid methods.

To allow the usage of the calculated key figure as data label, a normalised value between 0 and 1 is required. In this study, the value 0 represents that the part is "not feasible in AM" and 1 stands for "feasible in AM". Values between the two extrema indicate that a component is partially suitable for AM and some geometric areas may need to be redesigned.

The ABC dataset containing more than one million CAD files is used as dataset for validation. Its applicability regarding AI has already been proven in former studies [2, 21].

4.1. Relevant restrictive features for this study

As mentioned in table 2 there are several factors, which are important, when it comes to the manufacturability of AM parts. Almost all of the mentioned factors can be directly translated to restrictive and thus critical features for the additive manufacturability of a given part. The only two exceptions are the part orientation as well as the given build volume.

Although both factors in combination can easily lead to an exceedance of the given build volume of the printer and can thus be declared as critical, they do not represent a feature of the given part. Furthermore, the part orientation has an influence on the restrictive features, as their orientation in the layer wise build processes of AM influences their manufacturability. In order to distinguish manufacturability in the printing direction (Z-axis) and the printing plane (X-Y plane), the orientation of the component must first be defined. After the definition of the part orientation all other features can be directly derived by means of geometrical and mathematical analyses within the CAD software.

4.2. General Analysis

As mentioned earlier general restrictive features can be analysed equally across all relevant AM processes. Although not characterized as a critical feature of a given part, the first step in the analysis has to be the definition of the part orientation within the given build volume. Afterwards the analysis of the focused general restrictive features like the minimum wall thickness and gap width as well as the minimum bore and cylinder diameters will be discussed.

4.2.1. Part Orientation and Build Volume

In order to analyse all following restrictive features, the part orientation has to be determined first. There are several possible optimisation strategies regarding part orientation that pursue different goals. For example, parts can be oriented according to functional requirements such as surface quality and tolerances. If functional requirements are demanded for special surfaces, parts usually need to be post-processed. Therefore, this study follows an optimization strategy focusing on process reliability regarding FLM processes and thus identifies the orientation for the best adherence of the part on the print plate.

To achieve this, an algorithm was created, which first identifies all plane faces of the given part and checks if these are possible print bases by verifying if the planes are outer boundaries of the part. Afterwards all identified bases are analysed and coplanar faces are cumulated. Beginning from the largest base area, the part is then oriented using the normal vector of the face and the minimum bounding box of the part is derived. This bounding box is then compared with the given build volume of the printer and the largest possible base area is chosen as the basis for the orientation of the part. If no possible base can be identified or all bases result in an exceedance of the build volume, the minimum bounding box resulting from the orientation in the native z-direction of the part is checked. If this step does not result in a possible orientation it is declared, that the given part cannot be manufactured regarding the given orientation strategy. It is conceivable to integrate other orientation strategies in the procedure or to let the user specify an orientation.

4.2.2. Minimum Wall Thickness and Gap Widths

After aligning the part according to the identified orientation, the part is checked for deviations of minimum wall thicknesses and gap widths. In order to perform this analysis an algorithm is designed that slices the part in several equidistant planes parallel to the print base. From the resulting slices, the boundary curves are extracted and analysed via a comparison of pairs. In these comparisons it is checked whether the paired curves are the same curve, neighbouring curves or other curves. Afterwards the type of curve is checked. Depending on these two parameters the type of analytical comparison varies from a mere analysis of the contact angle of the curves (figure 3 a) through the analysis of the distance between the beginning and end point of the curve (figure 3 b) to the analysis of the distance between sets of equidistant points on the curves (figure 3 c).

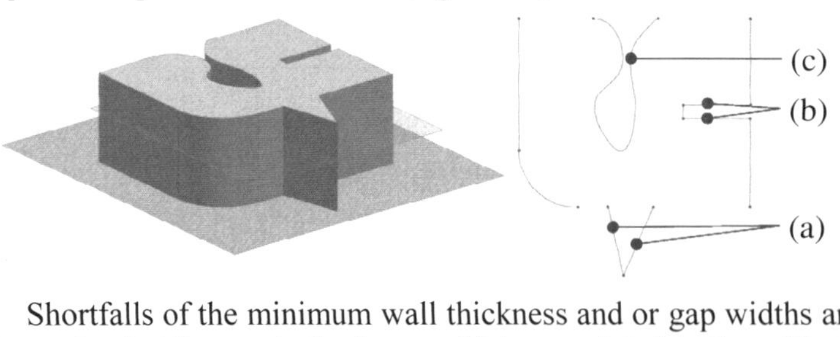

Figure 3. Exemplary results when checking wall thicknesses and gap widths

Shortfalls of the minimum wall thickness and or gap widths are recorded and the critical curves are determined. Afterwards the faces, which are related to the critical curves are derived and it is checked whether the faces achieve the minimum wall thickness or gap width.

IOP Conf. Series: Materials Science and Engineering 1097 (2021) 012020 doi:10.1088/1757-899X/1097/1/012020

An equivalent analysis is performed for the minimum wall thicknesses and gap width perpendicular to the print direction, although in this direction only the wall thicknesses and gap widths parallel to the print direction are considered.

4.2.3. Minimum Bore and Cylinder Diameters
Similar to the minimal wall thickness and gap width the minimum bore and cylinder diameter can also be checked in print direction and perpendicular to this direction. To find holes and cylinders which fail to achieve the minimum diameter, first all cylindrical faces of the part are identified. Afterwards it is determined if the face belongs to a bore hole or a cylinder and the direction of the central axis is identified.

With this direction and the print orientation the minimum radius of the section ellipsis is determined for the print direction as well as perpendicular to the print direction (see figure 4). If this radius falls short of the minimum bore or cylinder radius, the bore hole respectively cylinder is characterized as critical.

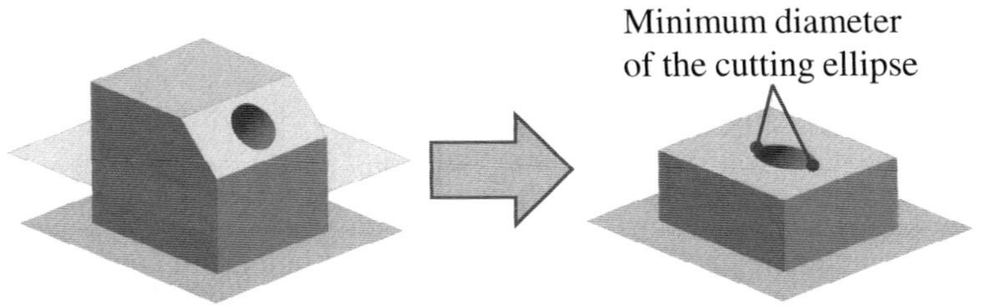

Figure 4. Section of an angled hole parallel to the building plane and creation of a sectional ellipse

4.3. Process-specific Analyses
As the name suggests, process-specific factors depend on the selected AM process. The focused AM processes are powder-based processes (SLS and SLM) and material extrusion processes (FLM) which both have different restrictive features. Whereas for powder-based processes it is important to ensure the removal of the powder from holes and cavities, material extrusion processes tend to overheat features with small cross-sectional areas and need support structures in order to print features with critical overhangs and islands.

4.3.1. Powder Based Processes
In order to ensure the powder removal from bore holes, the literature suggests to stay below a certain length to diameter ration R_{LD}. To assess this ratio all cylindrical faces, which form holes are extracted from the CAD model. Afterwards their lengths l_H and diameters d_H are extracted. With both values R_{LD} can be calculated and critical holes identified.

$$R_{LD} = \frac{l_H}{d_H} \tag{1}$$

Another important feature to consider, when talking about the removal of access powder are closed cavities. For cavities there is no specific limit value that can be derived from literature. Therefore, it is decisive to detect the occurrence of closed cavities in the given CAD model. To do so a cuboid with the measurements of an enlarged bounding box of the part is created and the model itself is subtracted from the cuboid. If the result of this operation creates n bodies with n larger than one, the part has a total of $n - 1$ closed cavities. If the resulting number of bodies is equal to one, there are no closed cavities in the model.

4.3.2. Material Extrusion
To determine whether features of the CAD model are prone to overheating during the manufacturing process, the minimum cross-sectional area was introduced. To determine the parts of the CAD model which fall short in regard to this area, the part is sliced in several equidistant heights parallel to the print

base. At every height the section faces are extracted and their area is measured. If a shortfall is detected, the associated feature of the section area is determined.

In order to minimize the usage of support structures and thus the need for postprocessing, the areas of the CAD model, which need such structures may also be detected by the algorithm. A critical value to identify the areas in need for support structures is the critical downskin angle of the faces. For regular faces like planes, cylinders and cones the determination can be achieved via the normal vector of the planes or the central axis of the cylinders and cones in combination with the cone angle respectively (figure 5 a). For free-form surfaces or other more complex faces, a net of test points is projected onto the face and the normal vector of the face at these points is extracted. With this vector the dowskin angle can be calculated and compared to the minimum downskin angle.

Other areas of the CAD part in need for support structures are islands. In order to detect such structures, the part is repeatedly sliced in several equidistant heights and the resulting section areas are extracted. Afterwards the number of areas is compared to the one in the previous section. If the number increases, potential islands are detected. To check whether the detected areas are actually part of an island, an equidistant net of test points is projected onto the areas of the previous slice and from there projected onto the next layer. The areas, which are free of any test points are then declared as part of an island and the related geometric features of the CAD part are extracted (figure 5 b).

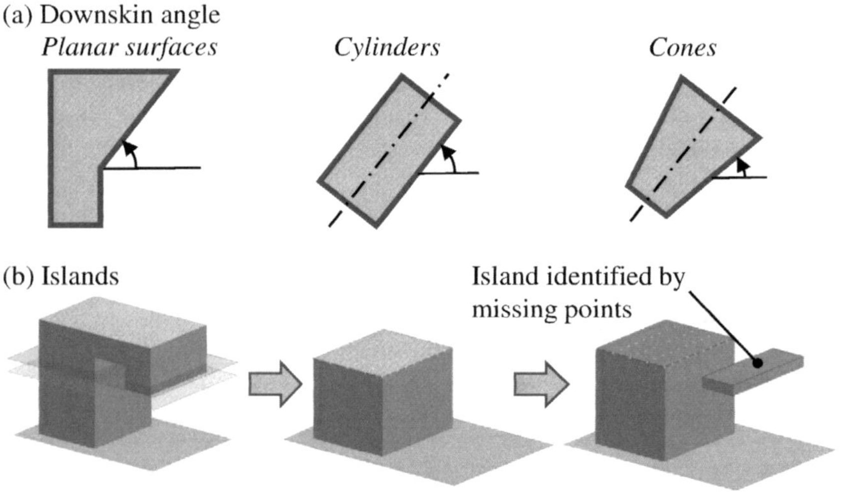

Figure 5. Evaluation of (a) critical downskin angle and (b) islands

4.4. Evaluation of manufacturability

To finally provide a measurable key figure representing the manufacturability of a component in a specific AM process the Additive Manufacturing Feasibility Indicator (AMFI) is developed. This measure allows to derive a statement regarding the manufacturability on the basis of the critical features described before.

Due to the current diversity of additive production plants and the different process capabilities, the effect of recognized critical features should be adjustable by user-specific weighting. For this purpose, the user is given the possibility to specify a maximum allowed number $q_{max,k}$ for each type of critical features identified, which may occur within a component. In addition, the user should be able to specify a value for each of these features for its integration i_k into the evaluation of manufacturability. This selection is made possible for both general and process-specific factors. Since the surfaces that require support structures do not directly influence the manufacturability of a component, they are not included in the evaluation at this point.

In order to specify a single key figure representing the manufacturability, the first step is to calculate the manufacturability of the individual critical features. For this purpose, it is checked whether the number of detected instances q_k of a critical feature exceeds the maximum number specified by the user.

IOP Conf. Series: Materials Science and Engineering 1097 (2021) 012020 doi:10.1088/1757-899X/1097/1/012020

If this is the case, then the manufacturability m_i of this feature is set to zero. If this is not the case, then the manufacturability of this feature is calculated as follows.

$$m_k = 1 - \frac{q_k}{q_{max,k} + 1} \tag{2}$$

Once the calculation of the manufacturability of the individual critical features has been completed, they still have to be combined to the total manufacturability. For this purpose, the weightings w_k provided by the users are calculated for the individual features using the following calculation rule.

$$w_k = \frac{i_k}{\sum_{j=1}^{n} i_j} \tag{3}$$

The AMFI of a part $P1$ can now be calculated as the sum of the products of the manufacturability of a feature m_j and the user-specific weighting w_j.

$$AMFI_{P1} = \sum_{j=1}^{n} m_j \times w_j \tag{4}$$

The overall approach of the quantification of manufacturability in form of the AMFI and the applied tools are illustrated in the following figure 6.

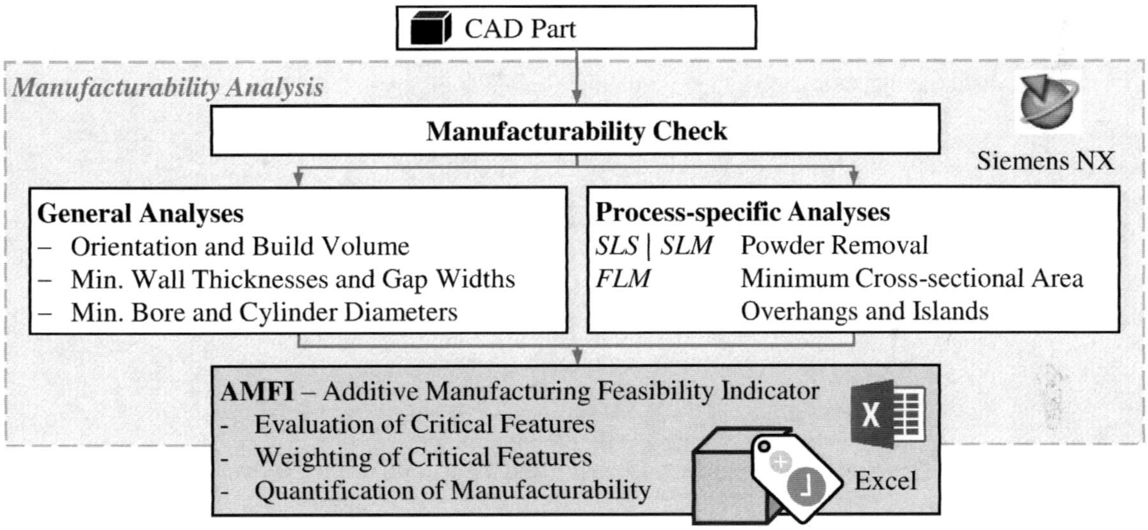

Figure 6. Process of the manufacturability analysis to calculate the AMFI

5. Results

The objective of this study was to develop and implement an analysis tool that evaluates restrictive part geometries in CAD files that impair AM and that generates a key figure that can be used as data label representing the manufacturability. The foundation of this analysis is given by the identification and evaluation of the most relevant geometric features that might impair a successful AM build. The developed algorithms described above evaluate the general factors build volume, wall thickness and gap widths as well as bore and cylinder diameters. Depending on the chosen AM process, process-specific parameters are analysed afterwards. For powder-based processes, channels and cavities in the part are specifically analysed to ensure successful powder removal. The cross-sectional area, overhang angles and islands are relevant factors in material extrusion and therefore analysed. Depending on the number of critical features identified in the analyses and their user-specific weighting, the AMFI of a part is calculated. The AMFI is designed to take values between 0 and 1 to meet the requirement for a suitable data label in the AI tool.

Therefore, it is proven that suitable characteristic values for data labelling can be derived from the part geometry depicted in CAD and that the manufacturability of the part geometry can be evaluated for specific AM processes. Further, the time for the automatic part analysis at approx. 30-40 seconds is significantly less than a manual analysis.

For the integration of the analyses in in Siemens NX a graphical user interface (GUI) has been developed in order to control the part analyses in a user-friendly way and to query necessary user-specific process limits. The GUI guides the user through a five-step process and allows the display of the results. Figure 7 illustrates the process steps that can be carried out for the analysis of a part. Step (a) allows the user to select one of the three mentioned AM processes to be used for part production. Afterwards, process limits can be adjusted individually for the general and process-specific analyses in step (b) depending on the focused AM plant. If the user does not wish to include individual factors in the calculation, this can be set in step (c) via the percentage weighting of the factors. Further the maximum number of critical instances of a specific feature can be specified. If more than the specified critical instances are detected in a part, the feature is indicated as not manufacturable (value 0). The analyses are carried out in the CAD software and an overview of recent analyses and the examined parts including their AMFI is prepared for the following step (d). The user is now able to choose one of the examined parts and in step (e) a table of the instances of each identified critical feature is presented for a user selection. Finally, the selected critical feature can be displayed directly in Siemens NX.

(a) Basic settings (b) Settings of process limits (c) Settings of weightings (d) Analyses and AMFI (e) Selection of critical features

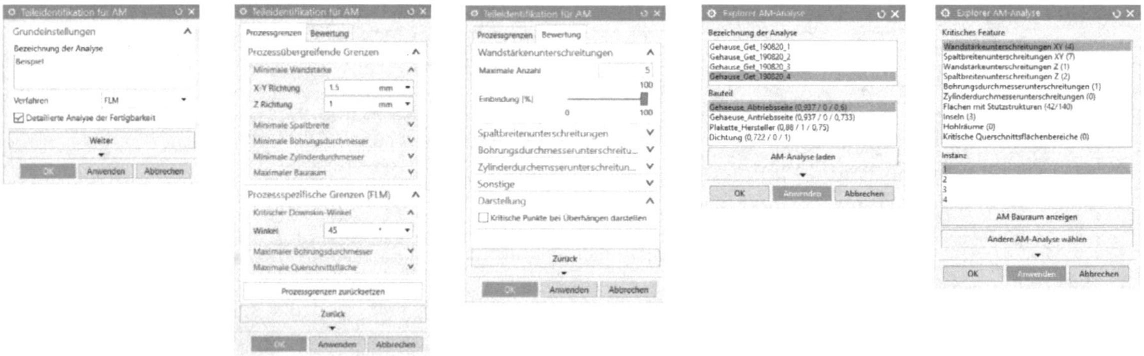

Figure 7: Developed user interface for setup and evaluation of the AMFI in Siemens NX

To illustrate the usage of the AMFI-tool figure 8 presents the analysis of an exemplary part. The calculated AMFI of this part results a value of 0.6. All weightings of the critical factors were rated equally for the calculation and each critical feature may be present a maximum of once. The evaluation reveals that the part contains areas that might be critical for AM as small bore diameters (a), cavities (b) and islands (c). No critical wall thicknesses, gap widths or cylinder diameters are detected. Therefore, the AMFI is calculated as shown in equation (5).

$$AMFI_{example} = \frac{1}{5} * \left(\left(1 - \frac{0}{1+1}\right) + \left(1 - \frac{0}{1+1}\right) + \left(1 - \frac{2}{1+1}\right) + \left(1 - \frac{0}{1+1}\right) + \left(1 - \frac{2}{1+1}\right) \right) = 0.6 \quad (5)$$

The AMFI value of 0.6 can be interpreted in such a way that the part is not fully proven to be successfully manufactured according to the specified process limits. Because of the identified critical features, it is advisable to check the highlighted areas (see figure 8) and rework the design.

(a) (b) (c)

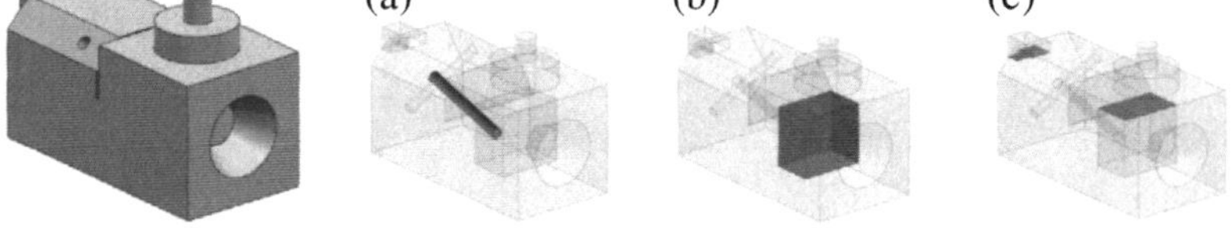

Figure 8. Indication of critical features in CAD

Validation

Finally, the presented method is validated with regard to its applicability for the automated data labelling of large CAD data sets. A subset of the ABC dataset of 3000 CAD-Files is analysed by the developed method and the calculated values for the AMFI are stored as data label. As an example, the process parameters and limits of the Prusa i3 MK3S printer were used as reference [22]. The diagram in figure 9 presents the distribution of the analysis results for the AMFI by plotting the number of parts on the logarithmic Y-axis over the AMFI values on the X-axis.

1280 parts are labelled with a value of 1, which means that they hold no geometric restrictions for AM. A sample inspection of these parts proves that all are suitable for AM. 1085 of the analysed parts are assigned with an AMFI value greater than 0 and less than 1. Critical geometric elements for manufacturing are identified in these parts. A total of 431 parts are assigned the AMFI value 0 because the dimensions of their bounding box exceeded the Prusa printers build space limitation of 250 x 210 x 210 mm (X x Y x Z).

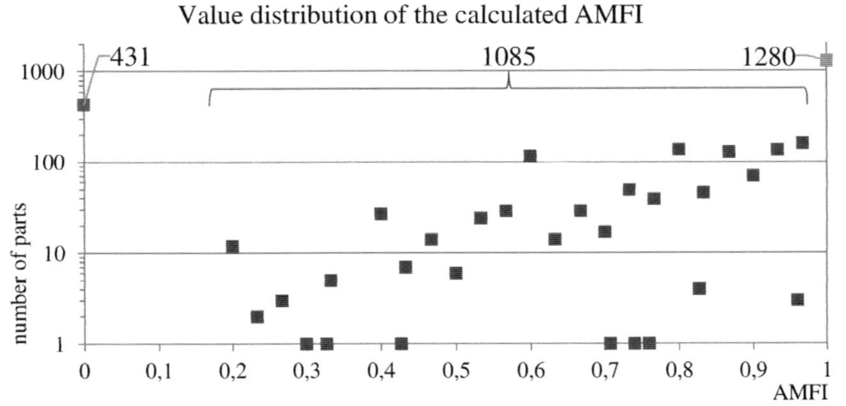

Figure 9. Results of the AMFI calculation of the validation data set

The AMFI calculation failed for the remaining 204 parts not indicated in figure 9. The analysis of the error rate of 6.8 % reveals that the analysed parts mostly show inconsistencies in the modelling, such as several intersecting bodies in one file (a) or consist of a high number of free-form surfaces with complex numerical definition (b) as shown exemplarily in figure 10. The presented analysis algorithms need to be further improved to intercept these kinds of errors and also provide an analysis of complex free-form surfaces.

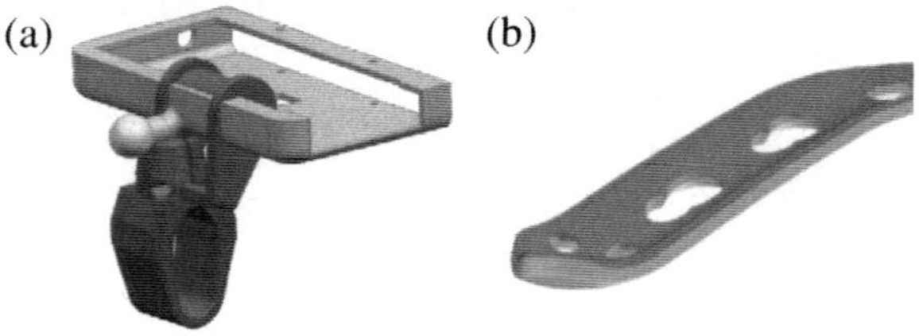

Figure 10. Examples of failed part analyses

6. Summary and Outlook

The developed toolchain realizes an automated evaluation of a large number of CAD data regarding their additive manufacturability. The presented indicator AMFI can be determined to quantify the manufacturability of a part using AM and thus make it accessible for further applications of data analysis. The challenge of a time-consuming manual data labelling described in the state of the art can be automated following this procedure. This allows an analysis of the necessary large amount of training data for AI applications and an unbiased labelling in a shorter period of time. In order to optimize the method, the analysis algorithms need to be further improved in order to be able to better evaluate components containing free-form surfaces. Since the AMFI is calculated in a value range between 0 and 1, it is suitable to serve as label for the training of AI algorithms like ANN. The application of the AMFI in combination with the existing AI toolchain will be conducted subsequently to this study.

References

[1] Redshift by Autodesk 2020 Generative Design: redefinining what's possible in the future of manufacturing

[2] Winkler M, Jacobs G, Stürmer S and Konrad C 2020 *International Conference on Intelligent Engineering and Management (ICIEM)* **281–88** London, UK

[3] Papp S, Weidinger W, Meir-Huber M, Ortner B, Langs G and Wazir R 2019 Handbuch Data Science: Mit Datenanalyse und Machine Learning Wert aus Daten generieren (München: Hanser, Carl)

[4] Kumke M 2018 Methodisches Konstruieren von additiv gefertigten Bauteilen (Wiesbaden: Springer Fachmedien Wiesbaden)

[5] Diegel O, Nordin A and Motte D 2019 A Practical Guide to Design for Additive Manufacturing (Singapore: Springer Singapore)

[6] Laverne F, Segonds F, Anwer N and Le Coq M 2015 Assembly Based Methods to Support Product Innovation in Design for Additive Manufacturing: An Exploratory Case Study *Journal of Mechanical Design 137* (12)

[7] Gebhardt A 2016 Generative Fertigungsverfahren: Additive Manufacturing und 3D-Drucken für Prototyping - Tooling - Produktion, 5th edn (München: Hanser)

[8] Lachmayer R, Lippert R B and Kaierle S 2018 Additive Serienfertigung (Berlin, Heidelberg: Springer Berlin Heidelberg)

[9] DIN Deutsches Institut für Normung e. V. 2016 Additive Fertigung – Grundlagen – Teil 2: Überblick über Prozesskategorien und Ausgangswerkstoffe (Berlin: Beuth Verlag GmbH)

[10] DIN Deutsches Institut für Normung e. V. 2016 Additive Fertigung – Grundlagen – Teil 3: Haupteigenschaften und entsprechende Prüfverfahren (Berlin: Beuth Verlag GmbH)

[11] Adam G A O and Zimmer D 2015 On design for additive manufacturing: evaluating geometrical limitations. *Rapid Prototyping Journal 21* **6**: 662–670

[12] VDI 3405 2014 Additive manufacturing processes, rapid manufacturing - Basics, definitions, processes

[13] Mani M, Witherell P and Jee H 2017 Design Rules for Additive Manufacturing: A Categorization *ASME 2017 International Design Engineering Technical Conferences and Computers and Information in Engineering Conference* **1**: 1–10 Cleveland, Ohio, USA

[14] Breuninger J, Becker R, Wolf A, Rommel S and Verl A 2013 Generative Fertigung mit Kunststoffen (Berlin, Heidelberg: Springer Berlin Heidelberg)

[15] Wohlers T, Campbell R I, Diegel O, Huff R and Kowen J 2020 Wohlers Report 2020: 3D Printing and Additive Manufacturing State of the Industry (Fort Collins: WOHLERS Associates)

[16] Paul H 2020 About the VDMA *https://www.vdma.org/en/ueber-den-vdma#1* accessed 12 November 2020

[17] VDMA Arbeitsgemeinschaft AM 2018 Additive Manufacturing: Metall Pulverbett Automatisierungs Roadmap: SMART FACTORY

[18] VDMA Arbeitsgemeinschaft AM 2020. Additive Manufacturing Kunststoff Pulverbett Automatisierungs Roadmap: SMART FACTORY

[19] Hauck C, Heering M and Gebhardt R 2019 Aktivitätenbericht 2018-2019: Arbeitsgemeinschaft Additive Manufacturing

[20] Langefeld B 2020 De-risking your supply chain in times of the coronavirus: How a digital inventory and additive manufacturing can mitigate risks

[21] Koch S, Matveev A, Jiang Z, Williams F, Artemov A, Burnaev E, Alexa M, Zorin D and Panozzo D 2019 ABC: A Big CAD Model Dataset For Geometric Deep Learning *The IEEE Conference on Computer Vision and Pattern Recognition (CVPR)*

[22] Prusa Research a.s. 2020 3D Printing Handbook

Design and manufacturing of a cylinder head by laser powder bed fusion

J Willkomm[1,3], K Kuhlbach[2], D Merget[1], M Wagner[1], S Reich[1], S Ziegler[1] and Prof J H Schleifenbaum[1]

[1] RWTH Aachen University – Digital Additive Production, 52074 Aachen, Germany
[2] Ford-Werke GmbH, 50735 Köln, Germany

johannes.willkomm@dap.rwth-aachen.de

Abstract. Nowadays the reduction of fuel consumption is the most important challenge for engine development. Laser Powder Bed Fusion (LPBF) as an additive manufacturing process with its implied design freedom makes it possible to reduce combustion emissions, increase engine efficiency and respond specifically to customer requirements. In this work, the outstanding degree of freedom of LPBF is exploited for a new design concept for a cylinder head of an engine in automotive industry in order to address further reduction of fuel consumption through improved cooling performance. Within the given installation space and under consideration of the existing mounting and connection positions, the design concept was developed, subsequently manufactured and reworked. In order to develop a design that meets the LPBF requirements, both the potential of the LPBF using the geometric degrees of freedom as effectively as possible and the limitations are considered. The opposing design drivers between exploitation of potentials and limitations in LPBF are achieved by an iterative design process supported by an own developed design tool. With the new design concept, a reduction of the metal temperature in the cylinder head of approx. 25 K is achieved, verified by simulation.

1. Introduction

Nowadays, the reduction of fuel consumption is the most important challenge of the engine development, and this determines most design decisions [1]. Additive manufacturing (AM) with its implied design freedom makes it possible to reduce combustion emissions, increase engine efficiency and respond specifically to customer requirements. Moreover, this technology has a high potential for future manufacturing, but the application is not yet standard in automotive engine components: Metal AM opens new development opportunities for powertrain components. The most widely used AM technology for metals is Laser Powder Bed Fusion (LPBF) [2]. For the LPBF process, CAD-information and a high-power laser beam are used to create three-dimensional metal parts by fusing fine metal powders. LPBF offers a great potential to add further functionality to the parts and to increase performance. As it can easily generate integrated hollow or lattice structures for the LPBF-fabricated parts, the volume and weight of the component can be reduced and/or simultaneously the surface can be increased (e.g. for highly-improved heat transfer). This technology opens the horizon to entirely new design approaches to reduce current production and lead times and to bring down production costs in the future. Using the new approaches, completely novel and multifunctional parts can be designed.

In this work, the outstanding degree of freedom of LPBF is exploited for a new design concept for a cylinder head of an engine in automotive industry to address further reduction of fuel consumption through various measures. Therefore, cooling performance (combustion knock avoidance) and manufacturing quality in parallel are addressed. Within the given installation space and under consideration of the existing mounting and connection constraints, the design concept was developed, subsequently manufactured and post-processed to be evaluated on a test bench simulating real operating conditions in the future.

2. Laser Powder Bed Fusion

The process steps of LPBF are shown in figure 1. The material as powder particles is processed and completely melted locally and selectively by a laser beam according to a digital slice model of the part. The layer forms a dense material after solidification. A relative density of over 99.5% is achieved if the correct process parameters are chosen, and the layer thickness is usually approx. between 30 and 60 µm. For a detailed description of the process see [3].

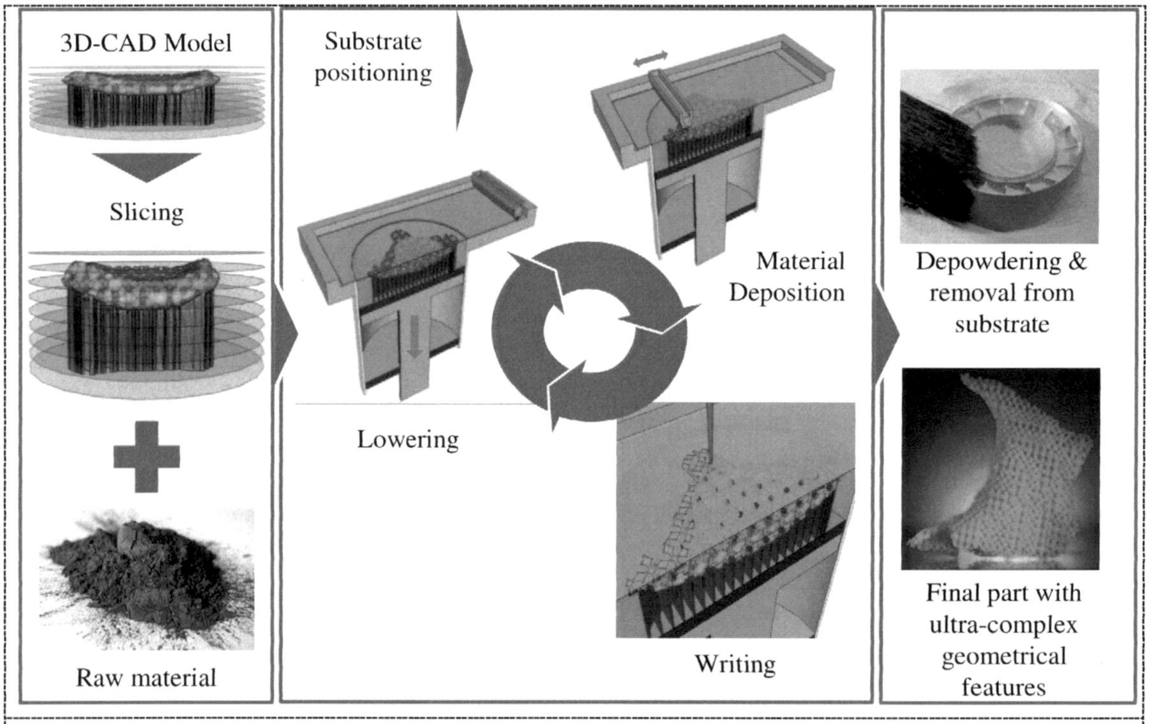

Figure 1. Schematic representation of the LPBF process [3]

Due to ongoing research, the number of processable materials increases. Nowadays, there exists a wide range of different commercially available materials, such as aluminium alloys (e.g., AlSi10Mg, AlSi12), stainless and tool steels (e.g., 1.4404, 1.4542, 1.2709), Nickel-based alloys (e.g., Inconel 718 and 625) as well as Titanium (e.g., Titanium Grade 2, Ti6AlV4) [4, 5].

2.1. The LPBF design restrictions

LPBF as an AM process offers a high degree of design freedom compared to conventional manufacturing processes such as milling or casting. However, there are also design restrictions for the parts produced by LPBF. The rapid heat input and cooling cause the shrinkage of the bonded consecutive layers leading to residual stresses in the part. For overhang features, these residual stresses result in deformation. When the overhang angle is below a critical angle α_{crit} this deformation can lead to failed built parts caused by high deformation and crash between recoater and part. To avoid it, support structures are used to counteract the stresses and to dissipate heat (figure 2) [6]. To prevent deformation of the parts after the removal of the support structures, the components might be heat treated after the process, depending on the material.

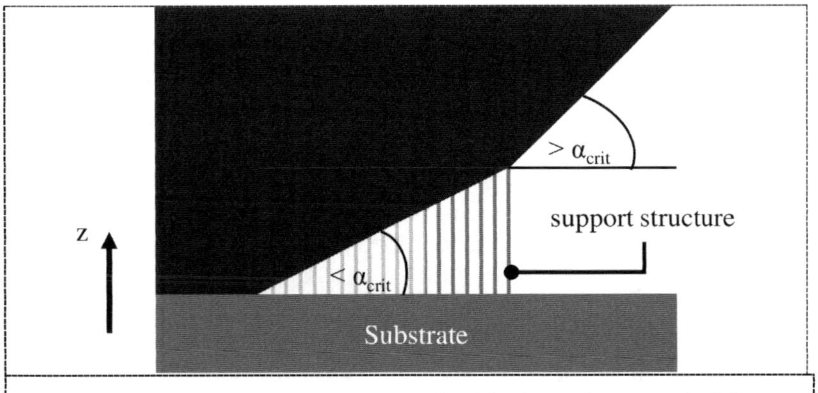

Figure 2. Support structures and critical overhang angle [6]

After the LPBF process, the post-processing of the parts on the substrate plate is required. On the one hand, the support structures must be removed mechanically, on the other hand functional surfaces with high surface requirements have to be post-processed by milling or drilling. In general, the LPBF components have a low surface quality (Ra = 10 -20 µm) due to sintered powder particles and/or the stair stepping effect which results from the layerwise nature of the manufacturing [6].

For applications where a cooling fluid is used, such as cylinder heads, it is important that no powder particles or support structure residues are left in the part, as these could damage the components in the cooling system. Therefore, cooling systems should be designed such that either no support structures are required or that the support structures can be removed after the process [7]. The concept in this work is to utilize the properties of the support structures and to employ them to improve the cooling properties of the cylinder head. For this purpose, internal cooling structures are designed, which on the one hand guarantee the processability of overhanging structures by LPBF and on the other hand, improve the cooling characteristics of the cylinder head.

The LPBF process has design restrictions, for example, the wall thickness and gaps are limited to a minimum (approx. 0.3 mm in the build direction). Bore diameters 90° to the build direction can be produced without support structures at a minimum of no less than approx. 1 mm and at a maximum of no more than 10 mm. However, these restrictions depend on the build direction, the processed material and the characteristics of the machine. A summary of the design restrictions for the LPBF process of different materials can be found in [7, 8, 9, 10, 11].

2.2. *Visual programming to exploit LPBF design potential*

Computer Aided Design (CAD) plays an integral role amongst various industries and most notably in mechanical engineering. Common CAD software products for mechanical engineering are built around the idea of a hierarchical design tree that allows the user to chronologically track modifications made to the model [12]. Although that structure is desired and well suitable for geometrically static parts within serial production, it does not provide the ability to exploit the design possibilities of AM. The geometric degree of freedom and the toolless production has led to unknown potentials in mass customization and creation of complex structures. Mass customized parts often consist of a main body design while additional features are customized and individually designed. With an increasing demand of mass customized parts, the manual workload to individualize the parts while preventing duplicates by exploiting the patterns exceeds the natural capabilities of creativity as well as boundaries of cost efficiency. The same problems apply for complex geometries. As an example, lattice structures consist of multiple struts that compose a unit cell which are, in the simplest case, repeated to fill a certain volume [13]. The explicit construction of every lattice strut would lead to an enormous workload and computational effort.

Visual programming environments such as Grasshopper 3D, which are originally designed for architecture, provide the user with the ability of designing 3D geometry using algorithms. These algorithmic modelling is not only well suited to automatize the aforementioned tasks, but further allows for efficient iterative design approaches that are often encountered in AM. One or multiple inputs in the form of geometries, parameter controls and others relate to a generative algorithmic model that produces outputs based on the connected functions within the algorithmic model. An integration of random number generators into the algorithmic model allow for the generation of a given quantity of geometries, that are individual and distinguishable concerning certain design aspects, while still sharing the general shape and connection points, which allows for an efficient customization [14]. As an input and output-based modelling approach, the algorithmic unit can be built in a manner, that the implemented modifications apply to inputs of different shapes which enables iterative modelling approaches. If implemented correctly, later changes in the base model are easy to handle due to the flexible input and require little to no adjustments to get the desired output model. For repetitive tasks on varying geometries, such as the generation of lattice structures inside of several different parts, the necessity of only a single algorithmic model greatly reduces the workload. This advantage also applies for other design adjustments such as offsets and fillets where the same design adjustment is going to be applied to several different base geometries.

3. Approach

To achieve the goal to design and manufacture the cylinder head, the three-step process shown in figure 3 is applied. In the first step, a material suitable for the specific application is selected and the process parameters as well as static and dynamic mechanical properties are developed for the selected material processed by LPBF. Once the material is selected and the mechanical properties are determined, the cylinder head is designed. To develop a design that meets the LPBF requirements, both the potential of the LPBF using the geometric degrees of freedom as effectively as possible and the design constraints are considered. The opposing design drivers between exploitation of potentials and constraints in LPBF are achieved by an iterative design process. The use of Grasshopper 3D allows for an efficient iterative design process in mainly two aspects. While the cylinder head base model was still undergoing geometric changes within the design space, the model was developed with an earlier version of the cylinder head while still being fully compatible with the final geometry.

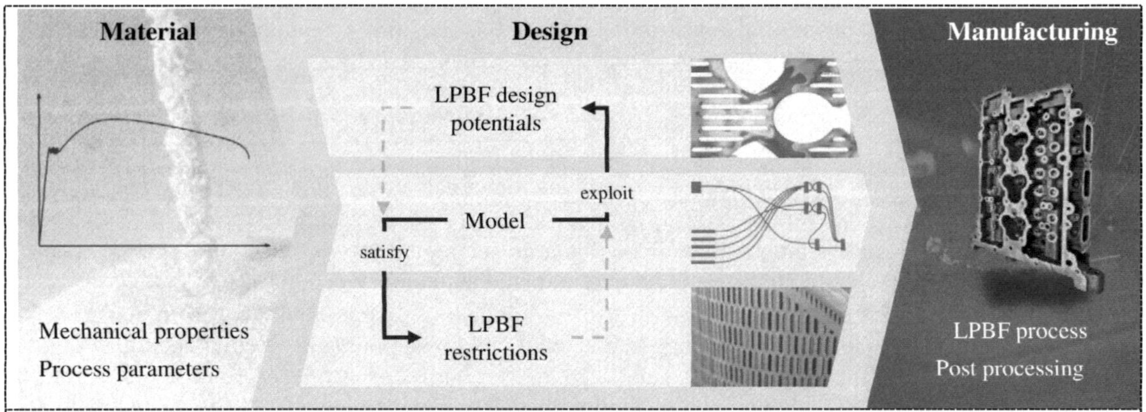

Figure 3. Schematic illustration for the design and manufacturing of an LPBF-suitable cylinder head design

After completion of the cylinder head design by the project partner the final model was used as the input to the Grasshopper 3D model and the rib structures were modelled according to the new geometry without further adjustments. Secondly, the use of slide controls for the ribs parameters enable intuitive and fast rib adjustments according to the simulation results of every iteration step. Finally, the component is manufactured by LPBF on a commercially available LPBF system, X-Line 2000, and post-processed by milling.

4. Material

In the first step, the potential alloys for the cylinder head are identified. For this purpose, aluminium alloys that can be processed with LPBF are investigated by a literature review. The following alloys are investigated: AlSi12, AlSi7Mg, AlSi10Mg, AlSi9Cu3, AlSi9Cu3, AlMgScZr, AlMg4,5Mn0,4, Al-12Si, and AlMg1SiCu. Based on the four selection criteria mechanical properties, processability, cost, and availability one alloy is selected. For mechanical properties Young´s modulus, tensile strength and ultimate elongation is considered. The objective is that the mechanical properties meet or exceed the target properties given for the casting alloy. From the alloys that achieve these target properties, AlSi10Mg is selected for the cylinder head because of its good processability, availability and thus low costs. For the selected material, process parameters for laser power up to 1kW are developed to fabricate specimen with a density greater than 99,5%. To achieve suitable material properties for the given use case and to reduce thermal stresses leading to the deformation of the part, a heat treatment process is investigated. Therefore, two different heat treatments are applied on the test specimens and mechanical properties are evaluated. Between the investigated heat treatments, stress relief annealing was identified as the most suitable one, as it showed the best balance between mechanical properties and low deviation over the specimen at different temperatures. The stress annealing consists of a heat-up, hold and cool-down phase. First, the part is heated up to 240° C in 1 hour. After a hold time of 6 hours, the component is cooled down to 100° C in the oven and then completely cooled in ambient air to room temperature. For the selected material under the selected heat treatment, the mechanical properties are determined. For the characterization of the material properties, the density, hardness, static mechanical properties (tensile strength, brake elongation, yield strength), thermal properties (thermal conductivity, specific thermal capacity) and dynamic mechanical properties (HCF) are investigated and simulated. Measured material properties from the tensile test are given in table 1. The measured thermal properties are given in table 2. All other properties for the simulation are taken from literature.

Table 1. Measurements from stress relief annealed tensile test specimens of AlSi10Mg.

Material property		RT	150°C	200°C	250°C
Tensile strength	Ø	311,00	247,75	210,75	171,13
[N/mm²]	s	27,22	13,89	27,12	7,90
Yield strength	Ø	163,00	152,00	138,75	127,63
[N/mm²]	s	7,55	6,48	7,89	4,10
Elongation	Ø	5,67	13,45	18,20	23,03
[%]	s	2,15	3,85	6,52	5,05

Table 2. Results from laser flash method measurement of stress relief annealed test specimens of AlSi10Mg

Temperature [°C]	Thermal conductivity Z-direction		Thermal conductivity XY-direction		Specific thermal capacity	
	Mean [W/(m K)]	Std. deviation [W/(m K)]	Mean [W/(m K)]	Std. deviation [W/(m K)]	Mean [J/(g K)]	Std. deviation [J/(g K)]
23	162,9	7,2	154,6	7,7	0,885	0,033
150	163,6	6,3	154,4	2,1	0,940	0,024
200	165,4	6,1	154,9	2,9	0,967	0,025
250	160,5	3,5	152,2	1,2	0,965	0,018

5. Design Process

The most promising approach to fully exploit the technological and economical potential of LPBF is to apply different design strategies like lightweight design, functional integration or improvement, monolithic design, or customized mechanical properties, which are feasible due to the great design freedom. While exploiting the advantages of the technology, the design restrictions that apply to LPBF parts must also be considered. As shown in figure 2, a main consideration when targeting complex parts as a cylinder head are cavities that can be found all over the part. In the first step, the main orientation of the cylinder head towards the build plate is determined, to find an optimal orientation where the number of critical surfaces in terms of LPBF is reduced. The remaining critically surfaces with an overhang angle below the critical angle, need to be addressed. Due to the complexity of the part, many of these overhang-violating surfaces can be found inside the water jackets of the cylinder head as well as in the combustion in-take and out-take. With high proximity to the combustion cavities, the highest potential for improved cooling performance lies in the lower water jacket. Therefore, the following work shows the design process for the cylinder head.

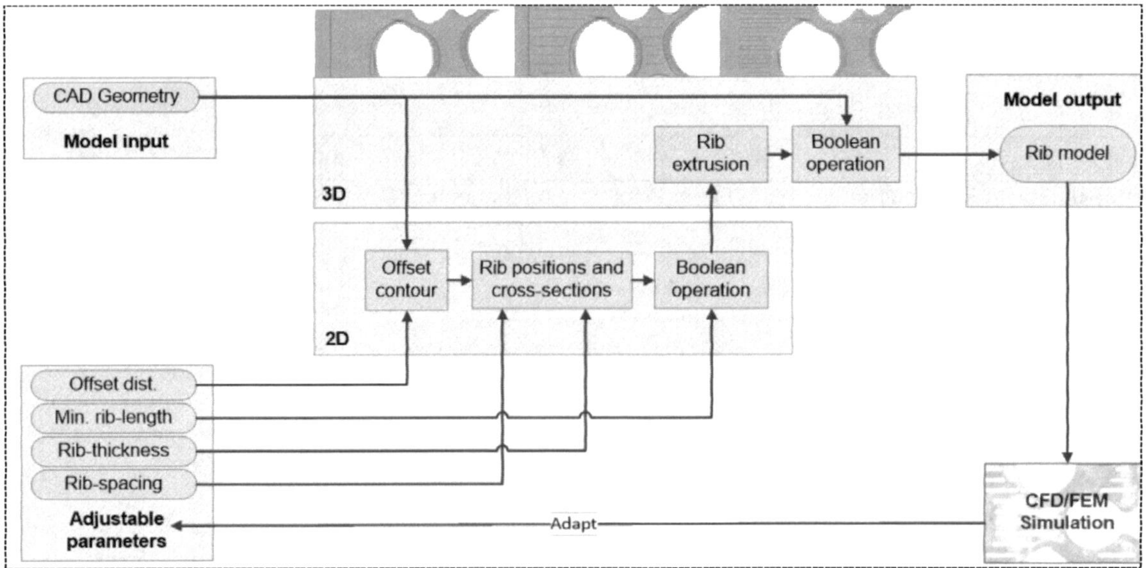

Figure 4. Schematic illustration of the parametric design model

While the design of the main body of the cylinder head is generated with conventional CAD software products, the design approach for the structures must meet the requirements of an iterative and LPBF-suitable approach. The used approach must provide the ability to easily adapt to changes that are made to the main body of the cylinder head during the development of the structures. In addition, the design approach must allow for a fast generation of differently parametrized structures and the ability to quickly adjust the structures according to the simulative validation results. The use of generative algorithms within the parametric visual programming environment of Grasshopper 3D was identified as suitable and likewise necessary to achieve the desired flexibility.

The CAD-file of the cylinder head serves as the input for the generative model. Before implementing the model, the main adjustable parameters need to be identified.

To allow the water inside the jacket to flow without problematic dead ends or strong narrowing areas, the offset of possible rib structures from the contour of the cooling jackets is a necessary parameter. The water flow is furthermore significantly affected by the cross section of free water flow, and thereby directly by thickness and spacing of possible rib structures. The named inputs are designed as slide controls within the Grasshopper 3D environment.

With the use of a non-planar intersection plane, the contour of the cooling jacket is derived from the CAD model of the cylinder head. The inner offset of the cooling jacket contour is now computed with the offset parameter, allowing easy modification later. The parameters of rib thickness and spacing are then used to generate an array inside the contour offset which represent the center line of individual ribs. With the use of the parameters of thickness and spacing, the cross section of each rib is defined. Starting with a rectangular cross section which would lead to high flow resistance, the generative model allows for a chained group of modifications such as fillets. The ribs fillet radius can also be modelled as a variable input for later modification. With the cross sections defined, the ribs can be extruded to exceed the length of the cooling jacket. The 3-dimensional cooling jacket is then used for a Boolean split operation, to cut the ribs to exact length. An example rib model is shown in figure 5.

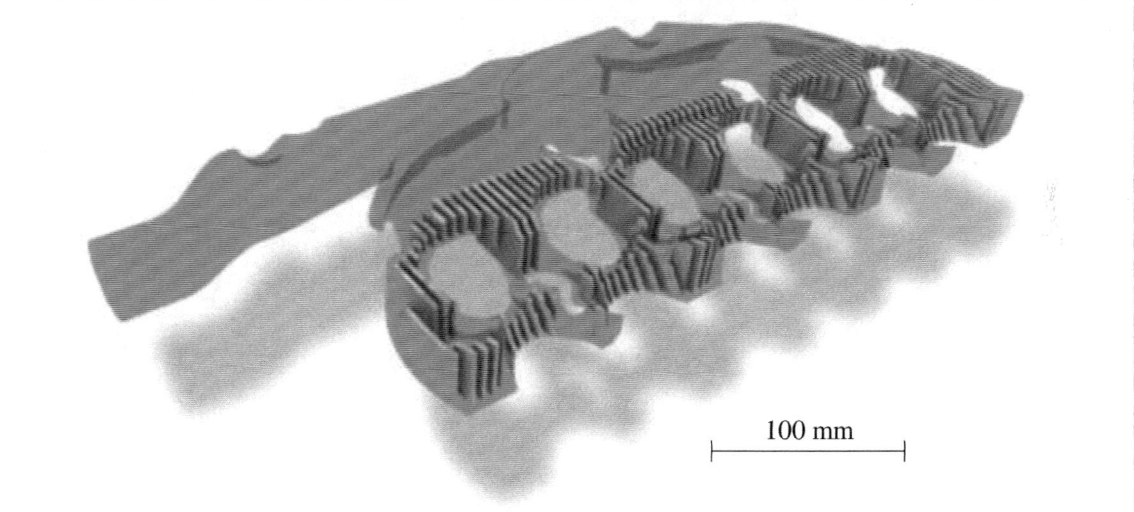

Figure 5. Cooling Struts inside the water jacket

With the described model, a variety of combinations in terms of rib thickness and spacing were derived just by adjusting the slide controls to the desired values. Moreover, the input-output nature of the parametric algorithm allowed the continuous development of the cylinder head without causing incompatibility. To meet the requirements concerning the LPBF process in terms of support structures, the rib parameters were chosen from a parameter grid that ensured that the remaining critical faces are sufficiently supported and producible with LPBF. With an initial set of rib variations, CFD simulations were performed to quantify the cooling performance and showed significant improvements in the temperature filed of the cylinder head. Furthermore, the FEM simulations were utilized to ensure sufficient stiffness to sustain the mechanical stresses caused by combustion. With multiple iterations of simulation and model adaptation, an ideal set of model parameters were determined, and the rib model was finalized.

6. Production and post-processing

In the final step, the cylinder head is manufactured by LPBF. Part of this manufacturing process is also the pre-processing, which means slicing and data preparation, and post processing. The post-processing includes heat treatment, removal of support structures and powder, and milling of functional surfaces and interfaces. Since the cylinder head is to run under real conditions, the same mechanical post-processing is applied as for conventional produced (casting) cylinder head, since the interfaces to the surrounding parts and functional surfaces are not adapted. The cylinder head is manufactured on a X-Line 2000 R system. The LPBF process time was around 12 days. To achieve the required geometric accuracies for the post-processing, the data, which is transferred to the LPBF system for the production, is scaled to compensate for expansion under the preheating temperature (200°C preheating). In addition, functional surfaces are provided with a post-processing offset. The cylinder head built in this project meets all the criteria for being installed on a real engine and used under real conditions. With the new design concept, a reduction of the maximum temperature in the critical areas of the cylinder head by approx. 25 K are achieved, verified by simulation of the coolant side metal temperature. In figure 6, the post-processed cylinder head is shown.

Figure 6. The LPBF manufactured and post-processed cylinder head

7. Summary

LPBF is an AM technology that offers new design freedom for metal parts to increase the performance of these parts. To exploit this potential, the design constraints of LPBF and the use of suitable design tools and methods are considered. These challenges and possible solutions are shown within this work by the example of a cylinder head. The visual programming language Grasshopper 3D was used to develop a generative and parametric in-house design tool for cooling structures of a cylinder head. Considering the material properties and production-oriented design, a cylinder head was produced that fulfills all functions required for the use under real conditions.

In the first step, only a specific feature, the water jacket, is considered and optimized to improve the function of the cylinder head. In order to exploit the full potential of the LPBF for this component, a complete redesign is necessary. This not only improves the manufacturability of the cylinder head, e.g. by reducing the number of support structures and thus reducing the process time, but also provides further functional improvements. Further functional improvements can be found, for example, in the lightweight construction by reducing the wall thickness compared to conventionally cast cylinder heads.

The process time for the LPBF cylinder head is 12 days. This significantly exceeds conventional production (casting). Even with current improvements in productivity (e.g. multi laser systems, shorter coating times), the production time cannot be reduced to a competitive level. One approach to reduce process time and costs is the use of hybrid manufacturing. The cooling struts are only needed near the combustion chamber, where the increased cooling performance is required. In hybrid production, a base body part is casted. The geometrically complex area is produced by direct buildup on this base body part using LPBF. This achieves a significant reduction in process time.

Acknowledgments
The studies for this work are carried out in cooperation with the Ford-Werke GmbH in Cologne, Germany. Thank you for your support and willingness to explore new paths for established technologies.

References

[1] Merker G P, Teichman R 2018 Grundlagen Verbrennungsmotoren Springe-Verlag p 73
[2] Wohlers Associates 2017 Wohlers Report 2017 – 3D Printing and Additive Manufacturing State of the Industry – Annual Worldwide Progress Report USA
[3] Meiners W 1999 Direktes Selektives Laser Sintern einkomponentiger metallischer Werstoffe Dissertation Aachen
[4] EOS GmbH acessed December 14, 2020 DMLS Metal Materials for EOS 3D printer https://www.eos.info/en/additive-manufacturing/3d-printing-metal/dmls-metal-materials
[5] GE Additive acessed December 14, 2020 Powders https://www.ge.com/additive/powders-overview
[6] VDI 3405 Part 3:2015:12 Additive manufacturing processes, rapid manufacturing - Design rules for part production using laser sintering and laser beam melting
[7] Matthiesen G, Daniel M, Pietrzyk T, Ziegler S, Schleifenbaum J H, Schmitz K 2020 12th International Fluid Power Conference (12. IFK) Dresden
[8] Hoeren K P J, Witt G, 2012 Konstruktive Möglichkeiten und verfahrenspezifische Grenzen in der additiven Fertigung (Ges. zur Förderung angewandter Informatik) p 127-136
[9] Adam G, Zimmer D 2014 Design for Additive Manufacturing—Element transitions and aggregated structures (CIRP Journal of Manufacturing Science and Technology 7)
[10] Adam G 2015 Systematische Erarbeitung von Konstruktionsregeln für die additiven Fertigungsverfahren Lasersintern, Laserschmelzen und Fused Deposition Modeling
[11] Thomas D 2009 The development of design rules for selective laser melting University of Wales
[12] Xu X 2009 Integrating Advanced Computer-Aided Design, Manufacturing, and Numerical Control: Principles and Implementations
[13] Panesar A 2018 Strategies for functionally graded lattice structures derived using topology optimisation for additive manufacturing (Additive Manufacturing 19) p81-94
[14] Bachman D 2017 Grashopper: Visual Scripting for Rhinoceros 3D

Evaluation of Parts for Additive Manufacturing utilizing System Models of AM Plants

Marcel Winkler, Georg Jacobs, Kathrin Spütz, Christian Konrad

Institute for Machine Elements and Systems Engineering
Schinkelstraße 10, 52062 Aachen, Germany

marcel.winkler@imse.rwth-aachen.de

Abstract. The objective of this study is to provide the design domain with knowledge about restrictive geometric parameters for AM-compatible part design. This knowledge is derived from the existing technical plant configuration and its component parameters. Parameter relationships between the AM plant configuration and the design parameters are derived from technical documentations and guidelines. The system model of an AM plant is built in SysML to formalise the complex relationships of the restrictive parameters relevant for part design depending on the AM system configuration. AM Manufacturing Models are introduced to calculate the technical constraints between AM plant components and restrictive design parameters. AM Plant Models are developed to formalise the configuration of a plant with the required parameters for the AM Manufacturing Models. A GUI is conceptualized for easy configuration setup of an AM plant.

1. Introduction and Objectives

Additive Manufacturing (AM) processes are a key technology in the development of innovative products. In an age of ever-increasing complexity of cyber-physical systems, AM can be applied to manufacture complex parts for which conventional methods are not sufficient. Due to its young age, the possibilities of the processes are constantly expanding. However, with every further development of the AM processes, the restrictions on part design must also be considered. For the successful application of these emerging AM technologies, manufacturing experts and designers need frequent communication and an exchange of expertise regarding Design for AM (DfAM) guidelines and dynamically changing process parameters must be assured. In order to produce high quality, functional parts, detailed knowledge about the existing AM plants, their configuration and the influence on design parameters is required already in early design stages.

Overcoming these hurdles, the support of Model-Based Systems Engineering (MBSE) is a promising approach. MBSE is a suitable framework for linking knowledge in a formalised way and making it tangible company-wide. It is obvious that bringing AM and MBSE together should be considered to clarify the relationships between the production domain and design parameters. This offers the potential to shorten product development phases and provide an economic benefit for companies.

2. State of the Art

In the following, the state of the art regarding part identification for AM by AI methods as well as the potentials of a MBSE approach will be discussed to disclose the need for research in this topic.

2.1. Artificial Intelligence in Part Identification for AM

A first study has shown that parts with high potential for AM can be automatically identified in a CAD database using AI algorithms that evaluate the part complexity [1]. In this supervised learning approach, the relationship between given CAD-files as input data and pre-assigned output values is learned by an Artificial Neural Network (ANN) as AI algorithm. The derived classifier is suitable for assigning parts to the clusters "AM potential" and "no AM potential" with an accuracy of over 90%. The procedure and used toolchain are presented in figure 1.

In order to train the AI algorithms for the part identification, a sufficient amount of data with known input and output measures (labels) is required. To reduce the manual effort for data labelling, the toolchain was extended by a method for the evaluation of the manufacturability of a part. The calculation of the Additive Manufacturing Feasibility Indicator (AMFI) allows a quantification of the manufacturability in a given AM process and the use of this value as a data label for the training of AI algorithms [2].

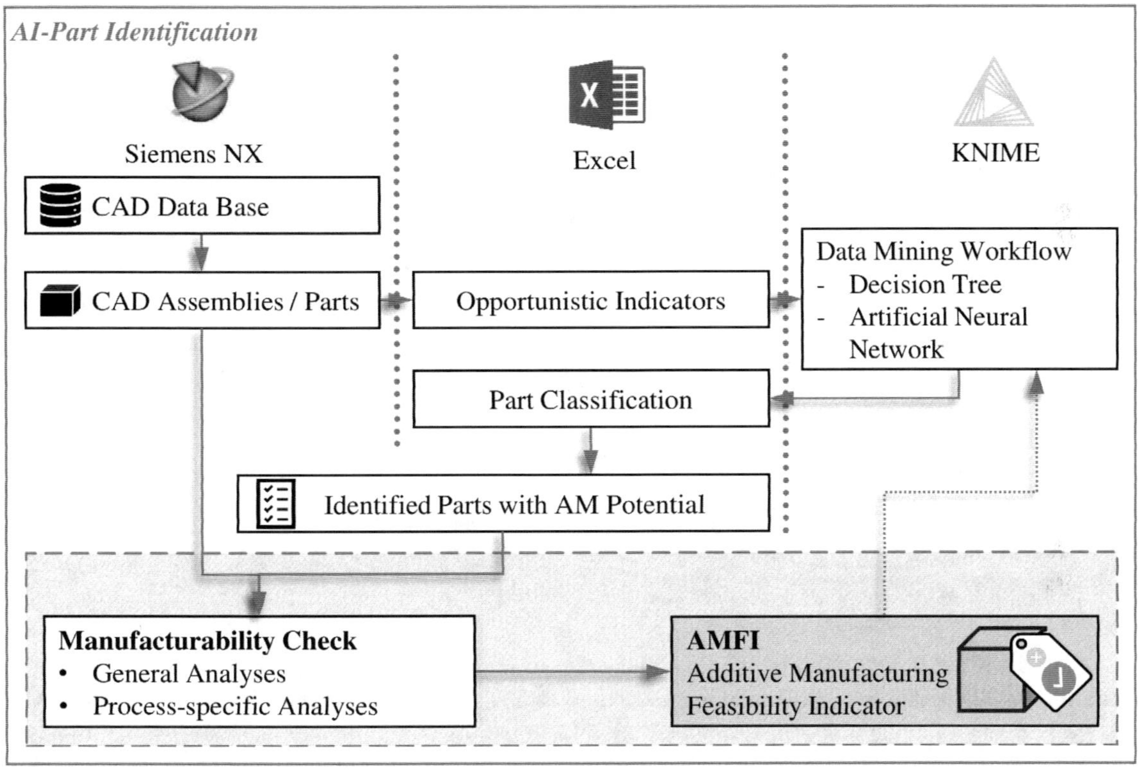

Figure 1. Toolchain for AI part identification for AM and manufacturability evaluation [2]

2.2. Additive Manufacturing Feasibility Indicator

The AMFI is developed to evaluate the manufacturability of a part in a specific AM process during the design process. By determining the AMFI of a part, the design engineer can already identify production-critical geometric elements of the part during the design process and adapt the design. The measure allows to derive a statement regarding the manufacturability on the basis of quantifiable design parameters. For the quantification of the manufacturability, critical features of a part are analysed regarding general and process-specific factors that promote or disqualify additive manufacturing (see table 1). Furthermore, the effect of recognized critical features can be adjusted by user-specific weighting [2].

IOP Conf. Series: Materials Science and Engineering 1097 (2021) 012022 doi:10.1088/1757-899X/1097/1/012022

Table 1. Overview of the most important restrictive design parameters determining the manufacturability of AM parts [3 - 8]

General factors	Process-specific factors
• Minimum wall thickness	Powder-based processes
• Minimum gap width	• Critical hole diameter
• Minimum bore diameter	• Cavities
• Minimum cylinder diameter	Material extrusion processes
• Part orientation	• Min. cross-sectional area
• Build volume	• Critical overhangs
	• Islands

General factors include criteria that can be analysed equally across all additive manufacturing processes. Focused parameters evaluated by the AMFI are part orientation and build volume, minimum wall thickness and gap widths as well as minimum bore and cylinder diameters. As the description suggests, process-specific factors depend on the selected AM process that is used for the manufacturability analysis.

Regarding the process-specific analysis three AM processes are focused to cover a wide range of industrial relevant applications. As representatives of powder-based processes, Selective Laser Sintering (SLS) and Selective Laser Melting (SLM or LPBM) are focused. Looking into powder-based processes, powder removal is one critical factor that needs to be considered. Holes with critical diameter for powder removal and cavities are identified during the manufacturability check. Fused Layer Modelling (FLM or FDM) is considered as material extrusion process using polymers. Parts with critical overhang angles or islands can only be manufactured in these processes by applying support structures (see figure 2).

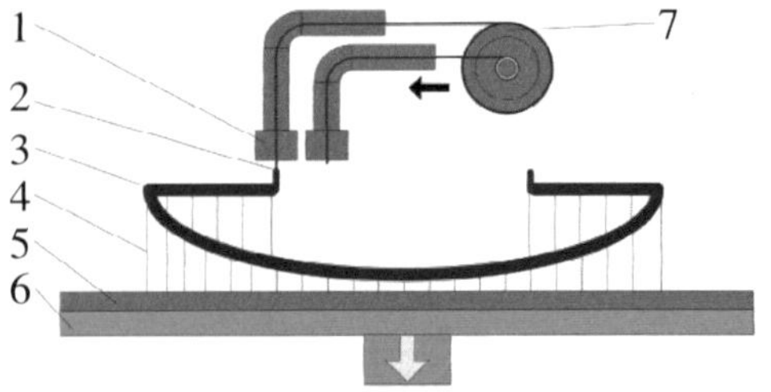

1 Heated nozzle
2 Layered build process
3 Built part
4 Support structures
5 Build plate
6 Build table
7 Material supply

Figure 2. Principle layout of an FLM plant [7]

The criticality of the individual general and process-specific features depends mostly on the process capability of the additive manufacturing system and thus on its configuration. The installed components of the AM plant have a direct influence on the design parameters that should be respected in the part design. A practical example is the diameter of the heated nozzle (1), which has a direct influence on the minimum wall thickness that can be printed. Furthermore, there are a large number of parameters which are not obviously connected to the design parameters but have a significant influence on the manufacturability due to complex interactions with other parameters in the AM plant [9]. The relationship between the components involved in the travel precision of the nozzle in the x-axis direction and their influence on the minimum gap width can be mentioned as an example of non-trivial relationships.

At the present time, the knowledge of interrelationships is restricted for production domain and can only be transferred to the design domain by an interdisciplinary exchange with a high communication effort.

2.3. Model-Based Systems Engineering

Today's increasing complexity of products due to the ongoing integration of electrical and software in technical products is leading to challenges in product development. To meet these challenges and reduce complexity, the use of a superordinate knowledge repository as a Single Point of Truth (SPoT) is a solution. The approach of Model-Based Systems Engineering (MBSE) offers great potential for the modelling of the SPoT, as it also enables the comprehensible modelling of the links. Further, it represents a centrally available repository that is used for development in the MBSE framework instead of decentralised document-based processes. The structure of the system model and the included relationships is supported by tool-independent modelling languages such as SysML (Systems Modelling Language). This graphical modelling language enables the modelling of product and design knowledge as well as the relationships between the individual knowledge elements. The advantage of the graphical modelling languages lies in their formalisation. Syntax and semantics are clearly defined and can be understood quickly and easily thanks to the graphical representation [10].

The foundation for the representation of different contexts is the possibility in SysML to allow different views on the same knowledge without duplicating information. Changes to the element of a product structure affect all other representations of this element, regardless of the viewpoint. The different views considerably improve the availability, use and communication of knowledge in development [10]. The use of SysML in a system modeler, e.g. the Cameo Systems Modeler™, offers further advantages. One of these advantages is the feasibility of the modelled rules and standards. The system model can thus be used as a design tool in development and construction, in addition to the pure visual representation of knowledge. This advantage is particularly effective in the integration in CAD environments, as a linked knowledge base for configuration setup [11, 12] and in the control of complexity [13].

In conclusion, the development of a system model is recommended wherever complex knowledge relationships from different company domains like manufacturing are to be formalised and made available globally. However, the MBSE approach is currently still on the threshold of broad industrial use. Especially in the field of knowledge engineering in manufacturing models further research is needed.

3. Need for Research

The preceding remarks on the state of the art reveal the need for research in the field of manufacturability assessment of parts for AM and the potentials of a connection with a Model-Based Systems Engineering approach in the AM production environment. In order to make a more precise statement about additively manufacturable features than is stated in general guidelines, a method needs to be developed to formalise the complex knowledge of the dependencies between restrictive design parameters and the AM plant configuration. The following research question and hypothesis are derived to address the white spot in the specific research landscape.

Research Question

How can restrictive design parameters be derived from the technical AM plant configuration and the knowledge be provided to the design domain to support AM-suitable part design?

- How can the complex knowledge about the relationships between AM plant configuration and restrictive parameters be formalized?
- How does the configuration of components in an AM plant influence restrictive manufacturing parameters for AM?
- How can the knowledge of the influences be made available to the design domain to develop an optimized part design?

Research Hypothesis

Restrictive geometric design parameters for AM can be derived from the technical AM plant configuration and formalised using SysML. The complex dependencies of the plant components on each other and their effects on the restrictive design parameters can be represented with clear syntax and semantics within a system model. The knowledge can be transferred to the design domain by linking the formalized models to a graphical user interface in CAD to support part design for AM.

Objective of the Study
The goal of the study is the development of a SysML model of an AM system in which the technical conditions are combined with restrictive design parameters for AM. The model serves the purpose of realising a knowledge transfer of the complex relationships from the AM domain into the design domain to support optimized design for AM. Based on this overall objective, the following sub-objectives are defined for this study:

1. Derived correlations between restrictive design parameters for additive production and plant technology (process, plant type, configuration)
2. Established *AM Manufacturing Model* for the calculation of the restrictive design parameters that depend on the components of the AM Plant
3. Representation of AM Plant in an *AM Plant Model* which displays the technical interdependencies in a configuration of an AM plant
4. Realisation of the connection of the system model to the toolchain for the calculation of additive manufacturability

4. Methodical Approach and Materials
To support the knowledge exchange between design and manufacturing, a system model is developed, which provides the knowledge about AM plant related manufacturing restrictions directly in the design process. The system model consists of the two parts *AM Manufacturing Models* deriving the restrictive design parameters and the *AM Plant Models* representing the configuration of an AM plant. The AM Plant Models provide the relevant parameters for the calculations in the *AM Manufacturing Models*. The methodical approach for the knowledge transfer between manufacturing and design domain is illustrated in figure 3.

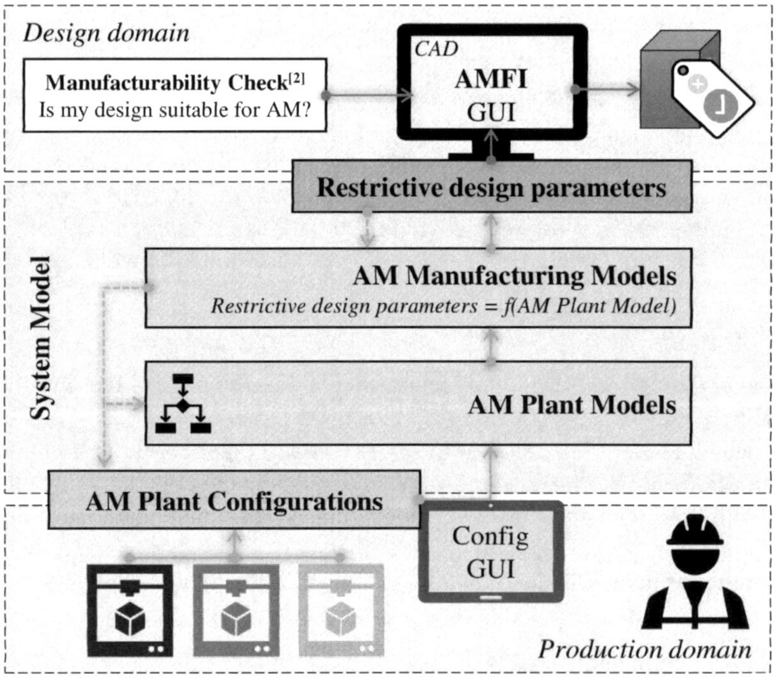

Figure 3. Methodical approach for the knowledge exchange by the system model and its interfaces

In order to achieve this goal, the interfaces and limitations observed for this study are defined. Looking into the design domain, the most relevant restrictive design parameters given in Table 1 are considered to connect the system model to the GUI developed for the CAD software Siemens NX to determine the AMFI. This ensures that the production knowledge is transferred into the design process to support AM-suitable part design by indicating geometrical elements that are critical for manufacturing. The knowledge about the relationships between restrictive design parameters and the determining components of the AM plant is extracted and formalised from technical documentations, guidelines, existing component portfolio and user experience. Once the identification of the correlations is done, the relevant modules of the AM plant can be defined which have to be represented in the system model. Furthermore, the specific configuration of these modules and the management of the configuration is considered in the design of the system model. An approach is taken to make the setting of the technical AM plant configuration user-friendly via a graphical user interface (GUI) for the production domain.

As example this study uses the Prusa i3 MK3S 3D-printer as representative of an FLM-process. Overall, the developed method aims at realising an easy extensibility in order to integrate additional AM procedures and further dependencies of restrictive design parameters, which are not considered in this study. The methodical approach of each individual step is described briefly in the following chapters.

4.1. Correlations of restrictive design parameters and plant technology

The relationships between the components of the AM plant with the restrictive design parameters are first identified and analysed. In a young manufacturing technology such as AM, a high degree of system complexity can be observed due to the level of integration of electrical and software elements in addition to the mechanics. In this study, the modelling of these dependencies is illustrated by the following idealised example of the AM Plant Prusa i3 MK3S shown in figure 4.

The components involved in the nozzle movement in x-axis direction are highlighted. A stepper motor (1) controlled by the system software is required to convert electrical energy into mechanical energy. The motor is equipped with an output shaft to which a toothed belt pulley (2) is attached. A toothed belt (3) is driven by this pulley, which moves a carriage in x-axis direction on which the print nozzle (4) is located. With the resolution of the precision of the traverse paths in x-axis direction, design parameters like the minimum realisable gap width in x-axis direction can be concluded.

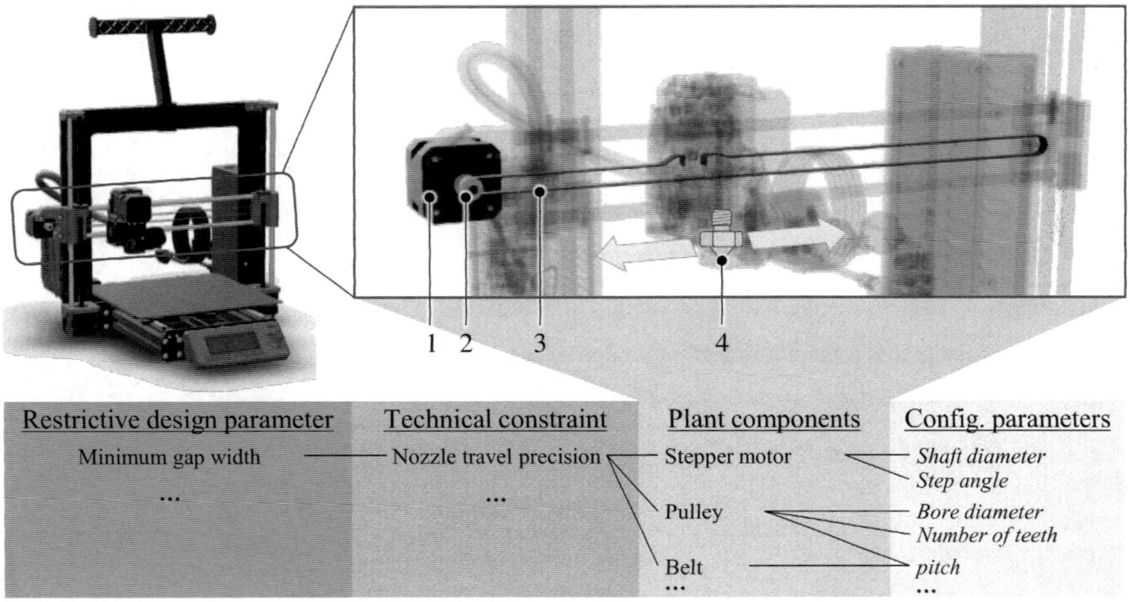

Figure 4. Prusa i3 MK3S with highlighted components relevant for the minimum gap width

The minimum realisable gap widths depend on the precision of the nozzle travel in the x-axis direction. The nozzle on the carriage is moved by a transmission of the rotary motion of the stepper motor via the pulley to a linear travel path of the belt. The travel thus depends on the angle of rotation of the stepper motor, also known as step angle and the diameter of the pulley on which the belt runs. This relationship can be described with mathematical calculations that are part of the *AM Manufacturing Models* described in the following chapter.

4.2. AM Manufacturing Models for the calculation of restrictive design parameters

As described above, the restrictive parameters depend on the plant configuration through technical constraints. Since considering all calculations of the restrictive design parameters exceeds the extent of this study, the development of a model describing the technical constraints are limited to the abovementioned idealised example.

In order to create the link between parameters of the AM plant components, technical constraints and restrictive design parameters, the necessary calculations are carried out in *AM Manufacturing Models* implemented in SysML. The basic structure of the *AM manufacturing models* is shown in figure 5 using the example of the *FLM Manufacturing Model* for the Prusa printer for the calculation of the minimum wall thickness as general restrictive design parameter.

In principle, different process-specific manufacturing models (e.g. FLM Manufacturing Model, SLM or SLS) can be integrated into the *AM Manufacturing Models*. Each of the specific manufacturing models contains constraints for the calculation of each restrictive design parameter depending on the respective *AM Plant Models*. General factors of the restrictive design parameters need to be calculated in each specific manufacturing model. Therefore, a connection to each specific manufacturing model is necessary. The calculation of the process-specific factors is only carried out in the respective manufacturing model. Figure 5 illustrates the calculation constraints to derive the restrictive design factors. The *Idealised Model Travel Precision* from the *Model Library* is represented in detail in the *FLM Manufacturing Model* as one of the calculation constraints. The other constraints integrated in the *Model Library* that are also indicated in the *FLM Manufacturing Model*, are used for the calculation of the other restrictive design parameters but will not be discussed in detail.

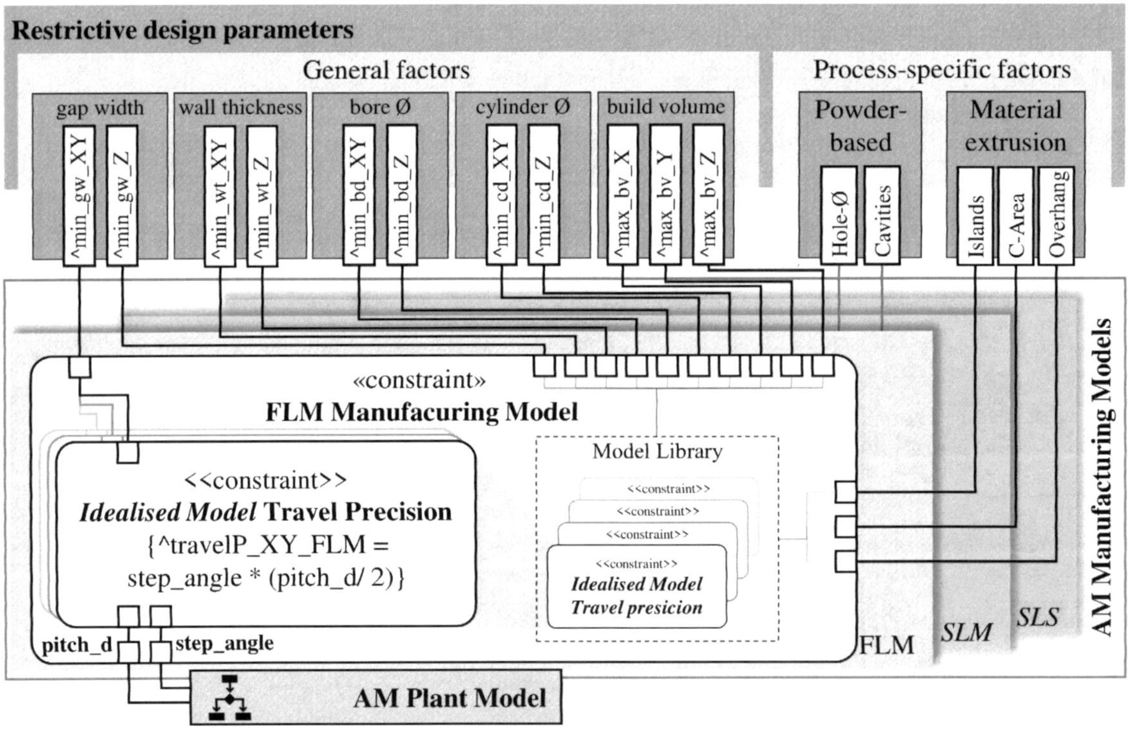

Figure 5. *AM Manufacturing Models* for the calculation of restrictive design parameters

All models used in the constraints can be integrated in different fidelity levels, depending on the goal of the system modelling. In the given example, the *Idealised Model Travel Precision* can be extended by additional parameters on higher model fidelity levels to increase the value accuracy if necessary. For example, the connection of the specific spring rate of the belt can be drawn from the component in the *AM Plant Model*, to be included into a more precise calculation. The constraint *Idealised Model Travel*

IOP Conf. Series: Materials Science and Engineering 1097 (2021) 012022 doi:10.1088/1757-899X/1097/1/012022

Precision can thereby be supplemented on a higher modelling fidelity by adding the specific spring rate of the belt and the acting force to define the positioning deviation of the nozzle.

For the modelling of high-precision industrial processes, the consideration of dynamic characteristic values is elementary and can be added to the *FLM Manufacturing Model* and be connected to other calculation constraints. However, due to the global tolerances of the considered AM process, they are not elaborated in further detail within the scope of this study.

The calculation models in the *AM Manufacturing Models* depend on values that are specified by the configured components in the AM plant. In the following chapter, the AM plant configuration and the determination of the required parameters derived from it are discussed.

4.3. AM Plant Models for the determination of relevant plant parameters

In the introduction of the *AM Manufacturing Models* shown above, parameters are used in the technical constraints that originate from a specific AM plant configuration. A plant configuration consists of a selection of components that are installed in the AM plant. These components possess the relevant parameters requested by the *AM Manufacturing Models*. If the configuration of the AM plant is changed, the resulting parameter values that are transferred to the *AM Manufacturing Models* must be adjusted accordingly. Furthermore, there are often technical dependencies between the components in the AM plant configuration that cannot be installed in any combination with each other.

With the development of *AM Plant Models* as digital twin of a physical AM plants, the configuration-specific parameters can be automatically updated and transferred to the *AM Manufacturing Models*. Furthermore, the representation of the *AM Plant Model* in SysML allows the definition of technical restrictions regarding the AM plant configuration. The following figure 6 illustrates the structure of the FLM plant as example for an *AM Plant Model*. It contains the derivation of the parameters required by the constraint *Idealized Model Travel Precision* in the *FLM Manufacturing Model*.

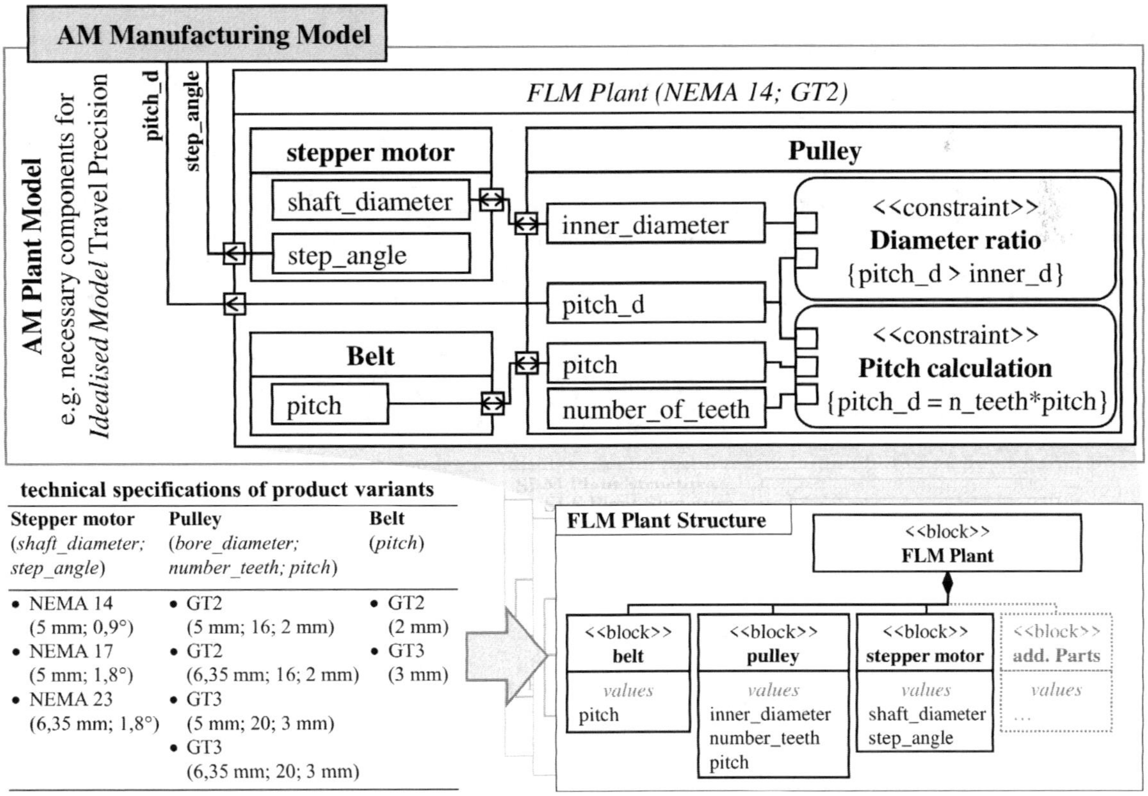

Figure 6. Instantiation of *AM Plant Models* and their linkage to the *AM Manufacturing Model*

IOP Conf. Series: Materials Science and Engineering 1097 (2021) 012022 doi:10.1088/1757-899X/1097/1/012022

In the FLM Plant Model the technical restrictions for the configuration of the components stepper motor, belt and pulley are illustrated. The free combination of the component's parameters for the calculation of the restrictive design factor is not reasonable for technical reasons. In the described example the pitch of the belt always depends on the chosen pitch of the pulley. The stepper motor is determined by the parameters step angle and shaft diameter. Similar to the pitch, the shaft diameter of the stepper motor has a direct link to the parameter of the inner diameter of the pulley to form a functional unit. The relevant pitch diameter (*pitch_d*) of the pulley now shows dependencies of parameters coming from both other components. On the one hand, the pitch diameter depends on the defined pitch and number of teeth. On the other hand, the pitch diameter depends on the shaft diameter of the stepper motor. These technical dependencies can be solved by creating two constraints. By the constraint *Diameter ratio* in the pulley it is defined that the pitch diameter always takes on a larger value than the inner diameter to ensure a realizable technical solution. Further the constraint *Pitch calculation* determines that the pitch diameter is calculated so that the required number of teeth with the appropriate pitch is achieved. In this way, the parameter values required by the *AM Manufacturing Models* can be determined and transferred by the *AM Plant Model*.

The specific configuration of the FLM Plant is created as instance of a structure of the FLM plant shown in the bottom of figure 6. In the structure, components are shown that are relevant for the calculation of the restrictive design parameter in the example. Their options for the realisation of different product configurations can be integrated following the method described in [11]. In order to ensure expandability with further components for the calculation of the other design parameters, the structure is modelled according to Weilkiens' VAMOS method (Variant Modeling with SysML) [15].

In practice, the AM plant components are replaced regularly due to signs of fatigue, wear or defects. Each of the components is available in different variants with different technical specifications. Table 2 shows an exemplary selection of technical configurations of the three components in the FLM plant relevant for the calculation of the travel precision with their specification values. The three variants of the stepper motor allow different choices of shaft diameters and step angles. The pulley comes in four different variants that are differentiated in the parameters bore diameter, number of teeth and pitch. With regard to the belt, only the pitch is considered for the example, which is available in two different variants.

When looking at the parameters, it becomes clear that dependencies between the components are relevant as they are integrated in the FLM Plant Model. For example, the shaft diameter of the stepper motor directly influences the selection of possible pulleys. If a stepper motor with a 5 mm shaft is selected, only the first or third variant of the pulleys shown in table 2 can be mounted. By mapping these dependencies in the structure of the *AM Plant Model*, the compatibility of the variants is checked directly and returned during configuration.

Table 2. Selection of technical specifications of product variants of the stepper motor, pulley and belt

Stepper motor *(shaft_diameter; step_angle)*	Pulley *(bore_diameter; number_teeth; pitch)*	Belt *(pitch)*
• NEMA 14 (5 mm; 0,9°)	• GT2 (5 mm; 16; 2 mm)	• GT2 (2 mm)
• NEMA 17 (5 mm; 1,8°)	• GT2 (6,35 mm; 16; 2 mm)	• GT3 (3 mm)
• NEMA 23 (6,35 mm; 1,8°)	• GT3 (5 mm; 20; 3 mm)	
	• GT3 (6,35 mm; 20; 3 mm)	

4.4. Interfaces for plant configuration setup

After the system structure and the determination of the restrictive design parameters are developed, the next step is to integrate the model into a usable tool chain. The fully automatic transfer of the calculated design parameters is realised using a Matlab code for the intermediate storage as comma-separated values file in Excel. This file is then automatically imported when the AMFI calculation is queried, thus loading the existing design parameters into the tool for manufacturability analysis in the design domain.

Due to the currently still rather low distribution of SysML software in industrial companies, the design of an intuitive GUI for the configuration of AM plants is recommended. The mockup of a suggested GUI (a) that can be used in the production domain is shown in figure 7 with its linkage to the AMFI in CAD (b). The GUI consists of three areas, the general settings, a display of the calculated design parameters and an area for adapting the system configuration and components. In the general area, important metadata such as the configuration name, the date and the employee editing the configuration can be entered. Further in the configuration area different modules for configuration can be selected for adaption. In the example below, under the tab "Travel X" the components involved in moving the nozzle in the X-axis direction are listed. The relevant parameter regarding the components pulley, stepper motor and belt can be defined in this view. To simplify the AM plant configuration using standard solutions, selection fields for preconfigured components are provided. This means that the user does not have to specify each parameter individually, but only has to make a component selection. Incompatibilities between the choice of standard configurations, for example an incorrect pulley for the stepper motor, can also indicated by highlighting the parameters. For those components that cannot be pre-configured, the user is provided with illustrations of the parameter relationships via principle drawings.

(a) AM plant configuration GUI (b) AMFI in CAD

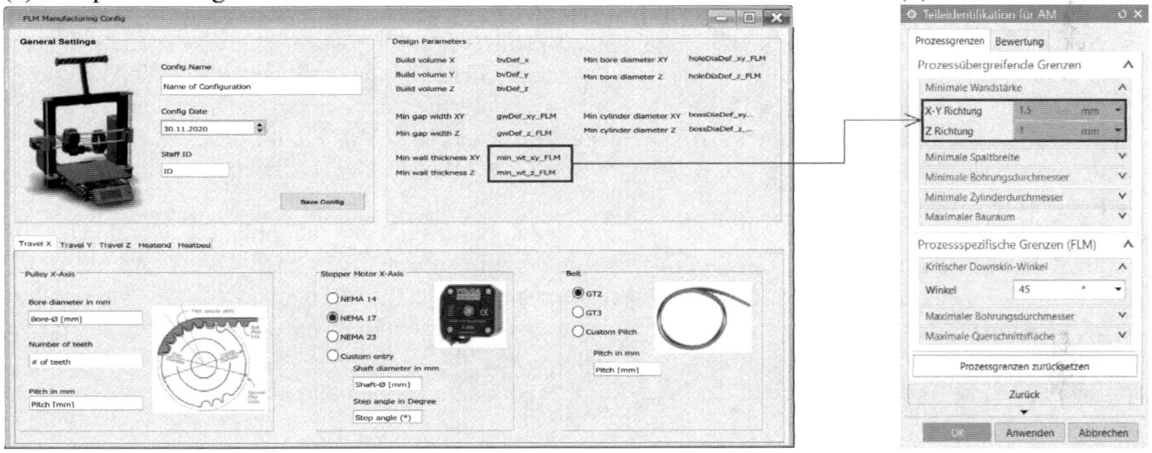

Figure 7. Mock-up of (a) the configuration GUI and its linkage to (b) the AMFI GUI in CAD

5. Results

The developed system model formalises the production knowledge of experts from the production domain about the AM plant configuration and derives restrictive design parameters from it. This knowledge is automatically integrated by an interface into a tool for a manufacturability check of parts in the design domain.

It is possible to derive correlations between restrictive design parameters for additive production and plant technology from technical documentations and guidelines. The knowledge about the interrelationships can be represented in *AM Manufacturing Models* using SysML. By the representation of the *AM Plant Models* it is possible to mind the technical influences of different plant configurations in the *AM Manufacturing Models*. Also, by the utilization of the software Matlab it is possible to transfer the derived design parameters into the existing toolchain for the evaluation of manufacturability in AM.

Setting up the system model using the presented structure makes it easy to connect further existing structural models of AM plants. Furthermore, modelled AM plants can be easily extended with various component configurations. In an assembled model of an AM plant, the existing plant variant and also the variants of installed components on lower levels of the reference product structure can be easily defined. The model structure is not only limited to the representation of additive manufacturing plants but can also be transferred to other manufacturing processes.

The structure of the *AM Manufacturing Model* allows an easy exchange of the interface between parameters of the production line and the design parameters. Furthermore, the automatic determination of the design parameters is ensured as well as the transfer of these parameters to interfaces to external tools such as the existing tool in CAD for the calculation of the AMFI. In this linkage, the currently existing AM plant can be set up automatically via the GUI of the AM Plant configuration and the resulting restrictive design parameters for the design domain can be considered in the component design for additive manufacturing.

6. Summary and Outlook

The developed *AM Manufacturing Models* enable the linkage of the structural and behavioural relationships of the AM plant with restrictive design parameters. In combination with the representation of the *AM Plant Models* an easy configuration of the AM plant components and parameters is allowed. The model structures can be successfully adapted for the other design parameters that are mentioned. By the combination of AM processes with the model-based product development steps are taken to overcome today's challenge of manual part and process evaluation. In addition, it is also possible to evaluate the manufacturability of a part already at early product design stages directly in the CAD software. In this way, the design engineer is supported in part design for AM, taking the company's specific production environment into account.

Nevertheless, the combination of the presented model structure with profound models for the behavioural description needs to be considered when applied to high-precise industrial AM plants. Based on the presented state, the system model can be further developed and supplemented with additional modules and components or by whole instances of AM Plants. The existing system complexity can thus be broken down step by step and the influences of the technical system configuration can be revealed.

The *AM Manufacturing Models* are also suitable for considering not only the technical plant parameters but also behaviour, process, material or cost models that can easily be connected to the system model. Thus, in addition to the components in the example, the extruder control, the temperature of the heatend or the material itself also have an influence on the width of the deposited polymer line in the FLM process and thus on the minimum realisable gap width in a part. These extensions of the system model need to be further explored in the future to further enhance the potential of this approach.

References

[1] Winkler M, Jacobs G, Stürmer S and Konrad C 2020 *International Conference on Intelligent Engineering and Management (ICIEM)*, London, United Kingdom, 281-288
[2] Winkler M, Stürmer S and Konrad C 2021 *Antriebstechnisches Kolloquium ATK 2021* (*Preprint*)
[3] Diegel O, Nordin A and Motte D 2019 A Practical Guide to Design for Additive Manufacturing. (Singapore: Springer Singapore)
[4] Gebhardt A 2016 Generative Fertigungsverfahren: Additive Manufacturing und 3D-Drucken für Prototyping - Tooling - Produktion, 5th edn. (München: Hanser)
[5] Lachmayer R, Lippert R B and Kaierle S 2018 Additive Serienfertigung (Berlin, Heidelberg: Springer Berlin Heidelberg)
[6] Adam G A O and Zimmer D 2015 *Rapid Prototyping Journal 21* **6**: 662–670
[7] VDI 3405 2014 Additive manufacturing processes, rapid manufacturing - Basics, definitions, processes
[8] Mani M, Witherell P and Jee H 2017 *37th Computers and Information in Engineering Conference*, **1**: 1–10
[9] Siebrecht J, Jacobs G, Beckers A, Löwer M, Schmidt W and Körber M 2016 *Design for X - Beiträge zum 27. DfX-Symposium Oktober 2016*
[10] Eigner M, Roubanov D and Zafirov R 2014 Modellbasierte virtuelle Produktentwicklung (Berlin,

Heidelberg: Springer Berlin Heidelberg)

[11] Wyrwich C, Jacobs G, Siebrecht J and Konrad C 2020 *Proceedings of the Design Society: DESIGN Conference* **1**: 2435–2444.

[12] Siebrecht J, Jacobs G, Konrad C, Wyrwich C and Schäfer W 2020 *Proceedings of the Design Society: DESIGN Conference* **1** :1105–1114.

[13] Konrad C, Jacobs G, Rasor R, Riedel R, Katzwinkel T and Siebrecht J 2019 *Procedia CIRP* **84**: 451–456.

[14] Drave I, Rumpe B, Wortmann A, Berroth J, Hoepfner G, Jacobs G, Spütz K, Zerwas T, Guist C and Kohl J 2020 *Proceedings of the 23rd ACM/IEEE International Conference on Model Driven Engineering Languages and Systems* 79-89

[15] Weilkiens T 2016 Variant Modeling with SysML (Victoria, BC: RTC)

AUTHOR INDEX

Institute of Physics
Dirac House, Temple Back
Bristol BS1 6BE UK

ISSN: 1757-8981
ISBN 978-1-7138-3352-9

CURRAN ASSOCIATES INC.
proceedings.com